THE CONNECTIVITY OF THINGS

Infrastructures Series

Edited by Paul N. Edwards and Janet Vertesi

A list of books in the series appears at the back of the book.

THE CONNECTIVITY OF THINGS

NETWORK CULTURES SINCE 1832

SEBASTIAN GIESSMANN

The MIT Press
Cambridge, Massachusetts
London, England

Originally published as *Die Verbundenheit der Dinge*, by Kulturverlag Kadmos Berlin, © 2016.

The translation of this work was funded by Geisteswissenschaften International—Translation Funding for Work in the Humanities and Social Sciences from Germany, a joint initiative of the Fritz Thyssen Foundation, the German Federal Foreign Office, the collecting society VG WORT and the Börsenverein des Deutschen Buchhandels (German Publishers & Booksellers Association). Work on this publication has been supported by the German Research Foundation (DFG)—Project Number 262513311—CRC 1187, Media of Cooperation, University of Siegen.

The MIT Press would like to thank the anonymous peer reviewers who provided comments on drafts of this book. The generous work of academic experts is essential for establishing the authority and quality of our publications. We acknowledge with gratitude the contributions of these otherwise uncredited readers.

This book was set in Stone Serif and Stone Sans by Westchester Publishing Services. Printed and bound in the United States of America.

Library of Congress Cataloging-in-Publication Data

Names: Giessmann, Sebastian, author.
Title: The connectivity of things : network cultures since 1832 / Sebastian Giessmann.
Other titles: Die Verbundenheit der Dinge. English
Description: Cambridge, Massachusetts : The MIT Press, [2024] | Series: Infrastructures series | Includes bibliographical references and index.
Identifiers: LCCN 2023057558 (print) | LCCN 2023057559 (ebook) | ISBN 9780262550741 (paperback) | ISBN 9780262381086 (epub) | ISBN 9780262381093 (pdf)
Subjects: LCSH: Civilization—History. | System analysis. | Social networks. | City planning.
Classification: LCC HM741 .G4513 2024 (print) | LCC HM741 (ebook) | DDC 302.3—dc23/eng/20240221
LC record available at https://lccn.loc.gov/2023057558
LC ebook record available at https://lccn.loc.gov/2023057559

10 9 8 7 6 5 4 3 2 1

CONTENTS

ACKNOWLEDGMENTS

My profound gratitude goes out to Steven Lindberg. Erik Butler, Laura Keeler, and Luane Hutchinson gave the translation its definitive shape. Wolfram Burkhardt of the Kulturverlag Kadmos in Berlin worked hard to make it possible for this book to cross the Atlantic to the MIT Press. This could not have been done without the support from Geisteswissenschaften International, the translation prize of the Börsenverein des Deutschen Buchhandels, awarded in 2020. I would like to thank all those involved for that honor. I am grateful to Geoffrey C. Bowker, Paul N. Edwards, and Janet Vertesi for including it in the Infrastructures series—something that this East German dissertation writer would never have dared to dream. Katie Helke, Justin Kehoe, and Suraiya Jetha made all of this possible at the MIT Press, and I am very grateful for their excellent work. The three anonymous reviewers knew precisely what remained to be done for an Anglo-American audience. I could not have wished for a better critical review of the German version. The University of Siegen Faculty of Arts and Humanities along with the Media of Cooperation collaborative research center in Siegen, and especially its subproject A01 *Digital Network Technologies between Specialization and Generalization*, provided the intellectual milieu to revise the text.

Before this text could set off on its trip over the Atlantic, it was made possible by a primarily European environment. Hartmut Böhme, Christian Kassung, and Erhard Schüttpelz deserve great thanks for their supervision of my dissertation. The Gerda-Henkel-Stiftung provided crucial support to the project in the form of a stipend and travel grant. The TOPOI Cluster of Excellence in Berlin and the Locating Media Research Training Group in

Siegen supported the writing of the text and the Topology of Technology Research Training Group in Darmstadt its revision.

I have been closely read, criticized, and encouraged by Anna Echterhölter, Thomas Haigh, Eva Johach, Dietmar Kammerer, Albert Kümmel-Schnur, Jasmin Mersmann, Christine Schnaithmann, Steffen Siegel, and Axel Volmar. I thank Jan-Frederik Bandel for careful copyediting, and David Sittler, Ursula Giessmann, and Katharina Zimmer-Lachmann for final corrections. Markus Karrer provided timely legal advice. Verena Wöbking capably organized the interlibrary loans necessary for the translation under pandemic conditions. Anna Katharina Büdenbender has worked brilliantly to process illustration permissions.

I had indispensable and inspiring exchanges with Jörn Adler, Inge Baxmann, Thomas Becker, Maximilian Benz, Dietrich Boschung, Geoffrey C. Bowker, Regine Buschauer, Mark Butler, Catarina Caetano da Rosa, Michel Callon, Chris Cheek, Heiko Christians, Gioia Dal Molin, Iris Därmann, Monika Dommann, Jörg Dünne, Marten Düring, Philipp Ekardt, Lorenz Engell, Christoph Engemann, Robert Felfe, Rainer Fischbach, Alexander Friedrich, Ursula Frohne, Alexander Galloway, Petra Gehring, Sokratis Georgiadis, Nacim Ghanbari, Lucas Marco Gisi, Melanie Gruss, Stephan Günzel, Wolfgang Hagen, Mikael Hård, Karin Harrasser, Eva Horn, Henry Jenkins, Michael Kerber, Namsee Kim, Sybille Krämer, Peter Krapp, Rebekka Ladewig, Mareike Layer, Gustav-Adolf Lehmann, Geert Lovink, Thomas Macho, Kirsten Mahlke, Dieter Mersch, Mara Mills, Lutz Musner, Stephan Packard, Jussi Parikka, James Purdon, Sebastian Quack, Bernhard Rieder, Detlef Rössler, Mathilde Roussat, Gabriele Schabacher, Valérie Schafer, Pascal Schillings, Hans-Dieter Schmidt, Jens Schröter, Sebastian Schwesinger, Oliver Lerone Schulz, Achim Spelten, Florian Sprenger, Markus Stauff, Nadine Taha, Eugene Thacker, William Uricchio, Erik van der Vleuten, Sebastian Vehlken, Joseph Vogl, Kirsten Wagner, Franziska Weber, Heike Weber, Niels Werber, and Daniel Werning; the research colloquia of Hartmut Böhme, Natascha Adamowsky, and Christian Kassung; and the students of the seminars "Was ist ein Netzwerk?" (Humboldt-Universität zu Berlin, summer semester 2007), "Am Apparat: Ein Telephonseminar" (winter semester 2008–2009), "Netzwerkanalysen" (Universität Siegen, winter semester 2014–2015), "Netzwerkkulturen" (summer semester 2021), and my lectures on "Network Cultures" (Humboldt-Universität zu Berlin, summer semester 2023 and Universität Siegen, winter semester 2023–2024).

I am grateful to numerous helpful people in libraries and archives: the staff of Berlin libraries, of the Universitätsbibliothek Siegen, and the Bibliothèque Nationale de France, especially the Bibliothèque de l'Arsenal; Kristine Clara at the International Telecommunications Union Library and Archive, Geneva; the staff of the Library of Congress; Laura Katz Smith at the Archives and Special Collections, University of Connecticut Library; Stephanie F. Krauss at the Countway Library of Medicine, Harvard; Helen Grove and Simon Murphy of the London Transport Museum; Jeff Mifflin and Lois Beattie of the MIT Archives and Special Collections, Cambridge, Massachusetts; and Christoph Eggenberger of the Zentralbibliothek Zürich. All errors that remain are my own.

Without the support of my wife, Ursula, work on the translation would have been impossible. This book is dedicated with love to her and our children, Elisabeth and Felicitas.

NOTE ON THE TEXT

The present book is a thoroughly revised translation of the second edition of *Die Verbundenheit der Dinge: Eine Kulturgeschichte der Netze und Netzwerke*, which was published in German in 2016. It is based on the author's eponymous dissertation, submitted to the Philosophische Fakultät III of Humboldt-Universität zu Berlin on May 7, 2012, and defended on October 29, 2012.

Unless otherwise indicated in the notes, all translations are by the translator and the author. Archival materials, including patent documents, are indicated in the relevant note. Online texts were last accessed on December 31, 2023. Selected important research literature has been integrated during the translation process. Parts of chapters 6 and 7 have been published previously in a translation by the author and Dawn Michelle d'Atri: "Drawing the Social: Jacob Levy Moreno, Sociometry and the Rise of Network Diagrammatics," *Media of Cooperation Working Paper Series* 2 (December 2017), http://dx.doi.org/10.25819/ubsi/7895; "Henry Charles Beck, Material Culture and the London Tube Map of 1933," in "Network Archaeologies," ed. Chris Cheek, Nicole Starosielski, and Braxton Soderman, special issue, *Amodern* 2 (October 2013), https://amodern.net/article/henry-c-beck-material-culture-and-the-london-tube-map-of-1933. Additional book materials can be accessed via http://www.sebastiangiessmann.de.

1 GETTING CAUGHT UP

Everything is given on pledge and a net is spread out for all that lives.
—Rabbi Akiva, *Pirke Avot*[1]

Language sets everyone the same traps: it is an immense network of easily accessible wrong turnings.
—Ludwig Wittgenstein[2]

What is the truth about the network? Would you like to know what it is? It is clear that networks have, first and foremost, come to serve a social function. Their material purpose receives hardly any attention at all—at least compared to "getting connected" on social media. This is striking, especially since the German word *Netzwerk*, English *network*, French *réseau*, and Italian *rete* all refer to something concrete. In response, this book explores objective points of reference that have gone missing. After all, "the semantic core of the 'net' is material." The concept and term can be traced back to spiderwebs and an age-old implement for fishing.[3] So, how was the net spun into a network? How much material reality is concealed by today's socioeconomic networks, and what are the media involved? What has led a seemingly simple device for catching dinner to become associated with an all-encompassing cultural technique?

The following pages attempt to answer these questions by way of historical episodes that illustrate the human, nonhuman, natural, and artificial qualities of networks.[4] Lines of division between nature and culture, on the one hand, and subject and object, on the other, don't disappear as much as never emerge in the first place. The physical object with its mesh and knots gives way to reticulations that consist of quasi-objects, hybrid practices, shifting relations, and various forms of exchange.[5] The difference between a net and a network—between a material object and something that is physical only in part—already represents an abstraction. Material practices of getting "connected" must be undertaken before one can make this kind of distinction at all, and the same holds for efforts to describe cultural uses, engage in theoretical reflection, or even provide a general account of conditions. For a society to see, describe, and calibrate itself in terms of networking, sundry spatial, social, economic, scientific, and artistic practices must already be in place and at work, "hooking up" things at a largely unconscious level.[6] To borrow a phrase from Vilém Flusser, they are "gestures of networking"—that is, "methods through which human beings try to give sense and meaning to their lives and to the world in which they live."[7]

Foregrounding the practices and gestures of networking does not presuppose a definition of what a network actually *is*. The principal concern is to understand how the status of being a network is achieved. Practices of networking become cultural techniques through discursive reflexivity.[8] But what networks should receive our attention? And what methods should we use?

MATERIALS AND METHOD

Since the material proves all but inexhaustible, it is vital to start with specific cases. To the extent possible, I put forward "primal scenes," turning points, and epistemologically and aesthetically significant examples to illustrate the material culture of sociotechnical networks and how they have been pictured both literally and figuratively. To examine the cultural memory of network societies, relevant material includes, among other things, the microscopic evidence mustered in anatomical research, handbooks on industrial supply, the setup of early telephone exchanges, a suddenly changing subway plan, curious developments such as speaking computers, and critical investigations of counterdiscourses such as conspiracy theories. These case studies represent starting points, not endpoints. It's only possible to write about networks and networking by exploring ramifications and lines of filiation.[9] *Realia* provide the analytical categories and concepts for cultural and historical investigation, not features of the field under examination that have been posited beforehand. Chapter 3 lays out the key polarities that inform our discussion: object reference–pictorial status, interstice (meshes)–heterarchy–density, organization–protocol, switching–mediation, fluctuation–synchronization, grafting–transformation, and disruption–breakdown.

The methodology follows from the material itself. The history of cultural techniques goes hand in hand with discourse analysis.[10] Its archive transforms knowledge embodied in objects and structures into language. It is not just written and visual documents that prove relevant here. It is just as important to examine the practices, techniques, and technologies that produce, operate, and maintain nets and networks. Approaches from the history of science, media theory, and visual studies play a corresponding role. I owe a great deal to studies by Walter Benjamin, Michel Callon, Michel Foucault, Donna Haraway, Bruno Latour, Michel Serres, Marylin Strathern, Cornelia Vismann, and Aby Warburg. It is difficult to imagine writing a cultural history of networks without the resources that post-structuralism offers for understanding the intricacies of decentralized actors and operations.[11] Significant portions of this book narrate genealogies of varying length and amplitude in keeping with historical chronology but without aiming for a linear order.[12]

While it would be gratuitous to list everything the book does *not* do, it seems prudent to indicate certain limits. Actor–network theory (ANT) is important for the discussion, but instead of applying interpretive schemes to historical phenomena after the fact, the task is to employ descriptive language and categories that emerge, as much as possible, from the sources themselves.[13] The same reservation applies to systems theory (which holds little promise anyway, inasmuch as networks can cut through social systems at an angle, so to speak). Nor is this study conceived along the lines of social network analysis (SNA). Instead of enlisting quantitative network analysis (with its graphs, diagrams, and charts) as a universal theory, I am interested in the interplay of differences and family resemblances across extremely varied networks.[14] The history of infrastructures and analyses of large technical systems (LTS) have provided crucial inspiration, but ultimately this book remains centered on matters of cultural and media history.[15] This book's translation into English and appearance in the MIT Press's Infrastructures series mark a further point of intersection between North American and European scholarship.

PASSAGES

Chapter 2 explores the *longue durée* of nets and networks, setting up the concept of cultural techniques and mapping out six historical strata that reach from textile materiality to mythology and theology, from the physical and reticular model of networking to its infrastructural and territorial implementation, and from the history of social networks to network science.[16] Starting with the actor's categories, chapter 3 outlines descriptive language for the cultural study of networks. Chapter 4 examines the nineteenth-century construction of canals and sewer systems as network building. Particular attention is paid to Saint-Simonianism, the utopian socialist movement whose program for integrated social life culminated in a colonialist project including the annexation of Algeria and the construction of the Suez Canal.

Chapter 5 turns to the early history of the telephone in the United States. Starting with Alexander Graham Bell's tours to promote his invention, a network unfolded on two complementary and interconnected registers: popularization and telephone exchanges. Chapter 6 sketches the first part of a visual history of the network diagram between drawing and writing. Examining eighteenth- and nineteenth-century microscopy, mathematics, and chemistry, it shows how semiotic practices originating in the natural sciences opened the way for the concept of the network to become established in the social sciences.

From chapter 7 on, my focus shifts from analyzing networks in spatial terms to exploring their modes of temporalization, which became increasingly important in the twentieth century. One key innovation illustrates this turn: the geometrized, topological subway map of London introduced in 1933. Henry Charles Beck's well-known Tube map and the clocks of London Transport were both means of coordinating and synchronizing circulation. Detailed timing, vital for networking projects across the board, also provides the subject of chapter 8, which takes up the transformation of production methods in an economy with increasingly interconnected parts. Here, returning to the visual history introduced in chapter 6, we look at the so-called network operations method adopted in the 1960s by Western and Eastern corporations alike. Globally networked production concerns not just administration and the workforce but also the very things that are manufactured.

Chapter 9 delves into early plans for computer networking and the emergence of the Advanced Research Projects Agency Network (ARPANET). Two elements provide focal points: the development of network protocols as a "language" or grammar of networking and the role still played by the system's hardware before the advent of digital communications technology relegated material components to a secondary position. Chapter 10 examines (from the outside, as it were) the evolution of networks into structures of social practice and symbolic order that seem to operate without reference to any object at all. In this context, the third part of our visual history of the network diagram treats the methods with which conspiracy theorists (Swiss Communist Georges Baehler and artist Mark Lombardi, in "real life," and the character Joseph Turner in the 1975 political thriller *Three Days of the Condor*) mapped and analyzed socioeconomic constellations of power.

Chapter 11 reviews the findings: the status of networks as quasi-objects, the long-term scope and shape of interconnectedness, and the increasingly few sectors of life, society, and the world that manage to resist being networked.

WHY NETWORK HISTORY?

This book, first published in German in 2014, arose from a long-standing interest in the spatial structures and visual schemes encountered on the Internet. During the 1990s, cultural theorists in the German-speaking world were fascinated, above all, by the artificial worlds of cyberspace. Around the turn of the millennium, interest shifted to "mixed realities," hybridity, and overlapping artificial and material spaces. In popular understanding, computer systems counted as networks almost as a matter of course. But if the archaeology of cyberspace and the history of cybernetics proceeded by leaps and bounds,[17] research on the history of networks had scarcely even begun.

Since then, the historical dimension of networks has become the object of considerable attention in cultural and social studies.[18] An international community is now dedicated to the history of networks. Historians of infrastructure and scholars in the field of science and technology studies have devised their own vocabulary for making material distinctions between systems, networks, and webs/internetworks.[19] Media theory still cannot get around the concept of the network, which Wendy Chun has aptly described as "belated too early."[20] Lori Emerson has been looking for "other networks"[21] outside the mainstream, and Kavita Philip is examining network cultures under postcolonial conditions.[22]

I began working on this book in 2006 after completing a brief, preliminary study on eighteenth- and nineteenth-century France oriented on the writings of Michel Foucault.[23] At the time, only scattered efforts had been made to investigate what has come to be known as the history, or archeology, of networks. Universal history and *histoire croisée* offered little to go on, but works such as Thomas Hughes's *Networks of Power*, Pierre Musso's *Telecommunications et philosophie des réseaux*, Laura Otis's *Networking*, and Mark Wigley's "Network Fever" sketched lines of inquiry and provided models of interpretation.[24] In spite of varied points of origin—the sociology of technology, philosophy, literary studies, and architectural history—these studies had the merit of emphasizing the materiality of nets and networks. This feature set them apart from the straightforward history of discourse, which is ill-equipped to handle the polysemous metaphors at issue.

Metaphorological reconstruction runs aground, at the latest, when terminology becomes as vague as its intended point of reference—what kind of net corresponds to what kind of network? Differences between languages only compound disparity and difficulty in attributions of meaning, and even within a single language, the range of possibilities for naming ("net," "fabric," "mesh," etc.) guarantees a diffuse semantic field.[25] Nevertheless, if only for heuristic purposes, certain distinctions will be helpful here.

In the following, "net" refers both to concrete, material artifacts and to connective structures that admit coded representation in topological diagrams. "Networking" primarily concerns spatial and visual ways of establishing links between identical elements. What is networked lends itself to cartographic depiction. "Network," in contrast, applies to fundamentally heterogeneous and open-ended practices that integrate people, things, institutions, signs, and spaces. As quasi-objects, networks display specific temporal regimes. As a rule, their operational chains defy graphic representation, but for this same reason, they encourage efforts to devise models and schemes of visualization on a massive scale. The three terms—"net," "networking," and "network"—should not be conceived independently. Especially in light of the varied material, modes of representation, and cultural practices examined here, they are constantly overlapping with, and spilling over into, each other.

Nets hold, connect, and catch; they also entangle, bind, and ensnare. This ambiguous feature is forgotten as readily as are the material and nonhuman components of networks. Indeed, it is only recently that the concept of the network has received a positive evaluation. For much, much longer, nets harbored a conflicted quality in the Mediterranean Basin and in Europe. Closer inspection reveals an array of highly different and contradictory meanings in prehistory and antiquity. It is warranted to take the status of nets in cultural history seriously. Under what historical conditions do nets' material and instrumental aspects slip through the cracks, as it were? How does a net turn into a "quasi-object" and "epistemic thing"[26]? What makes a relation a substance, and vice versa? And how is it that nets have come to signify networking, which brings actors, material connections, immaterial circulation, aesthetic design, social practices, and political techniques together in the framework of a cultural technique?

VERONESE'S NETWORK

An oil painting depicting a tidy net decorates the ceiling at the center of the Palazzo Ducale (Doge's Palace) in Venice. Stretching between the raised hands of a regally attired young woman, the net is a geometrically precise *symbolic form* (see Cassirer). In her right hand, the figure is holding a rod. She is pinching the net to the rod with her index finger, which is visible, and her thumb, which is not; her left hand is open, and two threads extend to her index and little finger. This part does not seem to be fastened to anything. So, the net and the woman's gestures demonstrate an elegant breeziness.

At first glance, the center of the net appears similar to a web spun by a spider. Indeed, such a creature occupies the middle, and the woman with reddish-blond hair has fastened her gaze upon it. Closer inspection reveals an extraordinary allegory. The threads stretching between rod and fingers frame the radial structure of the web in what is almost a square, so that an open triangle or parallelogram appears on the right-hand side. In other words, the pictorial elements signal a Euclidean system that calls to mind not just the natural artistry and geometric precision of the spider, already noted in antiquity,[27] but also the grid (*velum*) of Renaissance perspective.[28] It is difficult to make out the tiny spider on the towering ceiling of the Sala del Collegio, but easy to see the half-kneeling woman (figure 1.1).

FIGURE 1.1
Paolo Veronese, *L'industria*. Oil on canvas, 1575–1577. Palazzo Ducale, Sala del Collegio, Venice.

This colorful painting inserted in a ceiling coffer among depictions of the Virtues is by Paolo Veronese (1525–1588) and his workshop. In 1581, Francesco Sansovino called it the personification of applied geometry: "*L'industria di Archimede.*"[29] More commonly, the painting is known as *Dialettica o Industria* (Dialectic, or Industry). The work was created after a fire in the autumn of 1574 in the middle of a plague epidemic. Some fifty thousand of the city's residents lost their lives between 1575 and 1577. Redecorating damaged parts of the Palazzo Ducale took place at a time of general uncertainty. People celebrated life and prayed all the more that it would continue for the fact that the end seemed nigh. Against this backdrop, the display of political and artistic skill represents an act of affirmation to strengthen social coherence.

The title *Dialettica o Industria* also reflects the Venetian culture of manufacture, which flourished in the sixteenth century.[30] The allegorical presentation is of industry as a whole, whereby vigorous spinning stands for vigorous trade; a secular virtue is presented as a technique of pious, republican governance.[31] Polished hand gestures and sapient gazes are integral parts of the iconography.[32]

Another point of interest is the sewing basket standing to the left of the figure. Finely wrought and precisely executed, it seems to echo and double the net the woman is holding.[33] Along with a pair of scissors, the white and dark-green fabrics that evidently have made her over- and undergarments lie ready to hand. These everyday objects remain all but mute, with the scissors pointing in a direction different from the woman's gaze at the spider and web.[34] Delicate clouds enrich the picture by hinting at a divine presence. Although it foregrounds a political program colored by religion, the painting expresses almost immodest pride—as if the artist himself had brought forth, like the silk a spider spins, all the civic and artistic virtues on display.

The ancient topos of artful self-fulfillment is transformed into a trope for early modern culture and knowledge. Spiders don't imitate,[35] but people do. In light of the criticism voiced by Diogenes Laertius against the exaggerated rhetorical tricks of the Sophists, the finely woven net in the Sala del Collegio also represents an enjoinder to observe modesty: "Dialectical reasonings . . . are like spiders' webs, which, though they seem to display some artistic workmanship, are yet of no use."[36] But here, where struggles for political rank and influence played out among members of the Venetian aristocracy, verbal battles did serve a purpose. According to the long-standing tradition of seven liberal arts, dialectic encompasses persuasion and debate, including the resourceful invention of lines of argument.[37]

The very title of the painting—the inclusive "or" between "dialectic" and "industry"—generates a field of ambiguity.[38] Invoking texts and contexts that are not given heightens the tension and density of allegorical representation.[39] Veronese's *Industria* functions as a "dialectical image" as defined by Walter Benjamin: an image stretched between past and present and charged with remembrance and oblivion. Between the two poles, meaning flashes like lightning between associations. The similarities recognized are more a matter of unconscious correspondence, which remains invisible, than of the formal properties of what is depicted. As Sigrid Weigel puts it, "Recognizing, or reading, the points of similarity that flash into view activates an ancient skill, which was still in great demand among lettered men" of the day.[40]

Notwithstanding the premium placed on resemblance in the early modern period,[41] the allegorical depiction of nets possesses a generative character both in terms of memory traces captured by words and in terms of a "language of things" that becomes visible and sometimes discloses magical possibilities. The fact that Veronese's painting is located in the Venetian Palazzo Ducale concerns this second register of significance. The net depicted here stands for a host of cultural techniques. It at once represents a figure of geometry, provides a version of the grid used for central perspective, points to the economy of material culture, signifies (political) rhetoric, evokes and transforms the heritage of classical antiquity, forms part of a Catholic allegory of virtues, and marks a correspondence of similitude. . . . One could call it the first network. The image intensifies the manifold and its organization. As such, it betokens a new level of complexity for the relationship between material culture and discursive formations. It is ahead of its time and points to the future: as a material symbol, the net will become the quasi-object representing networks.

It is more remarkable and by no means self-evident that "net" has been adopted to refer to other, generally invisible aspects of culture. Why has "net" come to imply that something is "networked"? Why, for that matter, has "network" come to signify fluctuating, relational constellations of people, things, signs, and spaces? How has it come to stand for connectedness per se and, as a quasi-object, mean both the production of, and the precondition for, what is connected? And, if so, where does that leave the material quality of networks? The 1889 Grimm dictionary was quite precise in its vagueness. There, "network" was defined simply as "something netlike" (*Netzwerk: etwas netzartiges*)?[42]

These questions can only be answered—if one does not wish to rely on justifications after the fact or generalizations—by way of case studies. Beginning in chapter 4, specific networks are the focus of this book. Readers interested primarily in a historical narrative should jump ahead to that point. But if you are eager to know more about the methodological framework of this book, chapters 2 and 3 reflect upon cultural techniques and their role for network formations.

2 SIX STRATA OF NETWORK HISTORY: GENEALOGY OF A CULTURAL TECHNIQUE

> No human collectivity exists without things; human relations go through things, our relations to things go through men.
> —Michel Serres, *The Natural Contract*[1]

It's difficult to imagine, but networks are a relatively young phenomenon with *explicit* beginnings in eighteenth-century Europe.[2] Didn't ancient cultures—such as indigenous ones today—have an understanding of the world that included space defined by articulated points and connections? Weren't cells, bacteria, mycelia, and whole ecosystems networked in highly complex ways long before human culture existed? If *implicit* forms of knowledge are taken as the standard, nets and networks can be understood as very old indeed. Then, "silent" knowledge is embodied in actions and objects. This view is not interested in cognition so much as implementation, since "to know something is not to have to think about it."[3]

But networks—that is, social processes and structures tied to strategies of networking—are not a timeless, universal concept. Their origin can be dated historically and described as a cultural technique.[4] To be sure, both *implicit* knowledge involved in carrying out actions and *explicit* knowledge put down in writing or otherwise codified represent blurry categories. These terms should not be taken to be hard and fast so much as heuristic: there were nets and schemes of forging connections, objects, and practices long before networks became a social strategy and the driving force that orders knowledge.[5]

What implicit knowledge can be woven into fabrics and nets as material objects? And what is the relationship to cultural practices and symbolic orders? These questions seem easy to answer only at first glance. Weaving and knotting, protection from the elements, and catching animals are elementary, embodied cultural techniques. As such, they not only demand collective self-reflection but also open ways of inhabiting the world that include much more than clothing and food. Ethnographic accounts of how what seems long past remains present make as much clear.

For example, the act of speaking is likened to weaving in an African creation story recorded by Marcel Griaule (1898–1956), who studied ethnology with Marcel Mauss and linguistics with Marcel Cohen in Paris before performing fieldwork. The comparison will hardly surprise those familiar with recent philosophy of language[6] or post-structuralist theories of the text,[7] but among the Dogon of Mali it corresponds to a lively ensemble of myth, the doings of the gods, and everyday culture.[8] Griaule shares his conversations with the wise man Ogotemmêli. From him, one learns that the seventh of the Dogon's eight ancestor spirits (Nummo) consigned his directives to human beings in a very special, indirect way via an ant:

> At sunrise on the appointed day the seventh ancestor Spirit spat out eighty threads of cotton; these he distributed between his upper teeth which acted as the teeth of a weaver's reed. In this way he made the uneven threads of a warp. He did the same with the lower teeth to make the even threads. By opening and shutting his jaws the Spirit caused the threads of the warp to make the movements required in weaving. His whole face took part in the work, his nose studs serving as the block, while the stud in his lower lip was the shuttle.
>
> As the threads crossed and uncrossed, the two tips of the Spirit's forked tongue pushed the thread of the weft to and fro, and the web took shape from his mouth in the breath of the second revealed Word.
>
> For the Spirit was speaking while the work proceeded.[9]

The words of the spirit fill the interstices of the material, so that they are themselves the fabric. In turn, the fabric is the word itself. In this creation story, material is called "soy," which both means "it is the spoken word" and is itself the word for the number seven. Griaule observes that the seventh spirit imposes his word on the world by technical means. "By so doing he showed the identity of material actions and spiritual forces, or rather the need for their co-operation."[10] The ant has imparted this same technique to human beings. A privileged role is assigned to spoken language, which represents a synthesis of basic techniques for mastering nature, especially forging and weaving.

Among the Dogon, the drum is equally important, and its origin story is comparable. Various kinds of drum are derived from the Nummo playing a game with threads.[11] Drumming and weaving are closely related: to drum is to weave and speak simultaneously.[12] It is unsurprising, then, that the loom and the forge are central objects to the Dogon cosmology.[13] The collective imaginary deems weaving, which is associated with the daily work of women, to be the material condition of community. This circumstance illustrates Bruno Latour's thesis that technology is society made durable.[14] Indeed, technology produces social cohesion in the form of cultural techniques.

Dogon mythology is indeed complex, one might object, but the example above has little to do with mid-twentieth-century Western media culture. In fact, however, continuity appears in the midst of discontinuity: at the same time that Griaule was decoding the mental categories of Dogon culture, efforts were underway to catalog and index material techniques for mastering nature.[15] Cultural anthropologists and ethnologists had described the use of nets and slings as early as 1900 in some cases.[16] In contrast to their forebears, André Leroi-Gourhan, Claude Lévi-Strauss, and also the young Marshall McLuhan[17] were conducting research in full awareness of the revolution that cybernetics represented for the "human sciences," which, as a general science of control, was closely linked to the development of the (as yet unnetworked) computer. The oldest cultural techniques and the latest media stood side by side.[18] Ethnological interest in the material culture of nets and associated practices constitutes an integral layer in the history of networks.[19] In light of pullulating electrical and electronic connections, scholarly investigation of actual objects and instruments represented an act of preservation and rescue from oblivion. This was especially the case since researchers enlisted new storage media for recording sounds and moving and still images, especially French ethnologists from the 1920s onward.[20]

Against this backdrop, Leroi-Gourhan almost seems to be reflecting on his own cultural milieu when inventorying forms of nets in *Milieu et techniques* (1945).[21] Just picture his ethnographic overview (figure 2.1) alongside topological network diagrams made by contemporary scientists and engineers. The treatment that the trapping instruments receive is no less complex than the attention devoted to twentieth-century artifacts of

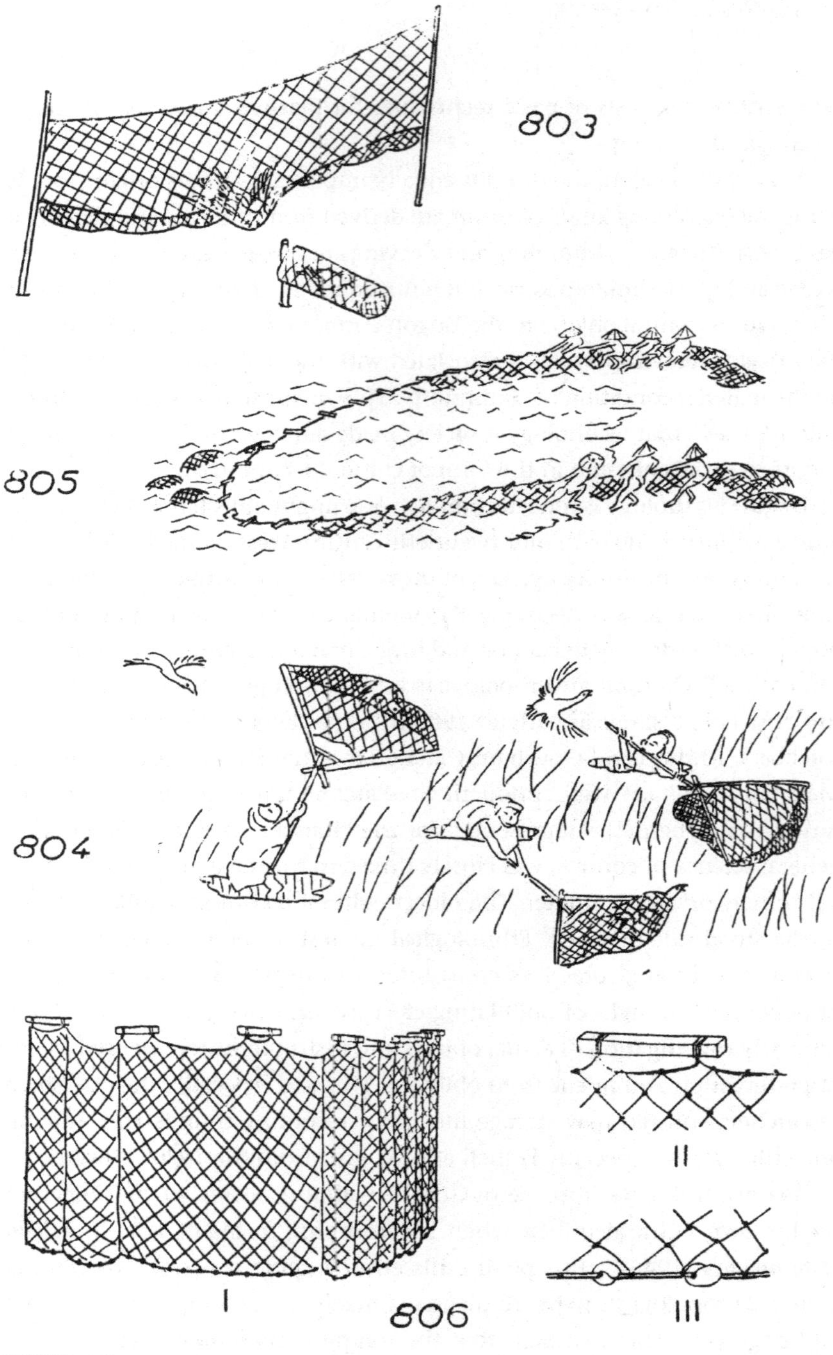

FIGURE 2.1
Overview of net forms in André Leroi-Gourhan, *Milieu et techniques*. Drawing, 1945.

knowledge. Large fishing nets already required the cooperation of whole villages. Leroi-Gourhan cites examples from Japan and the Mediterranean.[22] Specialized vocabulary for different kinds of nets displays an astonishing range across cultures; it points, in each case, to equally far-ranging forms of tacit knowledge ready to hand.

2.1 CULTURAL TECHNIQUES

It is scarcely possible to situate German-language scholarship on cultural techniques without reference to French anthropology and the implications it holds for actor–network theory. Geoffrey Winthrop-Young has described the concept of cultural techniques as unique to German-speaking media theory. This sets it apart from other international discourses on culture and technology.[23] Since the beginning of the twenty-first century, modes of investigation that do not consider culture and technique to be separate entities have become established, especially in Berlin, Weimar, Siegen, and Basel. One programmatic example is an essay by philosopher Sybille Krämer and art historian Horst Bredekamp, "Culture, Technology, Cultural Techniques: Moving beyond Text." At the end, the authors sum up their argument:

> Cultural techniques are (a) operative processes that enable work with things and symbols; (b) they are based on a separation between an implied 'know how' and an explicit 'know that'; (c) they can be understood as skills that habituate and regularize the body's movements and that express themselves in everyday fluid practices; (d) at the same time, such techniques can provide the aesthetic and material-technical foundation for scientific innovation and new theoretical objects; (e) the media innovations accruing in the wake of changing cultural techniques are located in a reciprocity of writing [*Schrift*] and image, sound and number, which, in turn; (f) opens up new exploratory spaces for perception, communication, and cognition; and (g) these exploratory spaces come into view where disciplinary boundaries become permeable and lay bare phenomena and relationships whose profile precisely does not coincide with the boundaries of specific disciplines.[24]

Cultural historian Thomas Macho offers a more concise approach by defining cultural techniques in terms of three specific features.[25] First, cultural techniques are distinguished by being capable of producing symbolic orders (which, in turn, exercise an effect on the cultural techniques themselves). Second, they are self-referential (i.e., recursive). Finally, they can be conceptualized only in intermedial configurations. Hereby, culture amounts to an operational field in which various modes of reading, writing, acting, exchanging, measuring, calculating, playing, and celebrating reveal criteria for intercultural comparison, on the one hand, and a way of writing cultural history, on the other. For Macho, cultural techniques are second-order phenomena that enable cultures to reflect on themselves by means of the symbolic work performed.[26] Attention should fall not on nouns—letters, images, numbers, architecture—so much as verbs—writing, drawing, calculating, building. The formative power of cultural techniques unfolds over long periods of time, which lends their study the coloration of quasi-universal history (or at least ensures commitment to the *longue durée*).

Macho's criteria have been much debated, beginning with his restriction of techniques to second-order, symbolically determined operations. This perspective, adopted from systems theory, has led to less emphasis being placed on the ability to teach and learn cultural techniques—their status as skills, in other words. Structure and agency do not balance each other out in this picture: cultural techniques function over long timeframes as civilizational media infrastructures. This definition has not only made the questions here possible but also revealed limitations—in contrast to Macho, I assume, following Leroi-Gourhan, that even basic actions have the potential to become second-order techniques with which cultures can be established, on the one hand, and reflect on their constitutive practices, on the other. Elementary cultural techniques—including agriculture, which is crucial to the historical semantics of *cultura*—should not be viewed separately from second-order cultural techniques. Instead, I advocate a relativistic and historicizing approach: techniques are always commensurate with culture in terms of their symbolic capacity, no matter whether they seem old or new. As Latour would say, networks admit both shorter and longer chains of mediation. Longer chains can be realized with modern techniques—which does not make them any less substantial than older, supposedly primitive techniques.[27]

Thanks to its overlap with ANT, research on cultural techniques has charted rapid progress in a spate of German-language articles.[28] Only parts of this discourse have been translated into English to date.[29] Like the remarks by Winthrop-Young quoted above, most contributions stress what sets the field apart—for example, when Bernhard Siegert situates his own works at the "end of the intellectual postwar in German Media Theory."[30]

Crucial texts remain inaccessible to an English-speaking audience—for example, Hartmut Böhme's groundbreaking essay on networks as a cultural technique, as well as Erhard Schüttpelz's discussion of the "media-anthropological turn of cultural techniques" and "elements of an actor–media theory."[31] The arbitrariness of what happens to get translated and what doesn't, although it always warps perspectives, is not the problem. Rather than the question "What's German about media theory?"[32] or "How post-Kittlerian is it?" what I regard as urgent is situating research in both German- and French-language traditions of anthropology and media theory. Friedrich Kittler already enlisted Marcel Mauss's concept of techniques of the body to ground his own approach to culture.[33] Likewise, the concept of operational chains (*chaînes opératoires*) is indebted to the works of Mauss's student Leroi-Gourhan. The connection to Mauss, Leroi-Gourhan, André-Georges Haudricourt, and Gilbert Simondon also explains affinities with ANT, which shifted the question of operational chains to the question of successful translation and transformation in sociotechnical networks. Finally, the combination of cultural techniques and ANT is particularly striking where they converge as actor–media theory.[34]

In keeping with considerations offered by Cornelia Vismann, the following reconstructs the "auto-praxis" of networks in historical context, emphasizing their grounding in material cultures. However, in contrast to Vismann, whose premature death deprived the field of perhaps its most important voice, I would like to stress the *agency* of human and nonhuman actors in symmetrical fashion. Anyone reading Vismann's "Cultural Techniques and Sovereignty" today will remark how the last offshoots of post-structuralist thinking foreground things and media, so that "man" (per Foucault's famous dictum) risks disappearing like a face drawn in sand at the edge of the sea:

> Cultural techniques describe what media do, what they produce, and what kinds of actions they prompt. Cultural techniques define the agency of media and things. If media theory were, or had, a grammar, that agency would find its expression in objects claiming the grammatical subject position and cultural techniques standing in for verbs. Grammatical persons (and human beings alike) would then assume the place assigned for objects in a given sentence.[35]

The above makes it clear how much studies of cultural techniques have benefited from the breakdown of the subject–object dichotomy enacted by actor–network theory. At the same time, Vismann rejects a symmetrical approach that also takes (distributed) human agency into account. In her eyes, what counts is not the specific execution of a practice but rather the rules for doing so.

In my own understanding, researching cultural techniques means bringing out hybrid constellations of bodies, things, and signs in distributed operational chains. The more one looks at practices, the better one understands operational chains as co-operational sequences.[36] Accordingly, networks are always already co-operative figures of what is more than human. They can also be understood as natural cultural techniques that bring forth ecologies, the environments in which we live. As a cultural technique, networking requires very specific skills. Taken in this sense, network archaeology is not a posthuman variant of media archaeology but rather its opposite. It is commonplace, in fact, for digital networks to remain tied to human agency. Reconstructing tacit knowledge and invisible work is more difficult than the infrastructural reading of mechanical protocols—and the latter, as a rule, are the result of multiple negotiations by human actors. In the best case, network archaeology yields an archaeology of practice. As Ben Peters summed up the matter at the Computer Networks Histories conference (Lugano, Switzerland, December 2017): "Networks are people!" All cultural techniques and operational chains ultimately depend on implementation by physical techniques and irreducible, embodied skills.[37] The question of the constellations in which human and nonhuman agents come together in networks anthropotechnically (and then separate again) cannot be answered ontologically; factors vary from one situation to the next.

This was long known to actor–network theory, with its focus on case studies of sociotechnical networks. All the same, ANT lacked a critical, historicizing perspective on its own foundations and mode of operation, which generalizes operational chains into networks. Programmatic initiatives were not wanting—for instance, Michel Serres's *History of Scientific Thought* or Bruno Latour's *We Have Never Been Modern*.[38] But for ANT, everything that could be described analytically as a network (or "worknet") qualified as an actual network.[39] Claims were bolstered by the self-evidence of the lifeworld (and academic practice) of the 1980s and 1990s. ANT could only come about because of the flourishing sociotechnical networks of the day. In contrast, subsequent, more historically oriented studies of cultural techniques—and of the history of infrastructure and science, technology, and society—demonstrated reserve by stressing the material grounding of networks, their metonymic character, situatedness, and specificity. Networks have genealogies, but they are not themselves genealogies. The book at hand is dedicated to examining this ambiguous, irreducible relationship.

2.2 SIX STRATA OF NETWORK HISTORY

Practices of networking are older than associated concepts or their crystallization into teachable and learnable cultural techniques. Furthermore, any networking cannot occur without metonymic reference to a physical net. Hereby, however, the point of reference can be an item of human manufacture, the geometrically precise web of a spider, a diagrammatic structure, a map documenting terrain as a space for navigation, or modern infrastructural innovations that organize supply, disposal, energy flow, and communication. Such diversity cannot be captured by means of intellectual history. It demands that one search for clues about past practices of networking that have not necessarily found expression in language. Thus, the goal is to provide an account of the back-and-forth between actors' categories and their adoption in practice or in reflection—as recorded, for example, in dictionaries.[40] Samuel Johnson's *Dictionary of the English Language* of 1755 defines a "network" as "anything reticulated or decussated, at equal distances, with interstices between the intersections."[41] This level of abstraction is unusual for early modern cultures of knowledge. Other large-scale, encyclopedic projects of the eighteenth century tend to define nets primarily as physical objects whose material design and repair requires great attention. Johann Georg Krünitz's *Oekonomische Encyklopädie*, for instance, provides detailed entries ranging from astronomical constellations to netted fences, but also includes geometric nets and grid constructions.[42] Johann Heinrich Zedler's encyclopedia proceeds in a similar manner and lists nets for animals from beavers to wolves.[43] Encyclopedias of the day do not yet contain a practically oriented use of the verb *vernetzen* (to network), but they devote space to patching and repairing physical nets,[44] as well as to practices of *netzen* (wetting, moistening). At the same time, anatomical descriptions of a *rete mirabile* inform entries on physiology.[45] Examples include Denis Diderot's unconventional discussions of perception and the brain in the context of the *Encyclopédie*.[46]

In fact, the eighteenth century in Europe is a period of key importance for the formation of knowledge cultures. The exchange of letters increased, and material uses of the concept of the net became established in natural history, physiology, medicine, and textile production. However, this development did not lead to any talk of "being networked" or engaging in "networking." Exceptional works such as Vitaliano Donati's *Della storia naturale marina dell' Adriatico* (1750)—in which the whole of nature is depicted as a net of species—confirm the rule.[47] The gradual emergence of the net as something understood not just in strictly material terms can be described, to borrow the language of Michel Serres, as the shift from an object to a quasi-object—an intermediate entity that exists in circulation, above all, with relationality trumping physical substance.[48] But in order to obtain such a quasi-object, whose loose ends form a network, as it were, a given society must be home to practices of netting and unnetting, binding and unbinding. Accordingly, it is only in the modern era, through building up industry and infrastructure, that networking came to achieve the prominence it enjoys now in the twenty-first century.

While this transition from reticular-material to infrastructural-functional semantics represents a crucial transformation in network history, it is not the only one. Six historical strata can be identified. Their borders are not hard and fast, and they should be thought of as continually shifting into one another. These levels correspond to (1) material understanding, (2) mythological use, (3) a physical-reticular perspective, (4) an infrastructural paradigm, (5) the dimension of social networks, and (6) the recent turn to "network science." These are described in turn below.

MATERIAL CULTURE

The material dimension of nets, along with their use as tools and traps located between culture and nature, goes back at least to the Neolithic period in Central Europe.[49] Here and there, collections in European museums include examples dating from the Mesolithic and Paleolithic.[50] These instruments for catching and holding prey were evidently standard among hunter-gatherers, who possessed the associated knowledge of how best to use them. As early human communities became more settled, techniques for making textiles evolved, notably the craft of weaving. However, it is difficult to date precisely the development of various net forms (creels, fences, weirs, trapping nets, dragnets, cast nets, scoop nets, etc.).[51] The invention of slings and hunting nets occurred among numerous groups in early history, and traces must be sought in practices of hunting. In Leroi-Gourhan's words, the net numbers among objects that the human body all but "exuded."[52] As a trap, it is more than just a tool. By nature, it is defined by a double absence: that of the hunter and that of the animal to be caught. Precisely because of this absence, the net could become a symbol of power.

Although setting traps is not a semiotic technique, associated practices signify the connection in space and time of hunter and hunted.[53]

MYTHOLOGY AND RELIGION

This use characterizes the mythological and theological dimension that became standard in the iconography of gods and rulers.[54] As river civilizations, Egypt and Mesopotamia had comparably advanced techniques for catching and controlling the bounty of nature. The latter, in particular, would never have come into being without its artificial irrigation networks. Both cosmologies were autonomous, even if common features predominated initially: enemies caught in an oversized net symbolized the might of Egyptian pharaohs and Mesopotamian deities. The surviving fragments of the so-called Stele of the Vultures—dated to around 2440 BCE and now exhibited in the Louvre in Paris[55]—offer a striking case in point. On the stele (figure 2.2), the victory of the Sumerian king Eannatum of Lagash over the rival king of the city-state of Umma is portrayed.[56] The defeated enemies of the city of Lagash are trapped in a net by the hunting god Ningirsu, who also personifies the local ruler.

Egyptian pharaohs used similar pictorial narratives—for example, on the front facade of the Temple of Edfu, which was built from 237 to 57 BCE. Even more than Ningirsu (or Eannatum), whose hand gesture signifies the ascendancy of Lagash, the pharaoh has power to command the death of enemies. The ruler's gesture of holding,[57] binding, and striking amounts to a magical death threat from far away. Comparable symbolism can be found in the Rigveda, the Atharvaveda, and the Old Testament. Indian mythology includes numerous gods who perform magic with slings and nets, such as Varuna, Indra (who likewise defeats enemy armies), Yama, and Nirrti.[58] The Old Testament radicalizes this negative charge into a theology of judgment. Both the Israelites and their enemies are caught like animals in God's net—hunted down and brought low for their misdeeds. Nets of divine rule constitute a space in which enemies are brought to heel and surrounded. The symbolism is the same in ancient Greece, albeit with an important difference. In the *Oresteia*, Aeschylus portrays Clytemnestra throwing a fishing net (*diktyon*) over her husband, Agamemnon, before she kills him.[59] Here, it signifies the reversal of the proper order, with a mortal woman prevailing over the lord of the house and land—a curse. Proximity is evident between the physical qualities of the net and magical practices of binding

FIGURE 2.2
Eannatum catches the enemies of the city of Lagash with a net. Fragment of the Sumerian Stele of Vultures, circa 2440 BCE. Limestone. Reconstructed original size: 130 cm × 180 cm; depth: 11 cm. Louvre, Paris, inv. no. AO 50.

and unbinding. In turn, the New Testament preserves but transforms the semantic register of material practices by making fishing a motif of the Christian mission. The proverbial "fishers of men" (Matthew 4:19; Mark 1:17) and the kingdom of heaven where the just and the wicked are held and separated (Matthew 13:47–49) open up a new theological dimension. Later, especially in the Middle Ages, the negative aspects—being caught as if in a spider's web—came to the fore.[60]

RETICULAR NETS

The beginnings of the physical-reticular view of the net can be discerned in ancient geometry, where admiration is evident for the natural precision of the spider's web and in anatomical writings, which describe tissue structures as nets. The two traditions could converge in surprising ways—for example, in iconography of the sense of touch (in which the spider and its web play a prominent role from the thirteenth century onward).[61] The discussion of the net (*réseau*) as part of the cerebral sensorium in Diderot's *Rêve de d'Alembert* (1769) can be understood as an offshoot of this tradition.[62] The Greek physician Galen already spoke of a *rete mirabile*—a complex, multilayered net—to describe organic functions.[63] Andreas Vesalius, in turn, searched the bodies he dissected to find the tissue in question. When he failed to do so, he correctly supposed that Galen had dissected a bull or a sheep.[64] René Descartes's *Traité de l'homme* (1633) already attributed a mediating function to anatomical tissues. Here, the philosopher discusses a *réseau* as a central part of the anatomy of the brain, from which individual nets (*filets*) extend as transport routes.[65]

Besides William Harvey's description of the circulation of blood, published in 1628, Marcello Malpighi's microscopic images of the frog lung, in 1661, were crucial for understanding the body as an ensemble of reticular networks. The anatomical description of the surface of the lung—which Malpighi also identified with plant structures—found a "miraculous net" in the form of a capillary system between the arteries and the veins. This model of a *rete mirabile* was part of a highly original aesthetic in the early modern period involving anatomical features and functions, especially where the flow of blood was concerned. In altered form, this living aspect of the reticular became the standard for forms of urban circulation and architecture modeled on the human body.[66] In the eighteenth century, physicians of the Montpellier school and, in their wake, Diderot developed a view of human anatomy in which tissue structures could scarcely be separated from biosocial networking energies.[67] Finally, Donati's *Della storia naturale marina dell' Adriatico* (mentioned earlier) made the net, as a diagram, a self-reflexive figure of knowledge for representing the order of living creatures.[68]

INFRASTRUCTURAL NETS

The infrastructural-territorial semantics of the net provides a touchstone for the cultural and sociotechnical dimension of networking. Fernand Braudel observed that the scheme can already be identified in the reorganization of European and Mediterranean space in the early modern period.[69] Significantly, new printing technology effected an analogous shift. Remarking parallels between typography and orders of knowledge, Michael Giesecke has observed, "It was by no means a foregone conclusion, if one considers world history, that communication techniques were employed to maximize the density of networking."[70]

In his history of the postal system in the early modern period, Wolfgang Behringer characterizes changing relationships between space and time in much the same way that Braudel describes effects, over the medium term, in the history of transportation: the regularity of the postal service changed habits and behavior. What counted was not the fastest connection but rather the most reliable one. The geographically shortest route was not always taken. Instead, itineraries were determined by the location of post offices and the connections they offered within a network of routes. The navigation of space and timeframes were arranged accordingly.[71] Until a veritable revolution in transportation occurred in the nineteenth century, operations were largely shaped by this combinatory logic. (Reconfiguration and reorientation had proceeded gradually as networks for transmitting news emerged in the sixteenth and seventeenth centuries, then as an increasing volume of letters, books, and ideas circulated in the eighteenth.[72])

Changes stemmed, above all, from efforts on the part of European nation-states to reorganize their infrastructure to mesh with industrialization and colonization. In France, after the revolution of 1789, for example, political projects and governmental programs repeatedly called for reorganizing collective existence. Projects abounded for building and connecting streets, canals, telegraphs, railroads, harbors, and shipping hubs, as well as overseeing banks and the circulation of money. The nineteenth century had no shortage of visionary schemes for improving infrastructure and global circulation, from first-, second-, and third-order networks for the management of streets[73] to plans for connecting East and West commercially such as Michel Chevalier's program for a "Système de la Méditerranée" (1832).[74] Modern systems of water supply were explicitly the model for designing networked infrastructures and organizing space under military control.[75] To supply Paris with drinking water, engineer Raymond Génieys developed the first example of a decentralized supply system that would equalize pressure throughout the city.[76] Alongside plans to curtail the shortcomings of optical telegraphs in radial arrangement by adding transverse lines of communication,[77] this innovation represents the beginning of nets as an infrastructural paradigm—with far-ranging effects for the design of cities, landscapes, and even the schemes of early utopian socialism (the Saint-Simonians).[78]

The program of continually securing new territory and expanding networked and networking infrastructures entailed increasing inequality of geopolitical power. Even egalitarian, emancipatory, Romantic movements such as Saint-Simonianism embraced colonialism and produced treatises on improving circumstances in Algeria by means of a *réseau de soumission*.[79] The infrastructural-functional paradigm of geographical networking along these lines was more or less established in full by 1880.[80] Its ascendancy was ratified in the wake of the electrification of Western societies by macrotechnical "networks of power" and system building.[81] Strikingly, in both Europe and North America, this development did not bring forth a new self-understanding of being immaterially "networked." Instead, the priorities of sociotechnical self-production continued to hold sway.

SOCIAL NETWORKS

The turn to a sociological, or simply pragmatic, diagnosis of the influence of social networks occurred relatively late. It was always implied, however—for instance, in Georg Simmel's investigations of reciprocal effects (*Wechselwirkung*) and the "sociological tie" underpinning human relations[82] or in Gabriel Tarde's sociology of imitation and innovation.[83] Examination did not necessarily occur in a formal, academic setting. The work of Jacob Levy Moreno and Helen Jennings is exemplary in this regard. In the early 1930s, Moreno, an immigrant and autodidact, combined group improvisation with social research at the New York State Training School for Girls in Hudson, New York. Jennings, for her part, contributed by conducting tests of the students. The social connections and emotional relations of girls who counted as "difficult to educate" provided the basis for large-scale mapping of social attitudes (attraction, rejection, and indifference).[84] Deriving immense benefit from the women with whom he worked, Moreno stood at the forefront of efforts to introduce network thinking to sociology, although his professional activity also included gestalt psychology and social therapy. Just as important as his own publications were the ethnographic studies they inspired—among others, numerous articles by sociologists and anthropologists in the newly founded journal *Sociometry*.[85]

Turning away from Moreno's interactionist approach, researchers after World War II adopted a sociostructural method. This orientation was evident both in the ways social graphs were analyzed and in the newfound use of computing software (following the lead of physics and chemistry) for digitally coding connection and nonconnection. Early network sociology, well into the 1970s, operated outside the mainstream of social science, especially in terms of its methods of analysis. The same is true of the study of social networks in anthropology, which continued to prioritize fieldwork and participant observation.[86] But as early as 1940, Alfred Radcliffe-Brown had gone so far as to declare, "A particular social relation between two persons . . . exists only as part of a wide network of social relations, involving many other persons, and it is this network which I regard as the object of our investigation."[87] The Manchester School of Anthropology, in particular, played a pioneering role by conducting a partially formalized network analysis of field research from Africa.[88] The intersection of infrastructural-territorial paradigms or macrotechnical nets with research into microsocial networks characterized research in the twentieth century. In the meantime, it has been joined by another development that continues to shape discourse. Mathematical graph theory, while dismissed as a parlor game since 1736, began, in the second half of the twentieth century, to establish itself as an autonomous and respected discipline.[89]

NETWORK SCIENCE

The application of graph theory to the natural, technical, and social sciences has led to the rise of network science, or what is known as the analysis of "complex networks."[90] The guiding assumption is that any phenomenon at all can be understood on the model of the network, a constellation of nodes and edges forming a geometrical figure. Connections that are predictable and admit visual representation take the place of specific, local practices. Cybernetics, systems theory, and structuralism live on in network science as pursued by prominent researchers such as Albert Lászlό-Barabási, and with United States government funding. Since the beginning of the twenty-first century, methods of computation and analysis that provide the foundation for real-time network analysis by secret services and social media platforms have become black boxes. Instead of retrospective analysis, efforts now center on anticipating the outcome of actions that follow patterns determined by "big data." Under these new conditions of media policy and economics, the practices and myths of open networking disappear into regulation—if not, indeed, outright control does prevail.[91]

2.3 NETWORK CULTURES

Since the 1960s, new social movements have made open networking out to be the measure of all things. The spread of the Californian alliance between "counterculture" and "cyberculture," with its preference for starting small, represents a case in point.[92] The ideals of sharing resources, grassroots politics, and environmentalism have flourished well beyond the Western, industrialized world. The rise of the World Wide Web was attended by open software such as Linux and projects for making information available free of charge such as Wikipedia. The goal was to promote egalitarianism by making networking tools universally accessible.

A critical assessment of this alliance between new, relatively accessible media technologies and social movements was not long in coming. Manuel Castells diagnosed the development as a symptom of a capitalist, globalized network society.[93] Luc Boltanski, Ève Chiapello, and others have also found fault with the model of horizontal management that is supposed to ensure hierarchy-free cooperation.[94] This double boom is the sign of a cultural technique: if science, industrial production, logistics, media innovation, management, counterculture, the ecology movement, and institutionalized politics all seek to operate in the same way, then it is clear that networking practices are ascendant—and unquestioned.

The other side of promiscuous "connectionism" is a critical view of ceaseless networking, which can focus on all that remains unconnected, interruptions in service, the need for repair, inefficiency, and systems coming undone. Anthropologist Marylin Strathern has pointed out that claims of ownership are liable to undermine, or at least reconfigure, network formations. Likewise, extra-network events—death, for example—transform relations and produce new arrangements.[95] Inasmuch as such breaks prove to be temporary, they do not fundamentally call into question the social, technical, economic, and aesthetic practices (or dependencies) at work.

The same is true for efforts to disconnect oneself, "get off the grid," or simply enjoy moments of silence and inaction. A networking imperative is taken for granted—say, when artistic projects thwart the rules of the market or contravene "relational aesthetics."[96] Establishing alternatives remains no easy matter, however. Even if the status of the network as an "absolute concept" has waned, it can scarcely be annulled entirely.[97] Freeing oneself from the net remains a task to be pursued in practical terms: infrastructurally, organizationally, artistically, and—especially in the case of a pandemic—through biopolitical initiative.[98]

Nor can practices simply be jettisoned once they have been adopted. In words recorded by Marcel Mauss, "But in fact I still do this: I cannot get rid of my technique."[99] The statement rings true, given ubiquitous references to the net that have become commonplace since the spread of the infrastructural-territorial paradigm at the latest. Metonymic references to at least one physical net—but usually more than one—continue to provide the infrastructural basis for networking practices. At the same time, the blackboxing of calculations and neural networks on digital platforms and the ongoing collection of information for economic and political purposes have made new analytical methods necessary. The latter would need to address the imbrication of tactical network analyses and military- and secret-service intelligence, as Donna Haraway noted earlier than other critics.[100] More than ever in the twenty-first century, the question is how to deal with data- and algorithm-driven networks. Broad swathes of network practices have been delegated to digital media agencies and platforms. The sluggishness and detail-oriented nature of nonprescriptive analyses makes them often seem obsolete when faced with machine learning.

To counter this trend, I would advocate critical, timely research that declares digital delegation of this kind to be merely provisional. Every network needs a qualitative assessment of its practices of connecting, repairing, disconnecting, and ceasing operation. It follows that there is no single method or vocabulary that will do the trick. Research on networks requires many approaches and languages, and there is always a need for active translation. Agency is too multifarious and intricate for any scheme of one-sided quantification and operationalization to work.

3 AN ARCHIVE OF NETWORKING

Does humanist discourse possess descriptive language for the practices of networking, the histories of nets and networks? Despite the interdisciplinary reception of actor–network theory, the question is difficult to answer. Given that graph theory (the formal analysis of networks), neural networks in machine learning, and theoretical perspectives along sociological lines are now firmly established, the matter does not necessarily seem urgent. These approaches also have the advantage of being predictable and operative. The boom in network theory during the second half of the twentieth century coincided with the "rise of network society" analyzed by sociologist and urbanist Manuel Castells.[1]

However, most of the theories that enjoy currency have the disadvantage of being ahistorical. They offer no self-reflection on their status as a modern form of magic by binding that promotes sociotechnical cohesion. They hardly provide a framework in which changes to networks over long spans of time as well as local and cultural variations can be addressed, but overall, they demonstrate the character of a universal theory. Their elementary vocabulary—nodes (points), edges (lines), directed/undirected graphs, matrices, numerical weighting of connections, algorithms to find the shortest path, and communication density—is geometrical in nature.

These semiotic practices for determining relationships—indeed, a distinct topology—proceed spatially. They do not contain a self-reflective moment. What counts is a form of analysis with which one can work, no matter whether it concerns traffic planning, social networks, molecular structures, the architecture of the Internet, or data-driven artificial intelligence. Quantitative network theory as promised by network science is a great leveler. It goes hunting with a diagrammatic formal net so that a multiplicity of phenomena, whether natural or cultural, will appear to constitute networks. Precisely because of the ahistoricity of epistemological instruments employed, mathematical network theory astounds us again and again with its capacity to produce true statements through model building.[2] Thus, it was possible to predict the spread of the severe acute respiratory syndrome (SARS) virus precisely in 2002 and 2003 by using simulations of global traffic. The accuracy of calculative network theory cannot be denied, it would seem. In establishing a constellation between biological phenomena, human bodies, political measures, transportation routes, and computer simulation, it positively shines as the driving force of a culture technique. But especially in the early years of the twenty-first century, research has been vigorous and become more highly differentiated.[3] Network simulations of the kind that worked for modeling SARS have only charted limited success for managing the coronavirus pandemic. Network science is clearly not equally suited for every specific situation.

To take an example somewhat at random, why should neural networks work in the same way as social networking on the World Wide Web? Because they can be described formally with the same set of methods? Does it matter whether the nodes are neurons or human actors? At least in its popular form, quantitative network theory would seem to suggest that it makes no difference. Advocates of mathematical approaches like to refer to John Guare's play *Six Degrees of Separation* (1990) as poetic evidence of a world in which everyone is connected via six people.[4]

In contrast, a materialist cultural history of nets and networks must insist on the differences between various objects—with full awareness that network diagrams and calculations constitute a crucial part of the picture. The following, then, attempts to outline a terminology based on the historical genealogy and concrete significance of networks. The terms are intended to establish a reflexive, historical descriptive vocabulary that does justice both to the materiality of what is networked and to the materiality of communication in and with networks. To employ the language of ANT—which now is also well established in fields of humanist endeavor—they mesh only in part, even though they are employed to related heuristic ends.[5]

3.1 OBJECT REFERENCE—PICTORIAL STATUS

Networks need a material foundation to operate. The more immaterial the "work" of networking, the more they seem to be heterogeneous, relational, quasi-objects in flux.[6] No human being has ever touched a network connecting banks to each other, as if something physical were extended between the tellers' counters and ATMs. All the same, the financial world is networked by means of shared protocols of communication, comparable economic practices, and cables and databases.

The more quasi-objects are produced, the more important the linguistic connection with material fact becomes. Is the communicative exchange between financial institutions best described in reference to a net for catching fish or birds, another manmade contrivance, or something entirely different? The objects invoked reveal the latent and unconscious aspects of networks. Networks derive their power and anchoring in the real world from metonymic ties to objects and infrastructures.

Michel Serres has elaborated the concept of the "quasi-object" on the basis of ludic situations (e.g., children's games such as "Button, button, who's got the button"?) and social interactions with a contractual element.[7] In this context, a network does not need to be a physical net; it emerges when a system of exchange is in place. Serres's discussion makes it clear that distinctions between subject, object, and collective stand at loose ends when a quasi-object is in circulation—hence common expressions about social bonds, fabrics, ties, and so on. Nor are nets the only quasi-objects that can symbolize connectedness. Quasi-objects mark subjects, form intersubjectivity, and create tenuous collectives. Their efficacy unfolds on an intermediate level corresponding to a crucial field of cultural history and theory, as well as media studies. In the game with a ball that Michel Serres cites as a classic example, the object that circulates remains in immediate, bodily proximity to the players.[8] The "net" of networks, in contrast, usually involves either very small spaces or larger territories. These spaces defy the grasp of the individual more than is the case with a ball game at close quarters. At the same time, their pull proves all the stronger.

It has become common practice among contemporary artists to give material form to disembodied structures.[9] The invisible strands of the network can be seen and touched in large installations. In this way, a metaphorical image is given physical reality and stripped of its figurative dimension. That said, networks are routinely made available to the senses through visual schemes: diagrams and maps. The pictorial form they are granted invariably brings the object being referred to into play. Material and visual culture overlap and condition each other. They can be read as signs of collective practices and desires. Is the action performed comparable to a spider's movements in its web? Or is a net cast in order to realize territorial ambition by dominating space, in competition with other nets but also connected to them?

Object reference and the visual depiction of networks raise questions. When networks and the practices, calculations, and schemes of representation involved are subject to observation and analysis, it is usually because a crisis, or at least a problem, has been perceived.[10] Almost always, the upshot concerns issues of power associated with political and/or economic hegemony. In the ancient iconography of power, the magical net signified equality inasmuch as everything and everyone was captured. Being caught up—usually with mixed feelings—in the social networks of today is a matter of maximal inclusion, but it also entails exclusion. The equality of actors is paired with gaps and spaces that usually are overlooked but constitute the material reality of the net.

3.2 INTERSTICE (MESHES)—HETERARCHY—DENSITY

Nets, as objects represented in diagrammatic form, and networks, as a relational cultural practice, are constituted by the interstice—the space in-between, which they leave open yet encompass.[11] The meshes of physical nets are what make them useful as instruments for hunting. In diagrams of networks, the interstice opens in unmarked areas that are neither point nor line. When territorial networks are built, they inevitably create spaces between points of connection and nodes.

The singular passion for making connections and forging ties that is characteristic of networking practices and theories draws attention away from what is surrounded but unconnected. Per Wolfgang Hagen's astute remarks on Aristotle, the in-between, the medial interstice, is a "non-concept" (*Unbegriff*). All the same, without the unmarked space of the in-between, perception of what does have consistency and definition is impossible.[12] Time and again, discussions of networks as cultural techniques gesture toward what

Aristotle calls "to metaxy,"[13] in which its environment becomes a decisive feature of a given place.[14] Networks make connections and thus environments but so do their nonconnections.

However, the permeable quality of networks does not lie entirely beyond our grasp. It owes its symbolic content to something material. A woven net needs the mesh between the tied knots. In medieval German and high Heideggerese, it is *lichtes Zeug*: both readily apparent and useful.[15] By the same token, lending networks visual form—and therefore making them available to perception—is inconceivable without planar, empty space between points. Designs that seem straightforward on paper, when implemented physically (in cities, landscapes, or airspace), take an articulated and striated form. Meshes come into being as sites of partial nonconnection. Depending on geographic distance, they create new modes of in- and exclusion, possible and impossible mediations. The interstices of networks are not empty in the electronic expanses of the Internet either; the space in-between is not non-space. Its meshes lend it a topographical dimension in geography and underpin a topological relation in the virtual realm.

Another key feature of networks is heterarchy. Unlike hierarchical order, heterarchy follows the principle of adjacency (*heteros*: "other," "neighboring"; *archein*: "rule," "govern")—always being slightly "off," as it were, and operating horizontally. Ever since the neurophysiologist Warren McCulloch introduced the term in 1945 to describe neural circuits, it has named phenomena of decentralized organization and self-organization. Heterarchy, which can also exist alongside and within hierarchical structures, represents a condition for the growth and condensation of networks. A wealth of interconnections yields unpredictable effects on the system as a whole.[16] Instead of ranked decisions or connections, a lateral path to another node is always available. The consolidation of neighboring points can produce highly resistant topologies.

Finally, a network's density results from the balance of sites and connections. Whereas interstitial space has no positive meaning, "density" is established terminology in quantitative network theory. It can be measured not only for individual nodes, groups, and clusters but also for the entire network, quantifying the ratio of actual connections to possible connections. The closer the quantity of links gets to the maximum possible number of connections, the greater the density. Beyond displaying this quantitative, "physical" aspect, density influences the organization and self-organization of a network.

3.3 ORGANIZATION—PROTOCOL

Networks are based on simple elements, the diverse interactions of which produce complex structures without being centrally controlled. The connection to biological concepts such as self-organization, autopoiesis, evolution, and emergence is self-evident—and made time and again along the lines described by systems theory. From the perspective of the history of science, this represents a legacy of cybernetics—in particular, the second-order cybernetics of the 1960s. The latter's radical equation of biological, social, and technological phenomena not only continues to echo in current media cultures but also has given rise to diagramming and calculation as universally applicable models for networks. The fascination of the "algorithms of the living world" (François Jacob)[17] proved so strong that "function," "organization," and "harmony" were taken over from molecular biology and neurophysiology.[18] More still, it made it possible that a computer scientist of Paul Baran's stature could ask his neurophysiologist colleague McCulloch, in all seriousness, how many nodes and links a net needs to be robust and reliable.[19]

A transfer of this sort might succeed on the level of modeling information theory. It is doubtful, however, whether sociotechnical networks, whose emergent qualities are undisputed, work in a self-organized and self-organizing way and are capable of compensating for failures, as the brain does. The need to conduct everyday repairs of disruptions and breakdowns (which are rarer) shows that self-organization represents a utopian ideal. No network society gets by without grounding, so to speak, without maintaining and repairing its structures and shoring up its foundations.[20]

Protocol is one way to understand how networks are created and organized, both pragmatically and in terms of power relations. A protocol provides instruction and a script according to which activity on networks can be oriented and directed. Anyone who wants to be part of a network must conform to its formal and informal requirements. Networks can be based on several protocols simultaneously. For the Internet, this is a technical requirement that makes transfers possible: the "p" in "https" stands for "protocol." Although its character as instruction is easily derived from its etymology—in Byzantium, *prōtokollon* referred to a page glued in at the beginning of official papyrus rolls with chronological information about its origins and author—protocols are always the result of social negotiations.

However informal their beginnings, protocols become binding codes. Exemplarily, they are the forms of ceremony and etiquette observed in a courtly or diplomatic context.[21] Protocols present a grammar that becomes the basis of the organization and possible behavior in a network. They permit checks and control by providing guidance in a heterogeneous material milieu where the power to perform actions is distributed.[22] Analyses of how power is apportioned and operates in a network often start with the social and technical codes inscribed in protocols. At the same time, no strict determinism prevails between instruction, controls, and other activities in a network.[23] Despite the formal nature of protocols, they enable a diverse range of operations.

3.4 SWITCHING—MEDIATION

Networks need places and actors with privileged status. If the power to perform actions is scattered and heterarchical, it does not mean that all actors have equal rights in the long term. On the contrary, central nodes almost always emerge. Manuel Castells uses the vivid term "nodality" to describe the phenomenon.[24] When concentrations entail the formation of quasi-monopolies, quantitative network analysis speaks of "scale-free networks." What remains decisive is that actors occupying strategic points are responsible for *switching*: shifting and channeling financial streams, data, traffic flows, and so on.[25] It concerns the style with which political, organizational, and technical apportionment occurs in networks.

Switching is hands-on—as when telephone connections are made manually.[26] Actors and actants—that is, human and nonhuman entities—perform the operation. Mediators, means, and ends exist simultaneously and in unpredictable arrangement. For this reason, they affect the social fabric.[27] Actor–network theory is dedicated primarily to the procedural steps by which mediation takes place. Mediation occurs continuously in operational chains, which in turn are distinguished by their specific mode of translation.[28] Mediation processes generate and, in a sense, modify the scripts that become the protocols of a network.

As an ongoing process of translating between material, media, and social givens, mediation is a principle that goes beyond the concrete switching performed by individual actors and actants and exceeds the conceptual purview of philosophy. It is through mediation that the collective so readily called a network assembles in the first place. Continuous mediation determines its morphology and conditions its organization and life span: "A network starts to form as soon as three actors are joined together by intermediaries."[29] A general theory of networks would have to integrate all intermediaries. As a rule, however, ANT restricts their number in keeping with the network described and its forms of translation. The latter admit description and accounting in the form of technical objects, bodies implicated, and money streams.[30]

Mediation does not occur automatically. It always takes place in the presence of fluctuation and disruption. Whether a network between actors and actants results depends on the actions and varies case by case: sometimes translation chains create a network; sometimes they do not.[31]

3.5 FLUCTUATION—SYNCHRONIZATION

Networks form specific temporal regimes. This circumstance does not concern only sociotechnical operational chains, from the superimposition and spatial and temporal arrangement of which stabilized forms of organization result. As dynamic entities and relational configurations, networks fluctuate constantly. The Polish physician and theorist of science Ludwik Fleck has given the matter a lapidary formulation. As ideas are linked, a fixed point is created that becomes the starting point for new lines of thought. New conjunctions are produced, and old ones change. "This network in *continuous fluctuation* is called reality or truth."[32] Fleck is referring, specifically, to linguistic, discursive phenomena and the association of ideas. His observations also hold, in general, for the nonuniform development of networks over time—a process that brings together multiple transformations.

The tendency to adopt a spatial view of networks has been criticized for good reason.[33] As a corrective measure, the aesthetic of digital networks has been linked to their temporality.[34] Even Michel Serres has expressed skepticism about the purely geometric view: "Networks, even if you add the idea of virtual modes of tracing, leave an image in space that is almost too stable. But, if you immerse it in time this network itself is going to fluctuate, become very unstable, and bifurcate endlessly."[35] The development of networks over time goes hand in hand with a tendency to lose cohesion. Does the reason lie with the fact that it is necessary to reposition oneself rapidly and repeatedly to adapt to changing situations? Or is it because of the forms that traffic flows and data streams assume? The kind of fluctuation in evidence varies from case to case. The upshot—whether constant calibrations and changes result in irreversible, lasting connections between actors—plays a key role in the ultimate organization and degree of self-reflection in sociotechnical networks.[36]

As "grammars," protocols are also responsible for temporally sequenced functions. The tendency toward heterochrony is countered by synchronizing circulation with clockwork precision. If the organization of a network in time is disturbed, it has tangible consequences: missed train connections, interrupted flows of data and goods, stalled cell growth, congested movement, and so on. In the long term, synchronization is an ideal state and difficult to achieve. Outside of purely biological systems, it requires strategic and sociotechnical work. Synchronization means finding suitable forms for the fluctuation inherent in networks.[37]

Synchronization can require considerable coordination efforts. Consider, for example, the countless telegraphic messages that were necessary for establishing a globally uniform regime of time and dividing it up into separate zones.[38] The whole project of industrial modernity can be understood as the all-encompassing effort to get synchrony and simultaneity to coincide with each other.[39] The twentieth century tells the story of coordinating time management and spatial connections, whether in transportation or in the networking of banks and stock markets.

Rhythms, simultaneity, and states of synchronization vary according to the specific character of a network. In general, synchronization adjusts the rhythms of networked operations to one another. To this end (in an ideal situation), relatively little interaction is required.[40] Successful synchronization is also successful mediation, which can represent an asset in sociotechnical networks. In contrast to strategies of synchronization, tactical operations make use of gaps and temporal intervals. Thus, asynchronous guerrilla warfare, while fluctuating continuously, demonstrates an unpredictable measure of self-organization that even high-tech network warfare cannot achieve. The temporal regimes of networks do not necessarily have to synchronize their rhythms. Some can unfold successfully by means of deliberate asynchronicity or the right balance of synchronization, asynchronization, and resynchronization.

3.6 GRAFTING—TRANSFORMATION

Networks succeed each other historically, but they also intersect and collide with each other. One network may also graft itself onto another. The telegraph followed the railroad, but railroads have also been built following telegraph lines. Grafting is a cultural technique for promoting growth and development that is also practiced outside an agricultural context.[41] Where sociotechnical networks are concerned, it involves turning a territory into a milieu of exchange. "One canal situates itself next to another, is taken up by it, grafts itself onto the preexisting one, to contribute casually to a denser networking."[42] In other words, no network stands alone. A given network usually emerges from existing elements and structures that lead to new relations. The coupling of one network with another can entail the fusion of individual parts. A complete takeover is also possible, whereby the original stratum disappears almost completely. But grafting is also a mode of parasitism in which other agents take over: excluded third parties, from hackers to terrorists, make use of others' protocols of connection to suit their own purposes.[43]

Grafting is one possible mode of transformation. How do networks change over time—historically, topologically, and in terms of relations of power? Transformations typically do not occur suddenly. They tend to develop over the medium and long term. Collective transition and passage—including *translatio imperii,* in the political sphere[44]—is packed with energy. Processes of transformation can take place in keeping with massive tectonic shifts or proceed gradually.[45] Their corollary can be the reconfiguration of a network, changes to protocol, or processes of asynchronization and resynchronization. Transformations often come to light as a result of changes in schemes of visualization: reading and comparing network mappings discloses middle- and long-term changes.

What are these changes in transformations? On an abstract level, the topology of a network is what changes—that is, the spatial relationship between the nodes. This implies the question of connectivity. What is connected and how? Who now occupies a place inside or outside of the network? The temporality of the network is also transformed by whether exchanges tend to be synchronous or asynchronous.

Describing transformations involves the effort to obtain a first-order standpoint of observation or, if necessary, a second-order one. Putting processes of change in perspective lessens the dramatic pathos of birth and death: "Seen as networks . . . the modern world, like revolutions, permits scarcely anything more than small extensions of practices, slight accelerations in the circulation of knowledge, a tiny extension of societies, miniscule increases in the number of actors, small modifications of old beliefs."[46] Nevertheless, the potential of transformation processes to cause rupture should not be underestimated: decisive events and crises often lead actors to reflect on, and alter, what they are doing.

3.7 DISRUPTION—BREAKDOWN

Networks also come to an end. They do so even though their distributed structure is supposed to compensate for failures, radical transformations, and breakdowns. Sometimes they are disrupted because protocol has not been followed, communication procedures went wrong, and temporal regimes could no longer be synchronized. Disruptions make the materiality of the underlying physical or communicative net palpably—and unpleasantly—real. They reveal the limits of opportunities for connection available to actors.

To put things in general terms, the reach of networks as a cultural technique becomes visible and audible precisely when their self-evidence is at stake and attention turns to instances of dysfunction.[47] But what would network-specific "signals of disruption" even look like?[48] Disruptions in the network are always local before spreading—for instance, one node fails to communicate, a certain connection is cut, or a flow interrupted. If the overall stability of the system becomes too shaky, breakdown can result. Network disruptions are always associated with a loss of connectivity. In "wired" societies, this generates specific forms of anxiety.[49] The rule holds not just where infrastructure stands at issue but also—and especially—where largely immaterial communication ceases. Consider the algorithmic trading that occurs in financial markets or the spread of computer viruses. The spatially distributed structure of networks can recoup intermittent disruption and the local failure of nodes to a certain extent, but it cannot prevent temporal dissociations and errors. In fact, negative dynamics such as unwanted feedback and cacophonic resonance are the rule. They just don't come to light to the extent that built-in resynchronization measures manage to ensure systemic stability.

Disruptions result especially from interaction. The temporary failure of one network leads to disruptions in the next one, which in turn holds consequences for a third. Overlapping and switched operational chains are susceptible to disruptions in succession, which affect the temporal development of networks system wide. Usually, small events interrupt flows more lastingly than spectacular phenomena. For example, crises in internal systems for baggage handling disrupt the flow of air travel over the long term more than the occasional eruption of volcanos.[50]

Disruptions, which are a regular occurrence, play a constitutive role in the emergence of new order.[51] A breakdown, in contrast, spells the end of a network. It is a question of time. In the 1980s, hackers coined the phrase "network meltdown," a phrase that combines cascading technical disruptions of networks with the image of a nuclear accident as the worst possible catastrophe.[52] Too much traffic—that is, too many data packets transmitted—causes this form of virtual collapse. The concept is paradoxical. It does not concern lasting radiation but only a temporary, absolute disruption that results from overload in the form of excessive network density. It also suggests the possibility of a total breakdown after which a network cannot continue to exist simply by transforming. The possibility of a breakdown does not surface as a purely technical problem. At issue stands the fate of human collectives entertaining the prospect of their own extinction.

What holds a network together? Its nonhuman properties represent just one part. According to Marilyn Strathern, at the end of their existence, networks are cut up according to socioeconomic property relations and the divergent interests of members.[53] A network is held together by more than faith in ongoing economic communication.[54] It stands as the narrative of a shared identity, which, at the beginning of the twenty-first century, apparently cannot get by without technological assistance.[55]

When and under what conditions can such a narrative begin? For a start, it means acknowledging that networks as cultural techniques require a self-reflexive and symbolic dimension. As described in chapter 2, this began in early modern cultures of knowledge and developed in eighteenth-century epistolary and salon cultures.[55] As far as the material culture of sociotechnical networks is concerned, the first explicit discourse can be identified precisely. It was nineteenth-century socialism in France that was attended by an explicit networking program whose profound effects on culture and history are the subject of the next chapter.

4 CHANNELS: THE POLITICS OF NETWORKING AROUND 1850

In his study of the Mediterranean, the French historian Fernand Braudel approvingly refers to the work of his colleague Lucien Febvre. The Mediterranean, Febvre has observed, is the sum of the traffic and transportation that occurs there.[1] Passages by land, sea, river, and along the coasts constitute "an immense network of regular and casual connections, the life-giving bloodstream of the Mediterranean region." The entire complex of the Mediterranean must be understood as dynamic space (*espace-mouvement*). This network (*réseau*) generates proximity via travel, promotes regular exchange, and corresponds to the region's urban cultures.[2] Apart from favorable geographical and climatic conditions, trade between cities stands at the origin of maritime and mainland connections.[3]

Braudel's geohistorical narrative concentrates on the reign of King Philip II of Spain in the sixteenth century. However, one is left with an impression of a networked space very much like today's. Industry follows mercantile capitalism, and finance follows industry—with corresponding effects on the intensity and directions of movement.[4] This ongoing transformation through networks that sometimes succeed each other and sometimes overlap is a point of reference for, yet different from, the network history told in this chapter. Braudel's *géohistoire* focuses on barely perceptible changes that slowly determine patterns and practices of exchange. His description of an evolving space of transportation maps the milieu of circulation with the same subtlety employed to discern the influence of the natural environment. Braudel observes that lively exchange continued seamlessly into the seventeenth century[5] and emphasizes the difference from the modern revolution of transportation that would occur in the nineteenth.[6]

The way connections are organized is a crucial historical difference between early modern and modern projects for canals and networks. The Mediterranean zone Braudel describes unfolded as a vast space, barely possible to survey, in which the power of individual cities balanced each other out.[7] In contrast, the part-nationalist and part-internationalist (and usually colonialist) logic of circulation in the nineteenth century aimed at continually expanding connectivity, with a view to political and economic control. As we will see below, it followed governmental strategies, which were often based on theories and treatises. More than ever, the mastery of physical space was a matter of networks of transportation and media. Political writings set the tone and the program for plans that all too often harbored hegemonic ambitions.

4.1 CANAL VIEWS: MAP AND PANORAMA

> But is that [a geographical expression] so insignificant?
> —Fernand Braudel, *The Mediterranean and the Mediterranean World*[8]

The biggest construction site on the planet! At the 1867 *Exposition universelle* in Paris, the Pavillion de l'Isthme de Suez regaled visitors with maps, photographs, and custom-built models documenting progress on the Suez Canal (figure 4.1). At the previous world's fair, in London in 1862, the Suez Canal Company had neglected to share news of ongoing efforts. Now, technological innovations were proudly illustrated as a "theater of works."[9] Examples included cargo ships, sand dredgers, and conveyor belts. The latter were especially admired. A diorama painted from photographs offered a synoptic view: excavations already made, building sites, machines, and the harbors of Port Said and Suez.[10]

Two years later, this public campaign of conquest through the media found an imperial continuation in celebrations attending the opening of the canal. No member of the assembled European royal houses was as delighted as Eugénie de Montijo, the wife of Napoleon III. Her exclamation on November 16, 1869—"De ma vie, je n'ai rien vu de plus beau!" (In my entire life, I've never seen anything more beautiful!)—expressed aesthetic astonishment and colonial ambition in one.[11] Viewing the scene from a ship united imperial pride and a hierarchized social status. As crowned heads towered in self-confidence, the territorial claims of nation-states extended below.

The pictorial status of the great canals of the nineteenth century was also consecrated by a sublime and edifying gaze from heavenly heights. The arrangement of models in the Pavillon de l'Isthme de Suez was hardly left to chance. It corresponded to the floating Montgolfier perspective already in evidence during preparations for building the Panama and Suez Canals (figure 4.2). Although such landscapes belong to the history of the panorama—especially synoptic views of the Alps[12]—they are not identical with the panorama or the diorama.[13] Instead, this hybrid combination of landscape and topographical map implies visual space over and above what Jonathan Crary has called "techniques of the observer." Building on Crary, we can assume that the way these views were perceived both publicly and individually blurred the distinction between internal feeling and external signs. The relocation of the viewer's standpoint from the 1820s onward called into question boundaries between inside and outside.[14] The imaginary journey by air or sea was meant to be immersive. Journals and world's fair spectacles rehearsed geopolitics and colonial rule on a domestic terrain. Merely indicating the course of canals on maps was already an anticipation of exercising power over foreign territory. I submit that the process of visual coding with an eye to material construction generated a specific form of media landscape without which nineteenth-century colonialism is scarcely conceivable.

As a cultural technique, maps not only generate representations of spaces to serve purposes of surveying and orientation but also belong to the sphere of political action.[15] As images with which one *does things*, they imply the question of who is acting and how these actors use them.[16] Additionally, they offer a shorthand version of the transition, described by Michel Foucault, from a disciplinary society to one in which colonialism and international communications are integrated into security apparatuses. In his lectures on the history of governmentality, before he took up a genealogy of the art of ruling (which can be understood as early techniques of control and regulation),[17] Foucault outlined a shift that is crucial to network history.[18]

Discipline functions centripetally. It concentrates, centers, and encloses, isolating spaces and regulating everything to the extent possible by means of prohibition. In contrast, security apparatuses are centrifugally organized,

FIGURE 4.1
Ferdinand de Lesseps presents the model of the Suez Canal at the world's fair in Paris. Engraving in *L'illustration*, 1867.

– Une Conférence de M. de Lesseps.

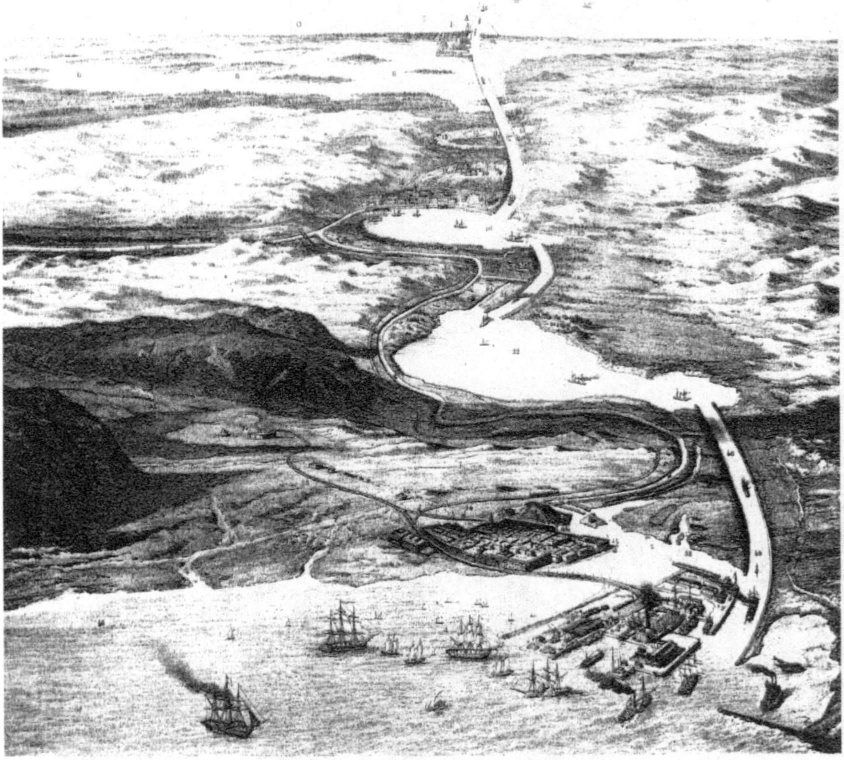

FIGURE 4.2
Vue à vol d'oiseau du canal de Suez. Lithographic reproduction of an illustration from the nineteenth century, 1935.

and they harbor expansive tendencies. "New elements are constantly being integrated: production, psychology, behavior, the ways of doing things of producers, buyers, consumers, importers, and exporters, and the world market. Security therefore involves organizing, or anyway allowing the development of ever-wider circuits."[19] In the nineteenth century, the art of producing a body politic included shaping a liberal milieu of circulation, often under the cover of geopolitical and colonial ambitions. "The milieu is a certain number of combined, overall effects bearing on all who live in it."[20] Centrifugal apparatuses of security that spread in the form of networks incorporated ever more people, things, and markets. The process was not limited to the sphere of national territory, whose internal security was organized by a police presence. As a matter of course, colonial powers acted in concert when it came to neutralizing straits, rivers, and canals for the purpose of international communication. We will return to this nineteenth-century development in international law below.

An apparatus of security ensures that activity can proceed in the sense of *laissez-faire*. As such, it numbers among the political techniques of liberalism. Hereby, the ability of people, discourses, and goods to circulate is equated with freedom. For Foucault, "circulation" means more than just connection to economic and media flows. Above all, the term refers to the political technique of encouraging desirable exchange and blocking undesirable exchange so that the "right kind" of exchange will proceed more intensively.[21] The concept of the network is apposite, since, in a prolonged process from the mid-eighteenth century onward, it became possible to see, symbolize, think, and write forms of circulation that were unrestricted by a closed circuit.[22] The changing morphology of cities and landscapes was closely tied to these partly material, partly immaterial networks pervading knowledge and goods.

Few historical phenomena reflect the contradictory nature of this development, which fused centripetal and centrifugal trends and sought to impose cohesion as a social technique, as well as the activity that became the talk of Paris (and beyond) around 1830. A group of young idealists who called themselves Saint-Simonians in honor of their teacher, the entrepreneur Henri de Saint-Simon (1760–1824),[23] had the presses working overtime. Journals such as *Le Globe* and *Le Producteur* proclaimed Saint-Simonianism as a new religion and prophesied the advent of social justice through technological progress and industrialization (Banks for all! Railroads! Canals! Steamships!). Advocating pacifism, the group set up a network spanning the Mediterranean region and, indeed, the whole world. Women who joined this early socialist movement published their own newspapers, such as *La femme libre* and *La voix des femmes*, to achieve social, legal, and economic emancipation.[24]

In light of these Saint-Simonian ventures, the history of nineteenth-century obsessions with canalization clearly involves much more than "just" securing a connection from point A to point B on geopolitically important terrain. Ferdinand de Lesseps's celebrated construction of the Suez Canal was a success. The Saint-Simonian project, while technically a failure, reveals awareness of the mediality of physical and social connections along lines similar to what Walter Benjamin remarked apropos of contemporaneous railroad projects.[25]

4.2 CITY: THE PARIS OF SAINT-SIMONIANS

> The Saint-Simonians: a salvation army in the midst of the bourgeoisie.
> —Walter Benjamin, *The Arcades Project*[26]

Specific ways of imagining urban life fueled the Saint-Simonian obsession with canals, streets, and railroads. At the same time, the grand-scale projects were hands-on and concrete. Paper was always circulating: voluminous correspondence, newspapers printed in large editions, and the assorted documents needed for bringing together banks and civic associations to fund projects.[27] The "association universelle" announced by the *Globe* sums up the will to network by means of industrial progress. Everything and everyone ought to be connected to everything and everyone else. "Association" is the word for "density" in urban communication networks.

The Saint-Simonians viewed Paris as a place of mobility. As Jewish and Christian sons of the bourgeoisie they aspired to careers in banking, engineering, medicine, the military, and politics. Institutions of higher learning—the École Polytechnique, École des Mines, and École des Ponts et Chaussées—were home to many sympathizers, especially in the time preceding the July Revolution of 1830. As these visionaries saw things, the society of the future was not something to be built so much as it had already been realized in their bold ideas. The dominant tense of Saint-Simonian thinking was future perfect: plans have always already been brought to fruition. The urban space of the day would yield a new world city, "La ville nouvelle" extolled by Charles Duveyrier as "Le Paris des Saints-Simoniens." Duveyrier composed his rhapsody against the backdrop of the cholera epidemic of 1832, lending poetic form to the practical arrangements for water supply and sewers outlined in articles that had appeared in the *Globe*.[28]

The rhetoric Duveyrier employs is borrowed from William Harvey's treatise on the circulation of blood, *De motu cordis* (1628).[29] The poet envisions the new Paris as a colossus to be built in keeping with providential design. Each part of the body is treated in turn, as in Harvey's book. This Leviathan is another version of the communal, social body of Judeo-Christian tradition: "The separated and mixed limbs that will compose [the city] are a monstrous mass, unformed, unanimated, dead. They are like the flesh, the nerves, the bones, the brain, and the entrails of man before the jolt of my will."[30] The author's own quasi-divine hand ensures that the city will come alive and be put in order. In the Louvre, which is to be expanded to untold dimensions, all the city's library collections, images, maps, and manuscripts will be stored. Here is the "brain" of the new Paris.

Duveyrier anticipates that Prussians, Englishmen, Russians, and Saxons will come running to see for themselves how the French people has finally become mighty and industrialized. The sect's Romantic aesthetic mixes with the poetry of advertising brochures and sets the way for the world's fairs: "Come! Here the earth is filled with the desire to live the life of man; here the earth gives itself to man, as a woman to her lover."[31] The water in the city's canals is crystal clear. The streets are curved to make rings inside of rings. The finished colossus breathes forth the fire of the steam engine. Finally, in the center of the new Paris, the Saint-Simonian "temple" will be constructed in the form of a woman,[32] a center of attraction and communicative density for the city's centrifugal expansion (figure 4.3).

The imaginative valorization of women and female bodies, however, contrasted with hierarchies and practices of exclusion. Although the movement was attractive to working women and clamored for their liberation, the voices of female Saint-Simonians remained in the minority and were drowned out by those of men. Women defended themselves against the one-sided, romantic projections of their male counterparts, insisting that they wanted to emancipate themselves.[33] In recent years, the contributions made by Pauline Roland, Élisa Lemonnier, Angélique Arnaud, Claire Bazard—granted the title of first "mother" of the Saint-Simonian cult—Cécile Fournel, Aglaé Saint-Hilaire, and Claire Démar have come to be better appreciated. A vital role was also played by women of a proletarian background, including Désirée Véret, Marie-Reine Guindorf, and Suzanne Voilquin, who published the newspaper *La femme libre*, which had an exclusively female staff.[34]

Duveyrier's industrialized literary treatment of the city and its feminine temple is still rooted in eighteenth-century traditions. The proper circulation of air, water, and (waste) products are supposed to promote general welfare.[35] His vision of a communal body does contain a new element, however, inasmuch as it displays a centrifugal orientation: the city is a whole world, and the whole world has become a city.

FIGURE 4.3
Joseph Machereau, *Le Temple-Femme*. Drawing, circa 1833. Courtesy of Bibliothèque Nationale de France.

In parallel to planet-wide networking, the nineteenth-century metropolis had to be rebuilt from the ground up. Saint-Simonianism sought to expand urban space to global dimensions and, in reverse symmetry, to compress the globe to a scale where it would be the "abode of humankind."[36] In other words, the aim was to maximize the possibility of circulation, or "milieu." ("The milieu is a set of natural givens—rivers, marshes, hills—and a set of artificial givens—an agglomeration of individuals, of houses, etcetera."[37]) Hereby, the borders of the nation-state would be suspended yet preserved. Constant expansion—or "association universelle"—would push outward, and the quickest means to this end was through the acquisition of new colonies. Everything and everyone should be connected. The apparatus of security in place in the motherland would extend to dependencies elsewhere. The metropolis remained the operative model of governance because it balanced centrifugal movement into new territories with a concentration of (controlled) communicative processes.

The *grands travaux* in Paris that were soon to come—above all, Georges Eugène Haussmann's remapping of city streets and Eugène Belgrand's redesign of its underground[38]—were part of a material colonization of domestic space. At the same time, they heralded the colonization of foreign realms. Branching canals, like rationalized thoroughfares, are the optimal medium for an apparatus of security in constant expansion. The canal epitomized smooth exchanges: telegraph lines, railways, water supply and disposal, roads, and financial transactions were its corollaries. The French state wanted networks to promote capitalist-liberal policies and practices of *laissez-faire*.[39] Nineteenth-century canals were not built on the logic of getting from point A to point B. What mattered were intermediate stops geared toward economic connectivity on a global scale. At the same time, they served domestic purposes of population control inasmuch as "good" and "bad" groups could be put in contact with or cut off from each other.

The construction of canals in France can be dated to a technological and epistemological shift in the 1820s. Planners faced the challenge of creating cross-connections within a ramified structure that would ease problems of traffic and transportation. In 1827, Raymond Génieys, chief engineer for water and streets at the office for *Travaux Publics* in Paris wrote of building a network (*réseau*) of canals for the first time.[40] Leaving aside military projects, water systems were the first large-scale technological structures designed to exhibit heterarchical network functionality.[41] The principle was applied to optical telegraphy two years later in a memorandum by the Chappe brothers, who were responsible for overseeing long-distance communications. Their idea of creating cross-connections between the lines running to Paris was also pursued by their successors in the Ministry of the Interior.[42]

Topographical maps proved invaluable for developing, constructing, and operating city water systems. The Seine was long the sole water source for Paris and also its largest sewer, until, at the instigation of Napoleon, the Canal de l'Ourcq opened up another supply from the northeast in 1804.[43] Génieys recognized that he would potentially be dealing with two overlapping networks (figure 4.4). Surveying the many ramifications, he set about calculating water routes and designed a single, subterranean network.

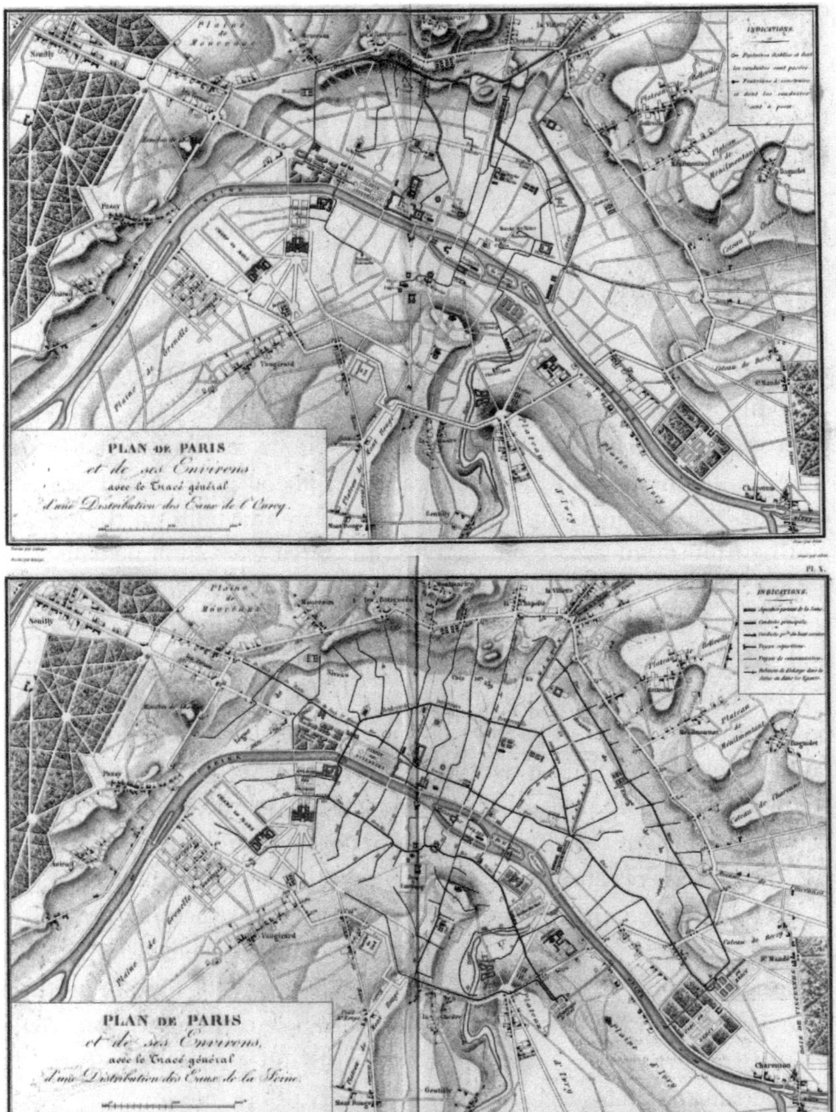

FIGURE 4.4

Raymond Génieys's plan for the Parisian water supply, *Plan de Paris et ses environs avec le tracé général d'une Distribution de Eaux de l'Ourcq / de la Seine*. Two engravings by Adam, 1829. Each around 22 cm × 32.5 cm. Courtesy of Bibliothèque Nationale de France.

Victor Hugo's *Les Misérables* features this top-down perspective in passages where the revolutionary Jean Valjean is wandering through the "intestine of Leviathan": "Imagine Paris taken off like a cover, a bird's-eye view of the subterranean network [*réseau*] of the sewers will represent upon either bank a sort of large branch engrafted upon the river."[44] When the manhole cover is lifted, as it were, another capital city is revealed—a dark double of the world above with its own streets, intersections, squares, cul-de-sacs, arteries of travel, and patterns of movement.[45] The sewers of Paris constitute a "polyp with a thousand antennæ, which grows beneath at the same time that the city grows above. Whenever the city opens a street, the sewer puts out an arm."[46]

4.3 LAND: NETWORKS FOR OCCIDENT AND ORIENT

> Les Colonies: En avant!
> —Michel Chevalier[47]

No one did more to prepare the Saint-Simonian push eastward than Charles Duveyrier's contemporary and coreligionist Michel Chevalier (1806–1879). No poet, banker, physician, or lawyer, Chevalier was an engineer with a strong interest in economics and, in many respects, a prime example of the Saint-Simonian movement. Trained as an expert in mining, he was appointed editor in chief of the *Globe* during the runup to the July Revolution of 1830. In 1832, he landed in prison alongside Prosper Enfantin, the "father" and head of the Saint-Simonian movement. Authorities had arrested both men after becoming suspicious about the large crowds flocking to the new cult's religious services in the forest of Ménilmontant, outside the gates of Paris. Later, after distancing himself from Saint-Simonianism, Chevalier enjoyed a remarkable career. He traveled to America, held a chair at the Collège de France, and negotiated free-trade agreements with Great Britain. Finally, in 1875, he founded the first Société d'études for planning a railway service under the English Channel.

In the months before the *Globe* went bankrupt and the sect was exiled to Ménilmontant, the number of large and small industrial projects proliferated. When a cholera epidemic descended on Paris, Chevalier made so bold as to call for a coup d'état by industrialists. He had already outlined his grand vision on an international scale on several pages of the journal on January 12, 1832.[48] Like Duveyrier's "ville nouvelle," his "Système de la Méditerranée" alternated between bodily metaphors and administrative precepts, setting forth a pacifistic economic project for the whole planet.

"La paix est aujourd'hui la condition de l'émancipation des peuples" (Peace is today the condition of emancipation for [the world's] people), read the epigraph. The geopolitical scope was no secret. Chevalier advocated a large-scale reordering of international transportation routes in the Eurasian realm, and he made the railroad the linchpin of his system. Widespread travel by rail and steamship would represent "not only an industrial but also a political revolution." Flanked by telegraph lines, the rails would afford access to the continents abutting the Mediterranean with the same uniformity and speed "that exists today in France." For Chevalier, these connections would be doubly productive: on the one hand, a material array of transportation routes and, on the other, a "spiritual connection" via a network of banks.

Disregarding competition between waterways and railroads, which was largely economically motivated, Chevalier counted on combining these two modes of transportation and travel. While people and light goods are well suited to railways for the sake of speed, the transfer of heavy and bulky materials is best done by water. Accordingly, Chevalier kept the large ports as points of reference and lent attention to facilities available in Spain, France, England, Italy, Germany, European Turkey, Russia, Asia, and Africa (running the last two together under a single heading).[49] Consideration of France's topography prompted the engineer to rhapsodize: no one could fail to see the expanse of *grande communication* linking Marseille, Le Havre, Lyon, and Paris through three "marvelous" valleys (the Rhône, the Loire, and the Seine).

Chevalier picked Venice, with its many canals, as the switching point for the Orient. The fragmented Iberian Peninsula (Spain and Portugal refuse to unite) and politically and religiously divided Germany represented ongoing sources of concern. According to the author's Système, the latter would play a central role connecting the Balkans and Turkey. Dresden, the "city of the French in Germany," was destined to be the hub of a "symmetrically distributed network." An Oriental-French railway would avail the "blunted" minds of Germanic scholars, who now could go and find inspiration under the skies of Turkey. In turn, "earnest" Berlin academicians and students from Göttingen could hop on a train and, in less than twenty-four hours, reach the Jardin des Plantes and Louvre in Paris to learn a thing or two about French elegance. Sleepy Russians all but buried in snow would benefit from the effects of the "vivifying network" (*réseau vivifiant*). Chevalier foresaw primary arrangements reaching completion with a railroad along the north coast of Africa. Later, *réseaux secondaires* would branch off from this line. The project would bring more than simple stopping and relay points. It promised to dominate space and divide it between as many as three tiers of major and minor routes.[50]

In a word, Chevalier's Système announced a program of universal connectivity. One might leave Le Havre in the morning, eat breakfast in Paris, enjoy dinner in Lyon, and then, the same evening, board a ship for Algiers or Alexandria in Toulon. This early example of universalist planning contained, under the camouflage of pacifist aims, the Saint-Simonian campaign to conquer the Orient.[51] Chevalier estimated the project would cost eighteen billion francs. His proposal for financing the endeavor was in line with the economistic logic of the article as a whole. For fifteen years, the budget for all European armies—one and a half billion francs annually—would be reallocated for construction. And inasmuch as all soldiers would be soldiers of industry, peace in Europe was guaranteed. Like a Jean Monnet *avant la lettre*, Chevalier had his plan culminate in the vision of a Mediterranean confederation.

The appeal to a communal body politic barely covered the emergent geopolitical strategy. Now, following a great historical battle between East and West, between matter and spirit, the Mediterranean was to become, at long last, the "nuptial couch of Orient and Occident." Chevalier's de facto Orientalism went far beyond all the land grabs his contemporaries envisioned.[52] It amounted to the cybernetics of state action under the sign of security, so to speak. The physicist André-Marie Ampère used the neologism "cybernétique" for the first time in his *Essai sur la philosophie des sciences* (1834).[53] In contrast to Chevalier, who outlined an international apparatus of security, Ampère offered reflections in the tradition of regulatory discipline. His thoughts on social economy (*économie sociale*), warfare, law, and politics are directed inward, toward shoring up and consolidating the nation-state. Chevalier's program, by contrast, aimed at continuous expansion—however peaceful it might claim to be.

The two models—internal political organization and the program for networking the world—can scarcely be separated from each other. In the nineteenth century, the question of governmentality concerned, above all, the viability of colonization—that is, practices based on political techniques for controlling the milieu. State control required strategic dominance of space that would permit transmission and reception to be channeled and strategically important hubs to be held firm. Only a network could maintain connections and encompass outlying territories.

The Saint-Simonians were fixated on physical transportation of all kinds because it pushed through and beyond national borders. The leader of the movement, Prosper Enfantin, was explicit, in his plans to settle Algeria, about building a colonial network (*réseau colonial*) that would enable a small number of colonists to subjugate many natives.[54] Two maps he included in the appendix to his *Colonisation de l'Algérie* (1843) make it clear how networked channels of communication would facilitate dominating the spaces in between cities (figure 4.5). While the upper map represents the actual topographical conditions of the country, the lower map indicates connections of strategic importance between important cities with straight lines. A detailed legend identifies civilian zones on the coast and military areas inland. The design of the network would not need to be especially dense: in order to subjugate the native population, mobility between nodal points and good (telegraphic) communication would be enough to ensure effective policing. In this way, ongoing intimidation and surveillance would be joined by the potential for quick and decisive action in the event of unrest.

Ships need ports, and trains need tracks: techniques of governing extend in a direction not yet envisioned by Ampère. Chevalier and Enfantin's term for the vital organization of currents and hubs of transportation is "association." But the Romantic notion according to which fuller connections would make possible a more equitable design for society was distorted in the course of political and economic action. In terms of power and markets, connectivity counts in proportion to the extent that local changes produce global effects.[55] In this context, the Saint-Simonians easily surpassed the genealogical ordering of knowledge that Ampère offered in his vast tableau of human knowledge. Whereas the latter divided sciences into first, second, and third order, they strove for *networks* of the first, second, and third order.

The same pattern is found in an explicitly geopolitical discourse, albeit under a different sign. At the end of the nineteenth century, the German cultural geographer Friedrich Ratzel made strategies of this kind the basis of his *Politische Geographie*, and he did so specifically with an eye to the "intensive domination of space." Hereby, traffic flows like water out of countless small- and medium-sized sources until it achieves massive dimensions.[56]

Ratzel also cast an imperial gaze that extended to all four corners of the entire globe: "The effects that traffic initiates culturally are, ultimately, the most powerful. From the highest viewpoint, routes seem to us to be the conductive tissue between large and small groups of humanity. Without them, millions of individuals and groups would not have grown to become the whole of humanity. Vigorous traffic is a symptom of culture, and it creates culture."[57] *Politische Geographie* camouflages colonial geopolitics under the cover of aesthetic enthusiasm and sublime contemplation of a vast, connected world. But in geopolitical discourse too, hegemony is a matter of who dominates the network.

Michel Chevalier, for his part, made it abundantly clear what was to be expected from the road eastward. "Les Colonies: En avant!" (The Colonies: Forward!) declares a long-winded hymn he wrote in April 1833, while incarcerated at Sainte-Pélagie Prison.[58] Another poem, titled "Géographie-Physiologie," ends with a straightforward proposition: "There are two isthmuses / the isthmus of the Orient, / Suez, / the isthmus of the Occident, / Panama. / They will be two holy routes."[59]

4.4 FLOW: EGYPT, SUEZ, AND THE NEUTRALITY OF THE CANAL

> Only from the geopolitical position of the canal can the thread be followed.
> —Gerhard Herrmann, *Der Suezkanal*[60]

Prosper Enfantin wrote on August 8, 1833, the Saint-Simonian "Year of the Mother," to his faithful apostle Émile Barrault:

> SUEZ
> Est le centre de notre vie de TRAVAIL;
> Là nous ferons *l'ACTE*
> Que le monde attend,
> Pour confesser que nous sommes
> MÂLES.[61]
> (SUEZ
> Is the center of our life of WORK;
> There we will perform the *ACT*
> That the world awaits,
> To prove that we are
> MEN.)

FIGURE 4.5

Prosper Enfantin, *Colonisation de l'Algérie*. Engraving by L. Bouffard, lithograph by Joseph Lemercier, 1843. 49.5 cm × 66.5 cm. Courtesy of Bibliothèque Nationale de France.

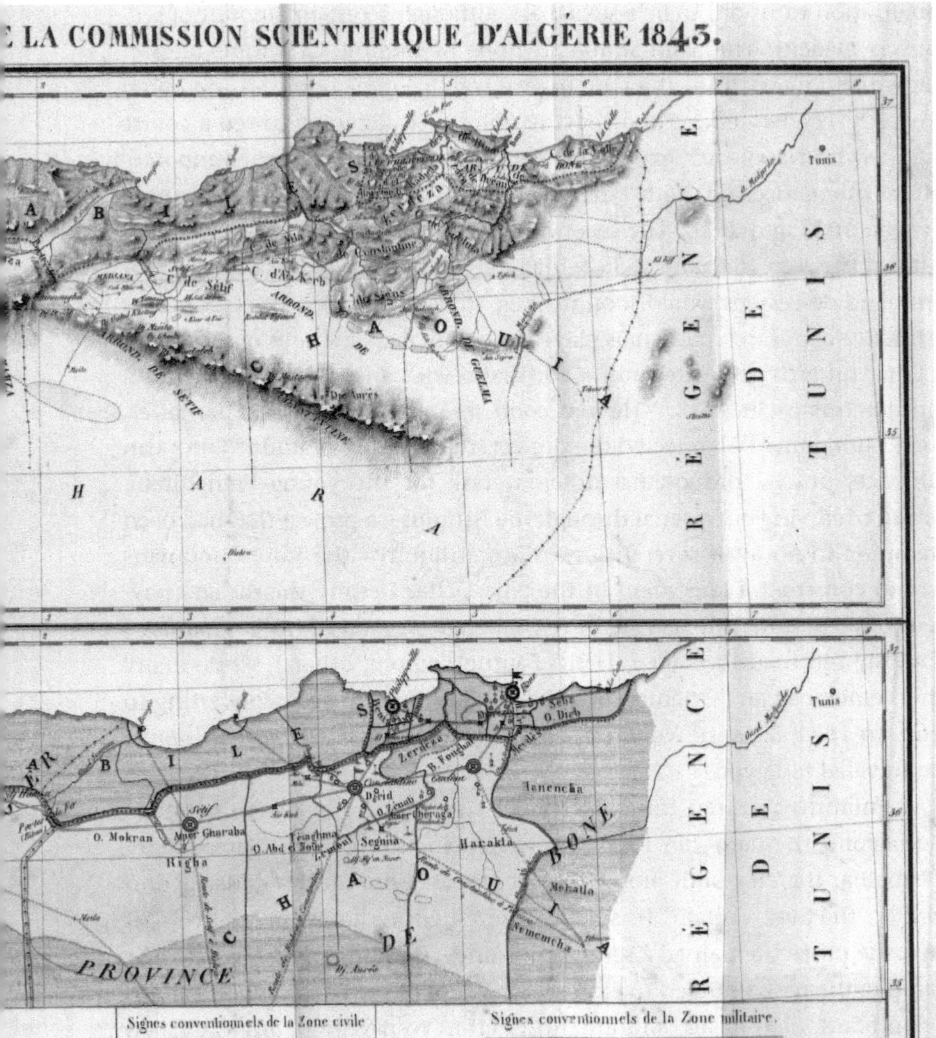

LA COMMISSION SCIENTIFIQUE D'ALGÉRIE 1843.
Tunis
RÉGENCE DE TUNIS
C. de Constantine
C. de Sétif
C. de Mila
ARROND. DE SETIF
ARROND. DE GUELMA
CHAOUIA
HARA
Tunis
Sétif
Constantine
O. Mokran
Righa
Amer Gharaba
O. Abd el Nour
Telaghma
Segnia
Zerdeza
B. Foughal
Hanencha
Marakla
Mehatla
Nememcha
O. Dieb
Sebt
BONE
PROVINCE DE
Signes conventionnels de la Zone civile
Chef-lieu d'Arrondissement
Fort
Colonie civile
Route civile
Limite de la Zone civile
Signes conventionnels de la Zone militaire.
Chef-lieu d'Arrondissement
Fort
Colonie militaire
Poste militaire
Makhzen d'Arabes
Makhzen de Kabiles
Route militaire
Route réservée
Limite d'Arrondissement
Limite de Cercle

Canalization turns out to be a sexual act, although a certain amount of self-irony is evident. The Saint-Simonian hope of finding "la femme-messie" (the woman-messiah) in the Orient gave way to other priorities soon after arriving.[62] The two efforts at achieving union with Egypt followed a course indicated by Napoleon's expedition decades earlier and the contemporary culture of engineering. Both attempts failed. The first expedition occurred between 1833 and 1836. The second initiative involved a Société d'études that met between 1846 and 1855, which did not manage to accomplish what Ferdinand de Lesseps would soon manage to do. The lack of success proves instructive about the role canals played in colonial apparatuses of security.

Enfantin arrived in Alexandria and Cairo with an entourage of faithful Saint-Simonians in 1833.[63] He was soon disabused of his Suez fantasies. Pasha Muhammad Ali rejected the project, rightly wary of undue European influence, but he proposed a different task for the young Frenchman. Instead of carving out a canal through the isthmus—a project that had been attempted to no avail several times since antiquity—the Saint-Simonians were to construct a large dam in the Nile Delta. Despite the meager payment offered, Enfantin accepted the challenge. Several of his companions promptly returned to France (Henri Fournel, among others), while many others embraced an "Oriental lifestyle"—a few of them even converting to Islam. In 1834, Suzanne Voilquin, editor in chief of *La tribune des femmes*, also traveled to Egypt.

As Enfantin pictured things, a model city would rise in the countryside beyond the dam—an ideal vision corresponding to the later reality of Ismailia, the city built along very European lines during construction work on the Suez Canal.[64] In spite of French efforts to assimilate to local ways, the project remained a series of misunderstandings and disasters. For example, the pasha offered to employ forced laborers. This was contrary to all Saint-Simonian ideals, and Enfantin's effort to pay them anyway failed miserably. The project halted altogether after a year of unremitting toil, when a plague epidemic decimated the local workforce and foreign adventurers in 1836. Enfantin returned to France, but he left behind the makings of a French elite. Saint-Simonians such as Charles Joseph Lambert, Louis Maurice Adolphe Linant de Bellefonds, Michel Bruneau, and Nicolas Perron would play important roles in the Egyptian educational and administrative systems. Ongoing correspondence with them enabled Enfantin to establish an international study group for building the Suez Canal.

Founded in Paris in November 1846, the Société d'études du Canal de Suez brought together British, French, German, and Austrian members interested in establishing a "kind of Bosporus in the desert of Suez."[65] Engineers outweighed the bankers and political economists. They included Robert Stephenson, a British expert on railroad construction, the brilliant Austrian engineer Alois Negrelli, and the brothers Paulin and Léon Talabot, who represented the French polytechnic tradition. From the outset, competition was on full display and not just between England and France. Articles placed in the major European newspapers kept the public abreast of developments and disagreements.

The course to be followed proved more than controversial. Linant de Bellefonds, who had accompanied Enfantin on the expedition and had remained in Egypt, proposed three alternatives based on his own observations and research: a railway across the isthmus, a canal connecting Alexandria and Suez, or, finally, a direct canal between the Mediterranean and the Red Sea. Linant himself preferred the third option and pointed out that it was shorter, took advantage of existing lakes, and would prove more economical in general (figure 4.6).[66]

The English liked the idea of a railroad connection, which would be unproblematic in geopolitical terms. But Enfantin and his banker friend François Barthélemy Arlès-Dufour wanted to combine a railway with a waterway from the outset. Paulin Talabot provided the design. In the spirit of the second of Linant's scenarios, it would connect Alexandria, Cairo, *and* Suez, including a section of the Nile to be dammed up to accommodate deep-sea shipping (figure 4.7).

The Saint-Simonians were counting on local connections redounding to the benefit of East and West alike. They envisioned a canal with three "stops" that would work like a railroad. The Egyptian rulers—Muhammad Ali until 1848, followed by Ibrahim Pasha and Abbas Pasha, then Muhammad Sa'id Pasha from 1854 onward—met the project with skepticism. It would be almost impossible to fund. This circumstance worked to the advantage of Ferdinand de Lesseps, who was the outspoken advocate of the plan proposed by Linant and Mougel. As late as 1855, Enfantin was still exchanging heated letters with Lesseps and seeking to persuade Napoleon III to have his Societé d'études du Canal de Suez recognized and integrated into the Suez Canal Company. Finally, when Negrelli took de Lesseps's side, Enfantin gave up.

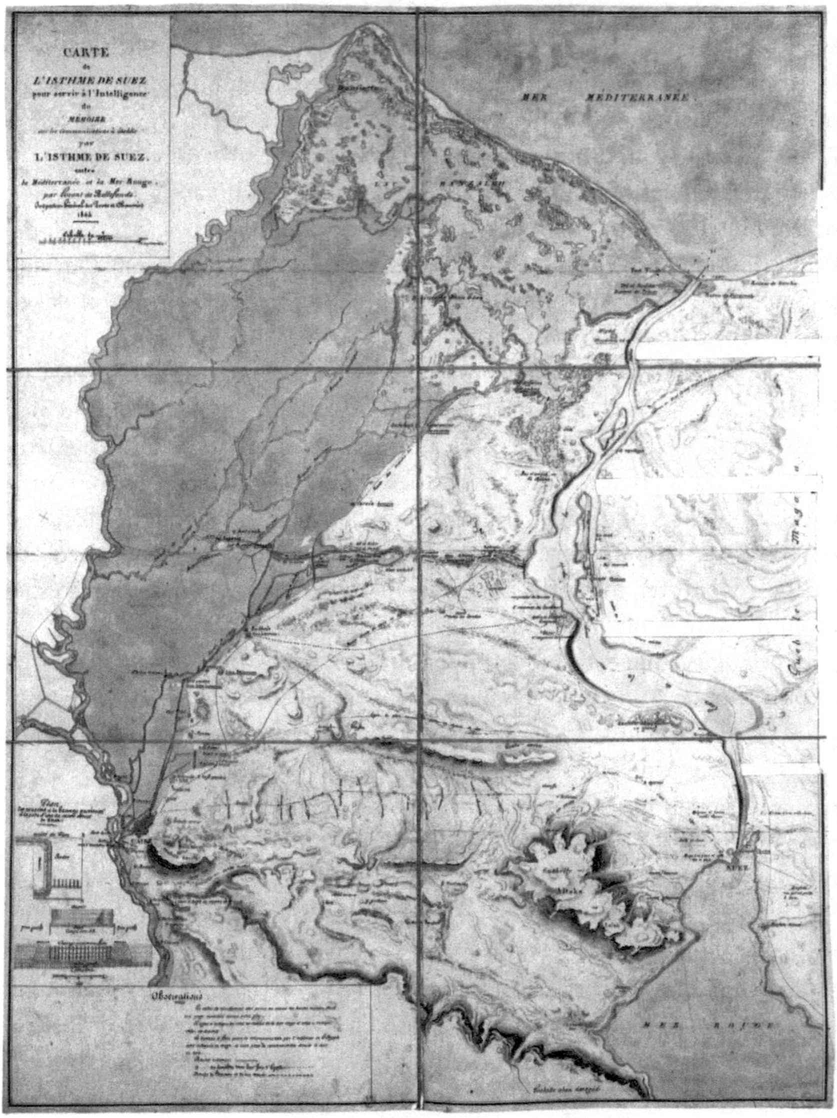

FIGURE 4.6
Linant de Bellefonds's design for the course of the Suez Canal, *Carte de l'Isthme de Suez pour servir à l'Intelligence du Memoire sur les communications à établir par l'Isthme de Suez entre la Méditerranée et la Mer Rouge*. Hand-drawn map, 1844. 92 cm × 67 cm. Courtesy of Bibliothèque Nationale de France.

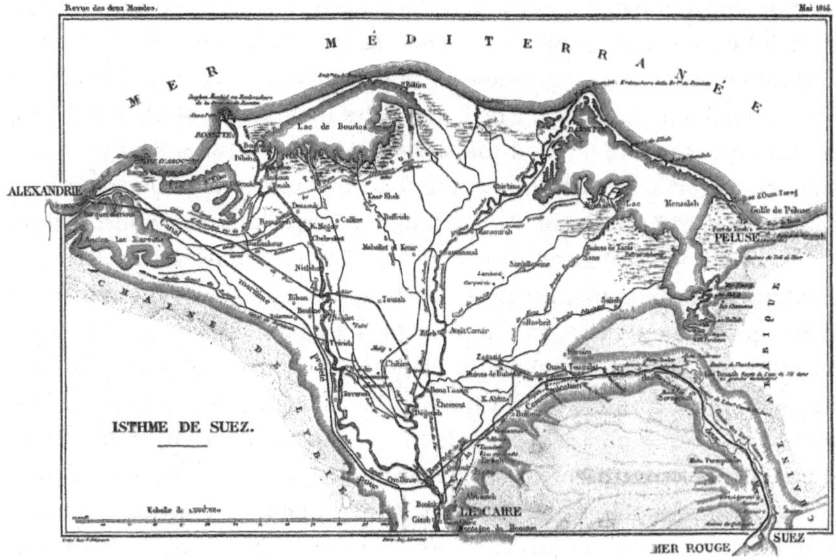

LE CANAL DE SUEZ, PROJET DE PAULIN TALABOT.

FIGURE 4.7
Paulin Talabot's design for the course of the Suez Canal, *Le canal de Suez, projet de Paulin Talabot*. Engraving by F. Delamare in *La revue des deux mondes*, May 1855.

One wonders how Egypt would have developed with a networked canal of local connections. From a geopolitical perspective, the Suez Canal ended up causing a bottleneck in international traffic, even though the idea had simply been to ensure passage from one point to another. The colonial transportation regime discouraged all local access and left Egypt as a protocolony. In contrast to the wishes of the Saint-Simonians, some of whom went native, France and England long kept Egypt from profiting in any way. Although commercial traffic nominally flowed through the land, the ninety-nine-year term of the agreement meant that it really passed Egypt by. There is no better proof to this than efforts to make the canal neutral under international law, which was supposed to ensure free shipping.

In Constantinople on October 29, 1888, the great powers of Europe first agreed on the neutrality of the canal zone in the event of war. There were to be no blockades, and warships on any side would be able to pass freely.[67] These provisions did not keep England from exploiting the control it had acquired through a stock coup in 1875 and consolidated with an invasion in

1882. In effect, nineteenth-century international treaties to ensure the neutrality of canals, rivers, seas, and straits turned entire stretches of physical land into media landscapes where expanding national security apparatuses and colonial ambitions flourished. Examples include the Black Sea (1856), the Danube (1865/1878), and the Strait of Magellan (1881).[68] At the same time, such profanations, insofar as they occurred under the rubric of universally networked "association," are ultimately Saint-Simonian in inspiration.

In "Philosophy of Landscape" (1913), Georg Simmel muses that the "subliminal formula that generates landscape as such cannot be evidenced in an equally simple way, and in principle may not be so at all."[69] To solve the dilemma, Simmel examines the "landscape as an art-form in painting." The subliminal formula would have to be different for the logic of networked space accompanying canals in the nineteenth century and later. The fact that water naturally resists being given a definitive form[70] gave rise to an international security apparatus: canalization meant colonization. Egypt was made a transfer station, existing only to speed along transport. In turn, the Saint-Simonian plan to settle Algeria involved a "colonial network" of cities for subjugating the territory. In political terms, colonial connections excluded other ones with a different social configuration and opposed to foreign rule.

Canals and straits became a media landscape at the intersection between colonialism, international traffic, and juridical neutrality. The legal encoding of standing bodies of water such as the Suez Canal as neutral already pointed ahead to the twentieth century, when canals were defined as a privileged site of symbolic operations. The rise of information theory from the 1920s onward inaugurated a new science based on neutralizing the *channel* for the purpose of transmitting signs. Hereby, disturbances and ensuring a frictionless stream of messages stood in productive conflict. "The channel. This is merely the medium used to transmit the signal from the transmitting to the receiving point. It may be a pair of wires, a coaxial cable, a band of radio frequencies, a beam of light, etc."[71] Water, ships, and colonial ambitions did not concern engineers such as Claude Elwood Shannon. Given the scope of the World Wars, the operative model was more commonly communication in code. Although international law and information theory both wanted to reduce the materiality of connections to a "mere channel," or noiseless conduit, different ways of dealing with interference were employed.

The material history of the Suez Canal, like that of any large technical system, is characterized by everyday interferences. Within a colonial imaginary, however, canals of transportation should harbor no noise. Interruptions, whether a popular uprising or a natural disaster, simply are not part of the plan, no matter how likely they may be. For Shannon and his peers, in contrast, operations for transmitting a message were oriented from the outset on the amount of noise.[72] The model presupposes an opponent of equal strength working at cross-purposes who must be taken seriously. To be sure, this leaves the question open of how signals pass from a material *canal* to a communicative *channel*. With a nod to Erhard Schüttpelz, we can say that Shannon's famous diagrams on the flow of messages are telegrams.[73] Transmission takes place thanks to, among other things, the alliance between shipping, telegraphy, and railroad that made places like the isthmus of Suez a medial milieu and a milieu for media. Or, in the words of a prophetic poem by Michel Chevalier:

> Au sommet des minarets
> le télégraphe agitait ses bras
> et de toutes parts apportait
> de bonnes nouvelles au peuple.[74]
> (Atop the minarets
> the telegraph waved its arms
> and brought good news
> to the people from all over.)

4.5 CHANNELS TO NETWORKS: GEOPOLITICS AND LIBERALISM

> Que les États contractants feront tout de leur possible pour constituer le réseau international.
>
> —Franz Alfons Desiderius de Chauvin, director of Prussian telegraphs[75]

Nineteenth-century fantasies of boundless associations and the material communication networks that resulted from them produced at least two conflicting bodies politic. In the first place, the actions implemented by the European nation-states aimed to generate ideal connections within their own domestic spheres, conceived along colonial lines. Territorial networking became a geopolitical imperative for the sovereign state. Continuous geographical expansion to other continents entailed border lines that were always changing. In turn, "imagined communities"[76] of a nationally and colonially networked industrial-political body met with the economic, logistical, and technological demands of international exchange. They formed a second body politic also traversed by networks. In a word, the space in-between posed a concrete problem, since railroads and telegraphs did not end where national borders did. Adaptation to the protocols of other national networks would prove costly and labor-intensive: changing railroad equipment to match different track widths or handing postal and telegraphic transmissions over to border officers.

Charles Joseph Minard's (1781–1870) statistical map of European railroad traffic promised almost frictionless movement across borders. Minard was an expert on canals, a graduate of the École Polytechnique, and former director and inspector of the École des Ponts et Chaussées. After retiring, he dedicated himself to thematic cartography. His tenure at the École des Ponts et Chaussées, from 1830 to 1836, had coincided with the crucial early years of Saint-Simonianism, which was then enjoying a heyday at the engineering schools of Paris. Minard's map, lithographed in 1865, shows the European continent to be full of arterial routes pulsing with life (figure 4.8).

The lithograph depicts the "vivifying network" that Michel Chevalier imagined. Because it is a picture, the connections are perfectly clear and vibrant. The map comprises the colonial apparatuses that secure territory and the national milieus where commerce and circulation are to thrive. However, a third element is not represented. Beyond national borders, regimes of connections between states were being set up, which led to the founding of international organizations. For goods, people, and information to flow, connectivity needed universal improvement. The goal was to create a new, largely political milieu in which neutrality would prevail, conflicts of interest notwithstanding. While the ambition was not explicitly pacifistic, measures did serve to secure intra-European peace for a while.

In 1865, diplomats met in Paris at the behest of Napoleon III for the first international conference on telegraphy.[77] Private British telegraph companies had long since established standards for electric telegraphy, which stretched well beyond the formal borders of the Victorian empire. But on the continent, only piecemeal agreements between states existed.[78] The dignitaries assembled in Paris had to contend with all the problems of standardization attending telegraphy across national borders. The conventions and social codes that underpinned in-person diplomatic negotiations set the tone and determined the form of international network protocol. The significance of the event is often overshadowed, in historical accounts, by the first transatlantic cable to connect England and the United States permanently in 1866.[79] Moreover, emphasis is usually placed on networking efforts by private companies instead of measures undertaken by governments.[80] The transatlantic cable was an object of controversy, especially in the United States, because of reservations about British ambitions. In contrast, European states did not view each other with as much suspicion. Debates did not concern *whether* to be connected so much as *how* to regulate things.

The first three articles of the document (of sixty-three in all) are devoted to the *réseau international*. How long do telegraph offices have to be open? In important cities, always; in summer, from 7 a.m. to 9 p.m.; in winter, from 8 a.m. to 9 p.m. How are messages to be sent? Initially, in Morse code.[81] Who is allowed to send them? Everyone, but governmental communications between states have priority.[82] What is the maximum word length? Seven syllables.[83] The cost of sending a message would be determined in keeping with a table of geographical distances. The standard currency was the French franc.[84] At the same time, signatories retained the sovereign right to withhold messages, interrupt service, and set local tariffs.[85] The regulations, which were based on postal regulations, even detailed how delivery was to be made.

On May 17, 1865, the agreement was signed by representatives from Austria, Baden, Bavaria, Belgium, Denmark, France, Greece, Hamburg, Italy, the Netherlands, Portugal, Prussia, Russia, Saxony, Spain, Sweden/Norway, Switzerland, the Ottoman Empire, and Württemberg. The treaty was set to begin, pending ratification, on January 1, 1866. Article 57 obliged participating lands to report all interruptions of operation and work on national exchanges. It also called for all countries to share, at the beginning of the year, message statistics and a map of areas served. This provision was renewed at the second conference in Vienna in 1868. The founding of the Union internationale du télégraphe in Bern on January 1, 1869, centralized the internal flow of information and established an administrative structure. The organization now had its own statistics, a comprehensive map, a conference schedule, and an official publication, the *Journal*

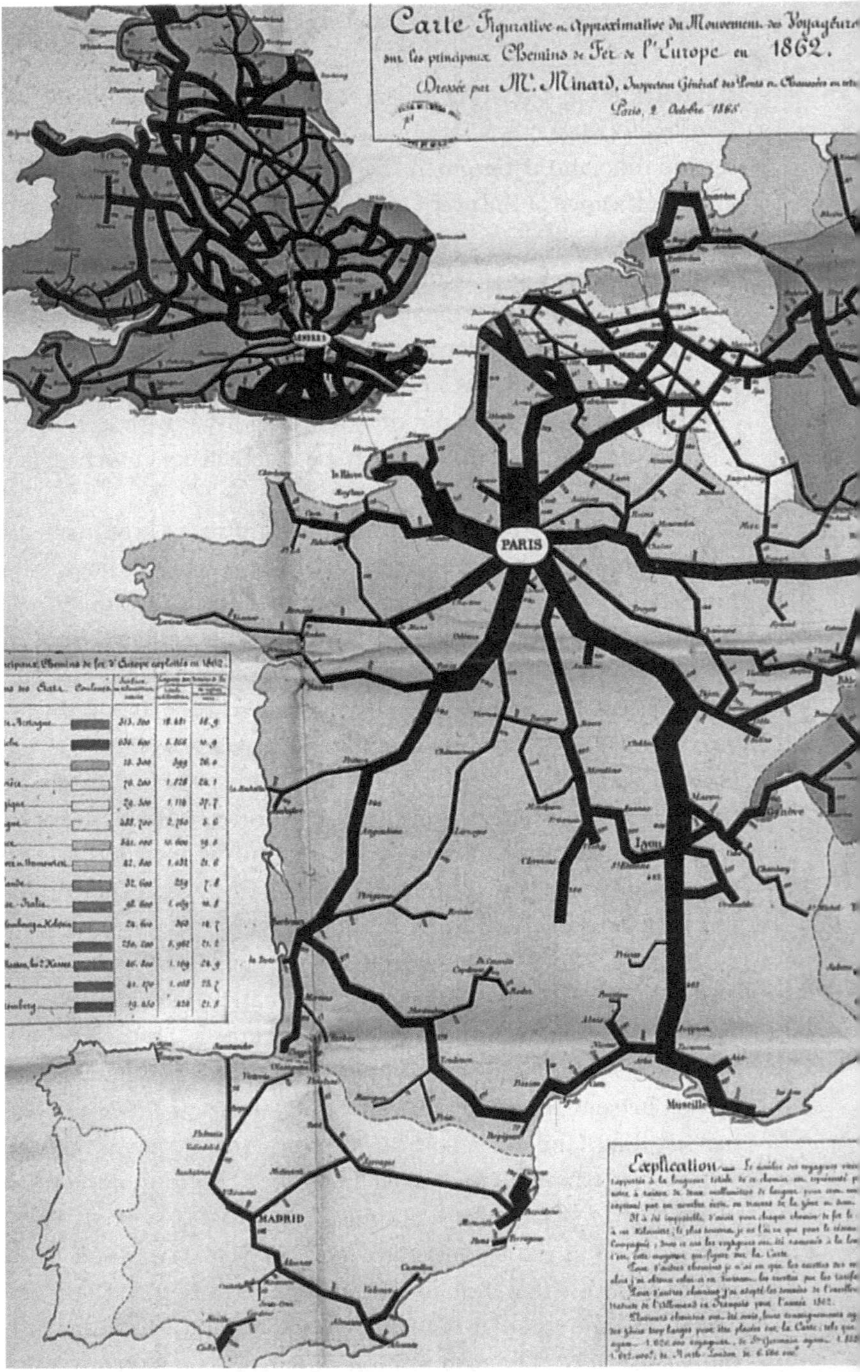

FIGURE 4.8

Charles Joseph Minard, Statistical map of European railroad traffic of 1862. Lithograph, 1865. 97.5 cm × 76 cm.

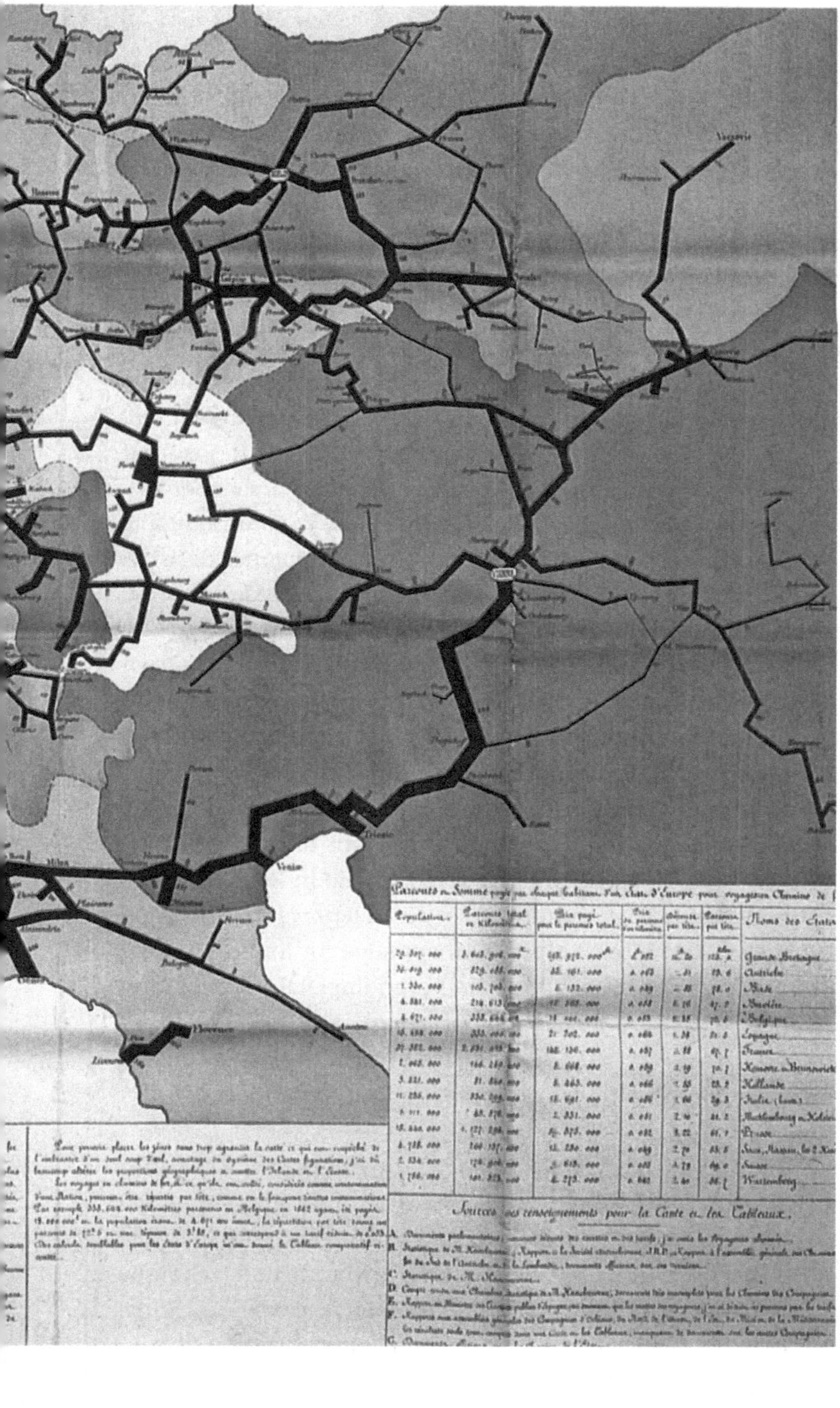

Varsovie
Venise
Trieste
Milan
Florence
Livourne
Bologne
Ancone
Plaisance
Alexandrie
Population.
Parcours total en Kilomètres.
Prix payé pour le parcours total.
Nom des États
Grande-Bretagne
Autriche
Bade
Bavière
Belgique
Espagne
France
Hanovre et Brunswick
Hollande
Italie (Lomb.)
Prusse
Saxe
Wurtemberg
Sources des renseignements pour la Carte et les Tableaux.

télégraphique.[86] Such elaborate measures were not unique. Similar international organizations followed, notably the Universal Postal Union in 1874. Weathering large-scale conflicts such as the Franco-Prussian War of 1870–1871, the Union internationale du télégraphe gradually evolved into the International Telecommunication Union (ITU) of today. Since 1948, its headquarters have been located in Geneva, opposite the Palace of Nations building of the United Nations.

Despite its unprepossessing appearance, already in the nineteenth century, this institution offered the site where the protocols of international politics converged. The Paris and Vienna conferences represent crucial moments in the history of networking. Here, diplomats hashed out how operations in different countries would be connected across borders. Regulating telegraphy, then, meant situating the activity in a transnational space. At the same time, at the level of tariffs, geography remained decisive. Unsurprisingly, fees in Russia were higher than elsewhere.[87] Messages traversing multiple territories yielded an interconnective network.

With the ITU, telegraphy became, in a sense, a political fact: an apparatus of spatial security negotiated by many governments requiring an office of its own for proper administration. It is no coincidence that headquarters were set up in Bern. Switzerland is neutral under international law.[88] There were, however, limits to making telegraphy politically neutral, and the liberal tendencies of centrifugally expanding connections did not take long to assert themselves. Thus, at the third conference, held in Rome in 1871–1872, European delegates gave a chilly reception to letters from telegraph pioneers Cyrus Field and Samuel Morse that appealed for lines to be kept neutral.[89] Especially in view of colonial—and even imperial—competition for better connectivity, the Americans' endorsement of *laisser faire—laisser passer* seems paradoxical. The trick was to strike a balance between consolidating domestic interests while remaining open to international connection and exchange.

So, what can we take away from the story of France and the Saint-Simonians?[90] Technical networks on the model of the canal embedded themselves in the spatial and temporal milieu and reorganized it from the ground up. On the one hand, they represented a governmental technique for controlling cities, nation-states, and colonial empires. At the same time, they were supposed to ensure security for the domestic population. If politics is the art of producing a body politic, then the liberal regime around

1850 was pursuing a specific goal: making "good" circulation and communication possible as the basis for the life of a growing collective body.[91] Saint-Simonian men exalted women, in the abstract, to an integral part of this collective body, while their "sisters" in the movement pushed for more concrete, legal equality.

At the same time, new bodies politic were emerging across borders, with geographical contours that were far less defined. Their anatomy was drawn by telegraph lines, shipping schedules, and technical requirements,[92] and their operations followed the demands of international political and economic exchange. This development led to paradoxical constellations. Regulating important routes of transportation and communication granted colonial powers legally neutral access to them. A political milieu ideal for imperial exploitation emerged. Even organizations such as the ITU contributed to this logic of colonialist networking. The latter's motto might have been, "Canalization means colonization"—or, in other words, "Create spaces of circulation that conjoin connectivity and self-interest." The formation of networks within the regulation of international transportation came at a price. The difference was paid by countries such as Algeria and Egypt, which were included geographically but remained politically and economically excluded.

The nineteenth century brought forth the colonial world order, including its political economies aimed at continual expansion but also networks that were not geared toward achieving or ensuring hegemony. Telephone networks were created overwhelmingly in times of peace, and they were intended for civilian use. What's more, the material culture of telephone communication was local at first. The next chapter turns away from globally networked milieus and focuses instead on developments on a smaller scale.

5 TELEPHONES, EXCHANGES, AND VOICES AROUND 1890

The material objects used to make a telephone call have become as foreign to us as the technology used to send an electric telegraph. Both types of technology can now be consigned to the same dusty corner of museums. But whereas it's still possible to turn the clicking of the telegraph into an engaging exhibit, the telephone switchboards and old dials remain silent and inert. The entire array of actions, places, and institutions behind those switchboards has vanished. Moreover, it is almost impossible to convey in the small amount of space given what can be called the knowledge of telephone exchange—that is, the real-world, embodied knowledge of all those who knew how to use a switchboard. This implicit knowledge has been transformed by the mechanical automation of connections.[1] Both the manual and mechanical operational chains of telephone exchange stand for specific styles of action that make speaking on a network possible in the first place.

At the same time, the embodied knowledge of telephone exchanges means the cross-references and intersections in the history of knowledge go hand in hand with networking. As an "epistemic thing," the telephone exchange has left traces in physiology, reflex theory, and association psychology.[2] Its popularizations, in turn, influence what can be known about the technology itself. And the telephone exchange, like telephoning in general, rehearses an eerie legacy of cultural history: speaking with the intangible, acoustically present ghosts of the dead.

The telephone conversation exemplifies the fact that connections have to be established and eventually ended in order to be effective. That said, in the early days, it was not clear whether a permanent or a temporary connection to synchronous communication was preferable. In the United States, for example, there were party lines that, instead of fostering intimate dialogue, let everyone join in the conversation.[3]

Franz Kafka's novel *The Castle*—written in 1922 and published posthumously in 1926—presents the scenario of multiple permanent connections. The village council chairman tells the surveyor, K.:

> At the Castle, the telephone seems to work extremely well; I've been told the telephones up there are in constant use, which of course greatly speeds up the work. Here on our local telephones we hear that constant telephoning as a murmuring and singing, you must have heard it too. Well, this murmuring and singing is the only true and reliable thing that the local telephone conveys to us, everything else is deceptive. There is no separate telephone connection to the Castle and no switchboard to forward our calls; when anyone here calls the Castle, all the telephones in the lowest-level departments ring, or all would ring if the ringing mechanism on nearly all of them were not, and I know this for certain, disconnected.[4]

Kafka's well-known scene does not just make interference and murmuring the foundation of all telephonic communication. It also offers a point of contrast to the principle of connecting telephones temporarily and as needed. The lack of connection between K and the Castle bureaucracy plays a vital role in the plot of the novel. The telephone is assigned the part of an uncanny and disturbing agent. "All those contacts are merely apparent," but they serve as an instrument for exercising power through disorientation.[5] Accordingly, the village chairman is quick to warn his interlocutor, "Of course it's best to run from the telephone before hearing a sound."[6]

Kafka's precise wording sets apart at least two important aspects of mediating communication. On the one hand, the telephone is introduced as an apparatus for transmitting language, very much in the sense of the "speaking telegraph" the device was said to be in the early days.[7] Inquiries are made to the castle via telephone.[8] On the other hand, Kafka creates a purgatory of permanent connection in which noise prevails because nothing ever interrupts it. The ongoing chorus of bureaucratic processes is only possible if there is no switchboard or operator regulating the interconnection of voices. Kafka's telephony is really a radiotelephony—not far removed, in fact, from the medium as it was in the first years after its invention.

Whereas telephones can be permanently connected, human beings require spaces and pauses. Urban areas played first host to arrangements in which signals and wires operated one-to-one: one voice, one line, a direct connection. Switchboard equipment and operations were tied to discrete addresses—as opposed to the permanent connection of party lines—with additional cables, besides the one for the voice, controlling who was speaking to whom. The busy signal became the strongest expression of this form of exchange.

The following account concentrates on North America, with a few asides on European developments. It focuses not on the parallel invention of the telephone apparatus itself but rather on its role for multiple inventions of networked communication technologies. Two transitions are particularly important in the American context: first, a heterarchy of small local networks formed; second, their ultimate merger under the monopoly of the national Bell System. One aspect of this development was a change in the style of networking brought by automated connections. Technical issues did not come first, however. What mattered was the value of the telephone as sound and sight in music halls, opera houses, and newspapers.

5.1 SPECTACULAR CALLS: THE TELEPHONE IN THE MUSIC HALL

> Telephony is a game of situating sites of discourse that can be from anywhere because the connection always exists. It's not that places are connected; connections create places.
>
> —Wolfgang Hagen[9]

On May 18, 1877, Thomas Augustus Watson, sweltering in the heat of the early summer, typed a brief letter to his wife, Elizabeth, from New York. He reported the difficulties of introducing the telephone to the city's inhabitants: "Alec's lecture was a failure in numbers. . . . Tonight, we shall have more because we have given away nearly 300 tickets."[10] The lack of popularity lamented by Alexander Graham Bell's assistant was unusual. In the music halls and opera houses of New England, the inventors enjoyed great success, but in New York, they had to compete with Broadway. Compared to the occasions with five hundred to two thousand attendees, an audience of three hundred people in New York's Chickering Music Hall was disappointing.

Watson's efforts to make a splash with the public were not without results. The New York newspapers reported from May 12 to 19 on the "speaking telephone" of the Scotsman Alexander Graham Bell.[11] But the response from magazines and journals between Boston, Salem, Providence, and New Haven was far more enthusiastic. The telephone was recognized not just as a device for conversation but also as a remarkable object full of potential and undisclosed horizons. It was "marvelous," "most wonderful," "one of the wonders of this age of marvels."[12] Additionally, it lent itself to spectacular demonstrations. In France and the United States especially, the early telephone appeared as a mass medium on the stages of opera houses, theaters, and music halls.[13] In both North America and Europe, it began as a "Theatrophone" that broadcast music and song from architectural auditoriums and showplaces.

Before he became a millionaire, Bell depended on the lectures he gave on both sides of the Atlantic for economic survival. His motivation to invent the telephone was not just scientific but also his love for Mabel Hubbard, the hearing-impaired daughter of his financer and lawyer, Gardiner Greene Hubbard.[14] Bell's courting of Mabel also led him to market his invention theatrically. As he noted in a letter to his beloved, five hundred attendees in the audience made him two hundred and fifty dollars a night, in contrast to the five dollars an hour he was otherwise paid for working on his invention.[15] That said, Bell did not come up with the idea of a telephone exchange: it owed much to the business models of telegraphy, hobbyists he inspired at his shows, and officials such as the German postmaster general, Ernst Heinrich Wilhelm Stephan, who declared the establishment of telephone connections to be an affair of state administration.[16]

Against this backdrop, Bell's visions for how his invention could be used in the future became part of a larger discourse, which he formulated on March 25, 1878, in Kensington, England, as follows:

> At the present time we have a perfect network of gas pipes and water pipes throughout our large cities. We have main pipes laid under the streets communicating by side pipes with the various dwellings, enabling the members to draw their supplies of gas and water from a common source.
>
> In a similar manner it is conceivable that cables of telephone wires could be laid underground, or suspended overhead, communicating by branch wires with private dwellings, country houses, shops, manufactories, etc., etc., uniting them through the main cable with a central office where the wire could be connected as desired, establishing direct communication between any two places in the city.[17]

Remarkably, Bell's proposal was not based on the telegraph network but rather imagined telephone connections on the model of urban infrastructure such as water supply. The spaces of early telephony remained local. It was a matter of conversations within buildings, between farms, and between different parts of town—not because the range of the signal was limited but because its use was fitted to existing social and economic practices.[18] In the 1880s, house telephones were probably spreading faster than networked telephones: it was not the equipment that was expensive but the network connection, which could be justified primarily for business use.[19] In luxury hotels, which quickly developed an interest in the telephone, internal networking and economic utility coincided.

Telephone networks in the American Northeast were less integrated in the metropolises than in the smaller towns between New York and Boston.[20] In part, this occurred because of electrical technology. The need to supply electricity for street cars and lighting led to many interruptions in telephone transmission.[21] An unexpected economic issue compounded the problem. Until the invention of a central battery supply, costs in large cities rose faster than revenue, as adding connections proved more expensive.[22] Above all, however, Bell's appearances in places such as Salem, Providence, and New Haven promoted geographical distribution. The cities in which overland connections led to opera houses and music halls evolved into the first hubs of the telephone network.

This overlapping of local connections and urban communication is nowhere more evident than in Bell's appearance at New Haven's Skiff Opera House on April 27, 1877. By his account, basic telephone services were tantamount to a bodily need: "In the future telegraphy and telephony will be in use generally in dwellings just as gas and water are supplied, not as luxuries but as necessities. Wires will be connected to a central office, when communication could be established with the butcher and the baker."[23]

George W. Coy, a Civil War veteran and manager of the Atlantic and Pacific Telegraph Company, assisted Bell that evening with setting up the equipment. The fact that Coy—who would later be the founder of a telephone company and engineer of the first switchboards—played the part instead of Watson is telling. In contrast to Bell's intense psychography, which has resulted in a kind of subgenre of telephone histories in cultural studies, scholars have barely lent attention to Coy's motives. He recalled his evening as follows: "I had quite a talk with him [Bell] about the possibilities of the telephone but could not get from him any information of how the telephone would come into commercial use."[24] The uncertainty over how to make the telephone profitable in competition with Edison and the telegraph companies (a problem Bell certainly struggled with) represented the other side of his spectacular calls. Coy's unremitting efforts to obtain a Bell license for New Haven, which he received in September 1877, were aimed at competing with the seemingly invincible Western Union Telegraph Company. The installation of individual direct connections, after some tinkering and experimentation, was followed by the inauguration of one of the first commercial telephone exchanges.[25]

The founding of the New Haven District Telephone Company is shrouded in legend but reveals something surprising nonetheless.[26] The first advertising campaign for subscriptions for a telephone line to the local network attracted exactly one customer willing to pay.[27] A local pastor, Reverend John E. Todd, sought a telephone connection to his parish. After another advertising round brought in twenty-one subscribers, his name was listed at the top of the first telephone book in the world. But this first telephone book was not a book so much as a "list of subscribers," with fifty entries for the residents of New Haven. The materiality of this object tells a different story than Bell's spectacular demonstrations of the "wonderful speaking telephone." It is a list dated February 21, 1878, on a kind of a plaque: the yellowish light-brown cardboard is three millimeters thick (figure 5.1). Several layers of paper have been combined, and the list itself has been glued on top on a separate sheet of paper. In its form, the list has a relatively robust, enduring look. Its nonalphabetical arrangement of private parties—including company owners George W. Coy, Herrick P. Frost, and Walter Lewis—does not suggest that anyone was expecting to update it. It has no telephone numbers.[28]

The following years' lists, as well as the revisions that became necessary in March and April 1878, take a different tack. The expansion of the network led to an instruction on the title page. "Destroy all previous lists," reads the document published in March 1881, now a veritable telephone book for New Haven, Hartford, Meriden, Bridgeport, New Britain, and Birmingham, with a paper index (figure 5.2). On each page, users are advised about proper telephone etiquette. George W. Coy's firm, now the Connecticut Telephone Company, presents the device that led to its founding in the *ex libris*. The frontispiece is Coy's self-built switchboard from 1878, stylized as an icon of the telephone company's identity.[29]

Switchboards and telephones functioned as complementary instruments and communication technologies. Just as much as the public spectacle

LIST OF SUBSCRIBERS.

New Haven District Telephone Company.

OFFICE 219 CHAPEL STREET.

February 21, 1878.

Residences.

Rev. JOHN E. TODD.
J. B. CARRINGTON.
H. B. BIGELOW.
C. W. SCRANTON.
GEORGE W. COY.
G. L. FERRIS.
H. P. FROST.
M. F. TYLER.
I. H. BROMLEY.
GEO. E. THOMPSON.
WALTER LEWIS.

Physicians.

Dr. E. L. R. THOMPSON.
Dr. A. E. WINCHELL.
Dr. C. S. THOMSON, Fair Haven.

Dentists.

Dr. E. S. GAYLORD.
Dr. R. F. BURWELL.

Miscellaneous.

REGISTER PUBLISHING CO.
POLICE OFFICE.
POST OFFICE.
MERCANTILE CLUB.
QUINNIPIAC CLUB.
F. V. McDONALD, Yale News.
SMEDLEY BROS. & CO.
M. F. TYLER, Law Chambers.

Stores, Factories, &c.

O. A. DORMAN.
STONE & CHIDSEY.
NEW HAVEN FLOUR CO. State St.
" " " " Cong. ave.
" " " " Grand St.
" " " Fair Haven.
ENGLISH & MERSICK.
New Haven FOLDING CHAIR CO.
H. HOOKER & CO.
W. A. ENSIGN & SON.
H. B. BIGELOW & CO.
C. COWLES & CO.
C. S. MERSICK & CO.
SPENCER & MATTHEWS.
PAUL ROESSLER.
E. S. WHEELER & CO.
ROLLING MILL CO.
APOTHECARIES HALL.
E. A. GESSNER.
AMERICAN TEA CO.

Meat & Fish Markets.

W. H. HITCHINGS, City Market.
GEO. E. LUM, " "
A. FOOTE & CO.
STRONG, HART & CO.

Hack and Boarding Stables.

CRUTTENDEN & CARTER.
BARKER & RANSOM.

Office open from 6 A. M. to 2 A. M.
After March 1st, this Office will be open all night.

FIGURE 5.1
First telephone book of the New Haven District Telephone Company. Paper on cardboard, February 21, 1878. 23.5 cm × 16.3 cm. Courtesy of Archives & Special Collections, University of Connecticut Library.

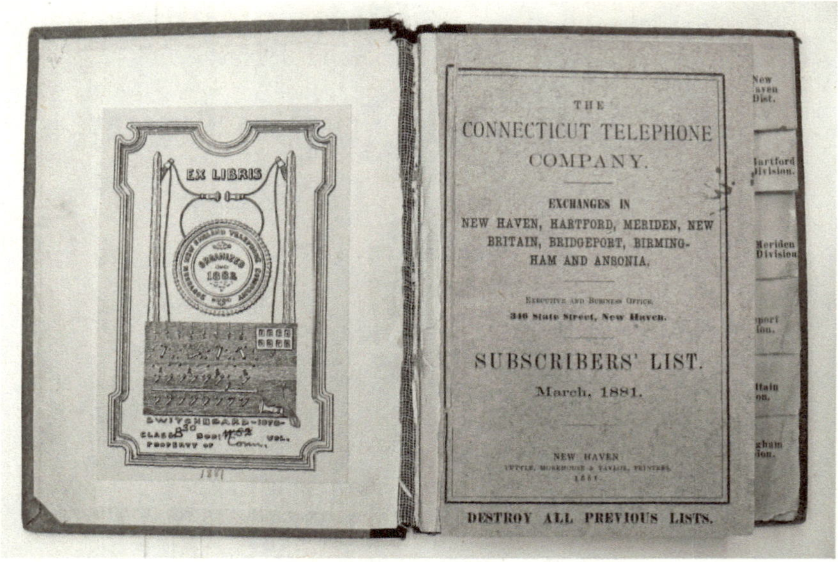

FIGURE 5.2
Telephone book from Connecticut Telephone Company. March 1881. 18.5 cm × 25 cm. Courtesy of Archives & Special Collections, University of Connecticut Library.

around the telephone, they embodied the places and agents of network formation that remained very local for years to come.[30] But distributed, decentralized organization eventually gave way to centralized offices, economic institutionalization in the form of telephone companies, and prominent sites of public presentation.

Bell broke off his lecture performances in May 1877, after a flop in Lawrence when his voice failed to be broadcasted. Other telephone pioneers took up the business of popularizing his invention. Stereophonic broadcast of operas, demonstrated by Clément Ader for the first time at an electricity exhibition in Paris in 1881, thrilled the public.[31] The founding of telephonic broadcasters such as the Théâtrophone, which began regular operation in Paris in 1890, or the Electrophone in London expanded the size of the telephonic audience. In Budapest, the telephone newspaper *Telefon Hírmondó* was introduced in 1893 by none other than Tivadar Puskás—the same engineer who was working on a telephone exchange for Edison in 1877, in parallel to Coy's efforts. It was Puskás who opened the first central office with a switchboard in Paris a year later and who also assisted Ader with opera broadcasts.[32]

No matter whether it was Bell, Coy, and the opera house, or Edison, Puskás, and *Telefon Hírmondó*, the laboratory conditions of the device's early years ensured that the telephone appeared as an open-ended object with a wide range of possible applications, not merely an instrument of business.[33] Without popularization, there would have been no telephone exchange as a concrete technological artifact. The mediation of mediation went hand in hand with the telephone being employed in local networks, which would have been unthinkable without performances in opera houses and music halls. The incalculability of technology that was a means and an end at the same time[34] corresponded to the unpredictability of staging events that demonstrated it in spectacular fashion and made it a cultural fact. The spectacle of the telephone belonged to public networking practices that proved decisive in the second half of the nineteenth century. Before the cables settled underground with the water, gas, and electricity, their presence in urban space was especially marked. The multiplying telephonic nodes, like the constant work on transportation networks and sewers of large cities, stitched together the city with new connections.[35]

5.2 BANK ROBBERS, GRAVE DIGGERS, AND NEW CONNECTIONS

> The idea of the enterprise appeared to be simple and convincing. Each buried coffin was to be linked via electric cable with the administration building. Telephones and electric bells would be connected up to the lines and each *customer* could not only report to the administration instantaneously if the occasion arose, but could also give the necessary orders to his doctor, his banker, and his family.
> —Walther Rathenau, "The Resurrection Co."[36]

The history of the telephone exchange begins with a service offered to protect against bank robberies. A little more than a year after Alexander Graham Bell registered his patent, during his publicity tours, Edwin Holmes, the owner of a burglar alarm company in Boston (Edwin Holmes Burglar Alarm Company, 342 Washington Street), offered a new service. Beginning May 17, 1877, his five customers (four were banks) could telephone at no cost over the existing alarm system lines. Another company grew out of its success: the Edwin Holmes Telephone Despatch Co. Earlier that year, the

FIGURE 5.3
Left: Replica of manual switchboard of the New Haven District Telephone Company. Right: Blueprint of the first model from 1878. 56 cm × 94 cm. Courtesy of Archives & Special Collections, University of Connecticut Library.

New England Telephone Co. had built the first local network with a telephone exchange for New England physicians in Hartford, Connecticut.[37] Based on its success, a new advertising campaign for commercial clients was launched on October 8, 1877, which attracted seventeen subscribers within a month.

SWITCHING BY HAND: MANUAL EXCHANGES AND LOCAL NETWORKS

At a comparatively modest rate, and without central planning, telephone exchanges were established in the following years in the eastern United States and then in Europe. The first manual switchboard, for twenty-one subscribers, inspired by one of Bell's public shows, went into operation in New Haven, Connecticut, on January 28, 1878.[38] Its engineer, George W. Coy, based his experimental design on the inexpensive cabling technologies common in telegraph offices (figure 5.3).[39] The telephone companies—in Europe, usually state funded; in the United States, nearly always privately

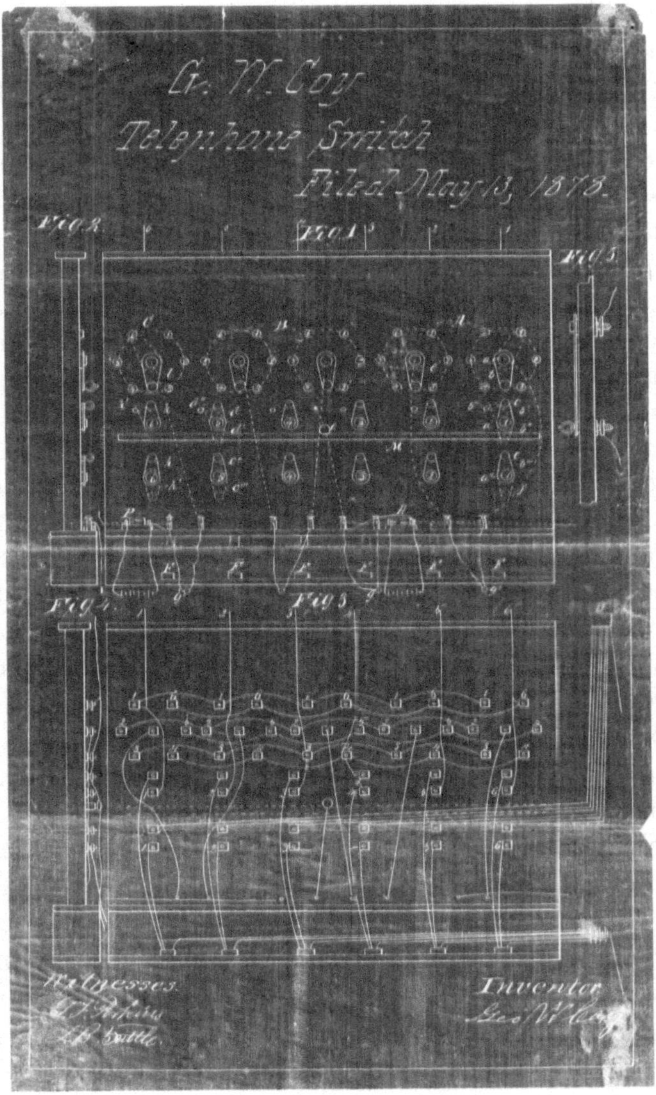

FIGURE 5.3 (continued)

financed—responded to the growing number of business customers by acquiring new switchboards. A dropping shutter signaled that the caller wished to be connected. After the number had been received by the operator, the connection was made directly by plugging in a wire.

In addition to the switchboards that were actually employed, many more were designed and patented but never built. They included, for example, devices with a ball mechanism that resembled a pinball machine to control the connection and a freight-loading station with small cars as switching elements.[40] The problem of a limited number of connections was solved with multiple switchboards (Leroy B. Firman, 1882).[41] Internal communications between different rooms were handled by messenger boys, and external connections to other telephone exchanges were set up. As the scope increased, however, so did the number and frequency of internal switching processes, which, in turn, required substantially more interaction between employees.[42] The suitability of men for complex switching work was doubted from the very start. Nevertheless, there continued to be strong resistance at first to hiring women as "the Danaïds of the unseen who incessantly empty and fill and transmit to one another the urns of sound," in Proust's words.[43] On both sides of the Atlantic, more and more women from the lower and middle classes were hired as cheap and competent labor—a development that went hand in hand with gendered attributions of communicative ability.[44] The work of telephone switching was characterized by both this class aspect and its gendered division of labor. While it was mainly men who built, administered, and repaired the infrastructure, women did the work of making connections.[45]

Herbert Casson's early *History of the Telephone* (1910) overflows with praise for the "telephone girl" responsible for producing synchronicity in the great flux:

> These telephone girls are the human part of a great communication machine. They are weaving a web of talk that changes into a new pattern every minute. How many possible combinations there are with the five million telephones of the Bell System, or what unthinkable mileage of conversation, no one has ever dared to guess. But whoever has once seen the long line of white arms waving back and forth in front of the switchboard lights must feel that he has looked upon the very pulse of the city's life.[46]

Casson's description does not mention the work of any nonwhite arms. Venus Green has noted that "as early as 1910, black women had sought entrance into the switchroom but met rejection by managers and the operators."[47] It took more than half a century for the "white lady" image to fade and for women of color to be integrated into the Bell corporate workforce.[48]

To get around the burdensome cranking of the phones in the 1890s, engineers such as Bell's J. J. O'Connell worked feverishly to provide current via central batteries, which would also make it possible to have light signals instead of mechanical shutters.[49] Their introduction eliminated numerous problems of manual telephone exchanges, which now had a network-based energy supply: there was no longer any need for recharging batteries or their on-site inspection by experts. The technical compatibility of the components of the telephone system improved, and—for the first time—operating costs decreased as the number of connections increased.[50]

From the outset, the equipment was prone to breakdown, and interference by ghostly voices and noises could not always be eliminated with better cabling. The English-language vocabulary of telephone experts for various noises was, not coincidentally, quite diverse. A connection could feature "spluttering, bubbling, jerking, rasping, whistling, screaming" that reached the ear as a rustling of leaves, croaking of frogs, hissing of steam, or flapping of birds' wings.[51] The systemic stability of the network was shakiest when many subscribers were connected. The solution to this came to be known as the "click busy test." Charles Ezra Scribner, head engineer at the Western Electric Company in Chicago and inventor of the telephone jack, came up with a practical procedure. By briefly tapping the jack, the person making the connection could hear whether the line was busy. With the click busy test, the two-person conversation increasingly prevailed as the central way of using the telephone network, but it did not eliminate other forms of its use, such as party lines. This operation gradually altered the actual character of the telephone: whereas once it was an open object also used as a mass medium, a network ruled out permanent connections.

While manual operation delayed the start of the person-to-person telephone conversation, projects beginning in the early 1880s focused on hiding the processes of connection.[52] At American telephone companies, it was already the philosophy for manual exchanges that callers understood they could rely on the protocols of the exchange service and had few steps to carry out themselves.[53] Designs for automatic telephone exchanges, in contrast, counted on the users being able to, and wanting to, select and address whom they called. Despite this difference, manual and electromechanical exchanges had a complementary and translational technical relationship. So, how were the operational chains of manual exchange and the knowledge embodied in (female) operators incorporated in the first working automatic telephone exchange in regular use, which was designed by the undertaker Almon Brown Strowger (1839–1902)?

SWITCHING BY MACHINES: ALMON BROWN STROWGER'S AUTOMATIC TELEPHONE EXCHANGE

The usual story goes like this: Like all aspiring entrepreneurs of his day, the Civil War veteran and former teacher Strowger hoped that installing a telephone would help his business. Despite a healthy number of deaths in the frontier town of Kansas City, Strowger's funeral parlor enjoyed little custom, while business was booming for his competitor, who had incomparably better connections to the local telephone exchange. Its director was a friend of Strowger's competitor. Moreover, his wife or girlfriend—sources disagree—worked there.[54] These relationships were so robust that even a good friend of Strowger's was buried by the competition when he died in 1887.

Within the next four years, Strowger developed, with the help of his nephew Walter S. Strowger, a device to automate the exchange process. The patent, which was filed on March 12, 1889 under the number 447 918 and granted on March 10, 1891, is for an automatic telephone exchange (figure 5.4): "The object is to provide means whereby a person at one station may make connection with any other station in the system, by the aid of electrical appliances, without the assistance of an operator at the central station."[55]

The invention was intended to improve "automatic, telephonic, telegraphic, and other electrical exchanges."[56] Perhaps strategically, the patent made no reference to precursors, although in the United States alone there were already twenty-seven patents claiming to have invented an automatic telephone exchange. For example, on September 6, 1879, Daniel and Thomas Connolly and Thomas J. McTighe proudly announced that they were the first to develop an apparatus to replace the system of manual work in telephone exchanges while permitting multiple connections of two parties each.[57] The Connollys and McTighe, who presented their apparatus in the summer of 1881 at the Exposition internationale d'Électricité in Paris, where stereophonic broadcast of operas also caused a sensation, never had

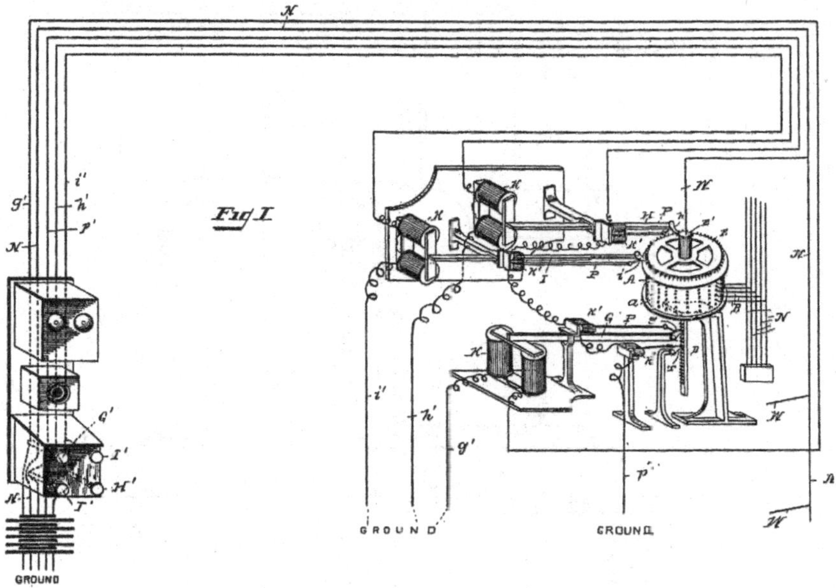

FIGURE 5.4
Almon Brown Strowger's automatic telephone exchange. Patent drawing, 1891.

any success with their automated device. It did, however, feature central mechanical ideas for the switching of circuits, including dials and rotating swivel arms, which returned in a different form in Strowger's design. Other developers such as Ezra Gilliland were also working on finding the way to make rapid electromechanical connections practical.[58] Likewise, there were experiments in the United Kingdom, Sweden, France, Italy, Russia, and—thanks to Tivadar Puskás—Hungary.[59]

Strowger's claims are not as original as the language of his patent application would have its readers believe. But unlike others, he was able to mobilize a network of apparatuses with a legal patent claim, economic feasibility, and popular communication that would fundamentally alter telephone networking. His proposal for a telephone exchange was based on the usual division of a central office and individual stations connected to one another "by and through line wires."[60] The technical heart of the first Strowger exchange was a switch cylinder. In the form originally patented, however, it was probably of limited suitability for long-term use because the attached wires had to move along with the cylinder itself (figure 5.5). The rotating cylinder as an information storage medium went back to

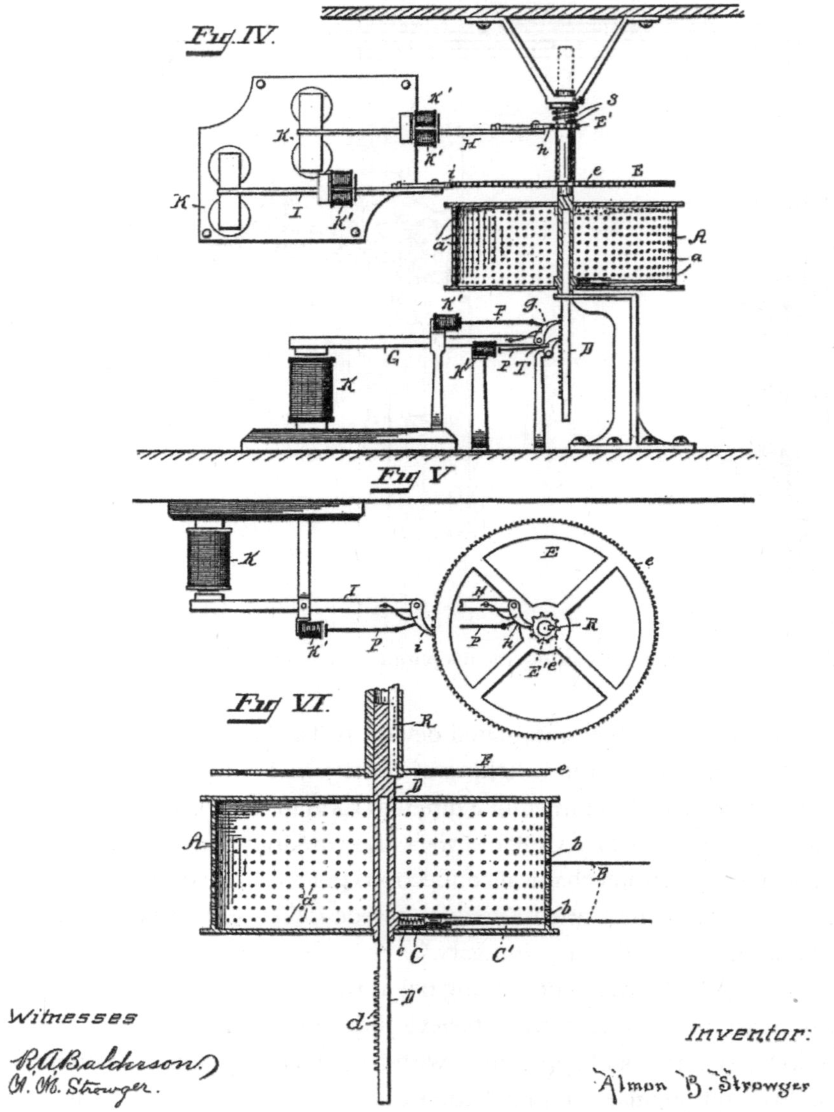

FIGURE 5.5
Switch cylinder from Almon Brown Strowger's automatic telephone exchange. Patent drawing, 1891.

the eighteenth century, with player pianos and Jacques de Vaucanson's design for an automatic loom. It was used in many fields in the nineteenth century, including telephotography.[61] Several early designs from the 1880s and 1890s clearly borrow from piano mechanisms.

The switch cylinder used for the first public demonstration of the automatic telephone exchange in La Porte, Indiana, about sixty miles from Chicago, was considerably different from the patent drawings. One important element of the apparatus had changed in the interim. The cylinder in the patent had become a rubber disk placed horizontally above the apparatus. All of the cables were now attached to the disk. The switching was still done by a contact arm, but now one that rotated (figure 5.6). The optimistic announcement of the company's electricians—that not just seventy-five people, as in La Porte, but as many as six thousand calls could be connected—was probably just commercial bluster.[62] An article in the *Chicago Tribune* on October 31, 1892, opened with the line "The telephone girl must go," but it showed the configuration in the original patent application.[63] Two days later, the front page announced the successful

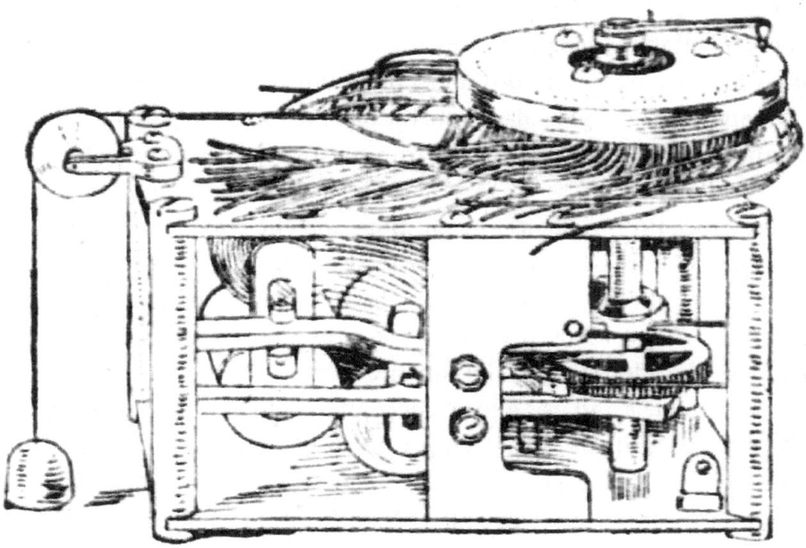

FIGURE 5.6
Construction of Strowger apparatuses from the first automatic telephone exchange in La Porte. Drawing from the *Chicago Daily Tribune*, November 4, 1892.

completion of the Indiana telephone system on November 1, once again with the headline that the automatic exchange would take the place of the "hello girls."[64] On November 4, a report on the front page recounted, with drawings, the official presentation and Strowger's opening speech of November 3. A special train had brought both experts and the curious from Chicago to La Porte, where a brass band accompanied a parade to the central telephone office. After opening remarks by the mayor, Strowger himself gave a speech. He alluded to the Connollys' and McTighe's patent of 1879, which he clearly knew, but also extolled his invention's many applications: in addition to telephone communication, it could also be employed as a universal switching instrument—for example, for detonating torpedoes or operating telegraphs along a railway line.[65]

Strowger's company also presented its innovation at the Chicago world's fair of 1893, having moved to the city two years earlier. Compared to the spectacular temple architecture of the Bell Company, the Strowger stand in the electricity pavilion looked quite modest, but it attracted attention nevertheless.[66] According to contemporary accounts, the invention was admired but considered only somewhat original. The Strowger exchange was said to be governed primarily by the principles of the old Morse telegraph, combining magnets, armatures, levers, and electric energy with other familiar elements, including pawls, ratchet wheels, and shafts—a comparatively unspectacular contrivance, especially compared to the long-distance telephony from Chicago to New York that Bell demonstrated at the world's fair.[67]

The crucial technological innovation, and the reason for the Strowger Company's future success, was the aforementioned stepping switch system, which could already be seen in preliminary form in Chicago in 1893. Over the course of the 1890s, it would replace the switch cylinder and addressing disk. The addressing of the subscribers was discretely separated by the separate contacts and was also arranged in a decimal system (ten times ten pawls per unit; figure 5.7). This permitted sequential switching of other "selectors"—so-called group selectors, which ultimately led to the "final selector"—a spatial arrangement proven to be practical in the complex installations of manual exchanges. Gradually, not just one hundred but one thousand or ten thousand numbers could be called. Whereas at first each key still had to be pressed several times—three times for three hundred,

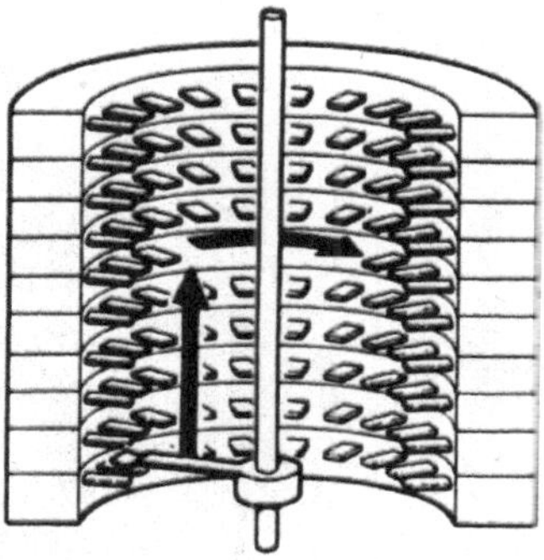

FIGURE 5.7
Schematic rendering of the step-by-step switching principle of the Strowger apparatus.

nine times for ninety, three times for three—beginning in 1896, rotary dials could transmit the electrical impulses directly as the mechanism returned to its initial position (figure 5.8).

Compared to the complicated key system, the rotary dial gave the automatic telephone exchange a form of "user transparency" that would become the standard for the Bell System's manual telephone exchanges. Independent telephone companies in particular, sensing an opportunity when the Bell patents expired in 1894, placed their bets on potentially cheaper mechanized exchange and lower prices for access to the network. Accordingly, the Strowger Company looked primarily to banks, department stores, factories, apartment buildings, schools, hotels, and offices as potential customers for its "immediate and efficient means of private communication."[68] All the while, however, the manual telephone exchange, which had become more efficient thanks to a central battery, continued to be used because it could cover much larger areas.[69]

The networking arms of the operators and the levers of the Strowger system did not function identically.[70] But they shared a method of address based on Western cultural techniques that made early telephone networks possible. The left–right schema of the Strowger switch follows the direction of writing but requires reversed searching from right to left within the rotational component. The spatial arrangement of the switchboard is based on the logic of a coordinate system. It also integrates a semiotic style of pointing and connecting that the letter grid of Charles Wheatstone's five-needle

FIGURE 5.8
Rotary dial by Erickson/Strowger, 1896.

telegraph had introduced to the electric communication technology of the nineteenth century.[71]

After 1910, the electromechanical Strowger principle spread more quickly in some European countries, including Germany, than it did in North America.[72] In the United States, it was used regularly only by independent providers (whereas the Bell System consistently relied on automation only in the 1920s).[73] As direct-dialing telephones gradually gained acceptance, the materiality of the telephone network formed part of a technological unconscious. In the case of automatic telephone exchanges such as Strowger's, that included a dead point in the exchange. Just as the purloined, misdelivered, or lost letter is constitutive of postal delivery, large direct-control systems are familiar with what is known as a technically "lost call" in the jargon of communications engineering.[74]

In Edgar Allan Poe's short story "The Purloined Letter," it is human actors who steal information, hide it, and use it for extortion.[75] But telephone operations from 1900 onward shifted, step by step, to the materiality of automatic exchange. Incommunicability and "lost calls" in the earlier generations of electromechanical exchanges occurred as follows. If, when moving from one selector to the next ("hunting")—for example, from the thousands digit to the hundreds digit—all ten reserved connections of the next Strowger apparatus were in use, there was only one sign of technological disturbance available: a "busy signal," whether the party being called was on the telephone or not. As a result, all switching procedures contained the technical possibility of generating lost calls, in which the numerical connection could not be established.[76] This was in part due to the economic requirements of networking. Both manual and electromechanical telephone exchange needed the opportunities for connection to be scarce and limited, which resulted in high prices: "In communications, bigger can easily mean more expensive, at least until the adoption of a comprehensive organizational system compensates for the disproportionate growth of complexity."[77]

This kind of telephone exchange, which survives today only in its patents and certain relics, is an artifact of the past. As a social principle, however, it remains in more recent transformations, such as when the party called being on hold leads to lost calls or when call centers take on the role of telephone exchanges as service agencies. Principles of telephone exchange and its specific networking styles have left distinctive traces in the history of culture and knowledge. In the next section, therefore, I reverse the narrative and explore some traces and doublings in cultural memory. What knowledge is set free by the telephone exchange, and how is it culturally embodied and materialized?

5.3 THE KNOWLEDGE OF TELEPHONE EXCHANGE

> It is nonsense to think that the history of cognition has as little to do with science as, for example, the history of the telephone with telephone conversations.
> —Ludwik Fleck[78]

Anecdotes and stories about the telephone—both the object and its inventors—are legion. The telephone exchange, however, is rarely the focus of attention. When it is, it is considered above all a paradigmatic "site of modernity," where white women (for the most part) do the work of making connections, or as a psychotechnical field of live experimentation for producing female nervous breakdowns.[79] Literary and media studies on the telephonic voice and noise often refer to the switchboard but rarely discuss it as an object of knowledge and a historical agent of economic and social networks.[80]

As late as the cyberpunk literature of the 1980s, console jockeys "jacking in" to the telephone network was a frequent motif, building on early hacker films such as *Three Days of the Condor*.[81] Reminiscences of these "good old days" can also be found in a brief scene in *The Matrix* (1999)—a film that Georg Seesslen rightly understood as a transition between the old alliance of cyberpunk and telephone and the new combination of cell-phone culture and computer networks.[82] Just before Keanu Reeves, as Neo, is liberated from the continuous simulation of a normal life, a rotary telephone prop is used to find the "real" location of his body, which was being used as a human battery for the machines of the Matrix.[83] This might be one of the many "lucky punches" the Wachowskis made while directing: the dramatization of automated dialing into the Matrix certainly works well with the odd ends of the material culture of the telephone.

More ambitiously and with greater precision, the artist Christian Marclay captured the operational chains of telephoning in the cinema. His seven-and-a-half-minute video *Telephones* (1995) collects and arranges a cross section of typical noises and gestures from the history of film.[84] By reducing the activity to serial operations, which contain exactly one element per sequence (dialing, ringing, picking up, greeting, brief conversation, saying farewell, and hanging up), Marclay outlines the anatomy of the audible and visible practices of phoning. The video delivers a punchline via the aesthetics of the medium. The editing makes it seem as if all the actors were connected in a single conversation. Its structure enables *Telephones* itself to function as a synchronous network across time, space, and social boundaries.

The Matrix and Marclay's video are not the only works in film history in which connection via telephone has left profound traces. One can see as much from a kind of subgenre of telephone films as well as in the diversity of media aesthetics and dramaturgy with which telephoning is presented.[85] One of the most prominent cinematographic examples of the aesthetics of the telephone exchange is *Grand Hotel*, the American screen adaptation of the German bestseller *Menschen im Hotel* (People at the Hotel) by Vicki Baum of 1929. Edmund Goulding's star-studded MGM film—its extravagantly high-class cast with Greta Garbo, John Barrymore, Joan Crawford,

Wallace Beery, and others was an innovation for Hollywood—did not just win the Academy Award for Best Picture in 1932.[86] It also established a rhythm of the telephone exchange and a rhythm of the whirring revolving doors that structure the misadventures of the carousel of people in Berlin: ballet diva, enterprising thief, outsider white-collar employee, and secretary. The very first scene is dedicated to the work of making connections at the hotel switchboard (figure 5.9).

The film's sudden final transformation into a tragedy, a homicide by technology, employs the telephone and switching as central elements.[87] Wallace Beery, playing the hot-tempered Preysing, kills the charming hotel thief "Baron" Felix von Gaigern over a stolen wallet by bludgeoning him with the device. Director Goulding then cuts directly to the telephone switchboard, where the confused operators are trying to make the desired connection: "Hello. Operator. Operator." Shortly thereafter the women comment soberly that someone is probably monkeying around: "They're having a nice little game up there with that telephone."[88]

FIGURE 5.9
Telephone switchboard in *Grand Hotel*, 1932.

In *Grand Hotel*, the switchboard is still operated manually, even though in the years between 1920 and 1935, automatic electromechanical systems could be found in cities such as La Porte and Hildesheim and in hotels, company offices, and other buildings—not just in cosmopolitan cities such as New York, London, Berlin, and Paris.[89]

THE TELEPHONE EXCHANGE AND ITS PHYSIOLOGICAL DOUBLINGS: BLOOD CIRCULATION AND REFLEX

Even with the installation of the so-called Strowger switch, the proverbial "telephone girl"[90] and her work remained part of the iconography of the telephone—at least as a media shadow (figure 5.10). Despite progressive automation, manual switching—not only a media technique but also a social technique of communicative exchange—was by no means unusual in the final years of the Weimar Republic. The automation of manual switchboards was interrupted in Europe by World War I, whereas in America, the Bell System was mechanized only once the firm had a monopoly.

The 1927 cover of the French journal *La science et la vie* featured a color lithograph of the telephonic-symbolic order of gender. In a wild mix of organic metaphor and technical display, it evokes a connection with blood circulation.[91] The blue and red wires correspond to the usual medical color coding for arteries and veins, which here become conveyors of language, just like Strowger switches. The electromechanical switches on the left and right lower edges of the image represent the rotary selectors introduced as standard equipment in Paris in 1926. The phone with letters in the center of the image points to an unusual feature of Paris telephone switchboards of the time. The first three letters of a telephone code indicated the telephone exchange. They were followed by four digits representing the person being called, which correspond here to the four dials of the "brain."[92] The seven-digit number could be entered using dials on which the twenty-five letters were printed.[93] While the "brain" of the automatic telephone switching system is superimposed on the head of the female shadow, the hands are noticeably free. Gone are the waving arms, their skillfully woven connections, and the helpful voice of the operator. At the same time, the mechanical form of address is still depicted as a feminine activity. The woman being called, in contrast, has the attributes of an operator of the day and uses a headset instead of the usual telephone handset. Both representations of women are eroticized here, be it in an actual telephone conversation or as

FIGURE 5.10
Title page of the journal *La science et la vie*. Color lithograph, May 1927.

a symbol of automatic infrastructural mediation. "The origin times of the telephone are characterized by its imagination as a female network," as Virginie Julliard has aptly put it.[94]

Forms of object reference that synchronize blood circulation, nerves, and gendered technology into a network should not be understood exclusively as "organ projections" in the sense of Ernst Kapp's *Grundlinien einer Philosophie der Technik* (Principles of a Philosophy of Technology) of 1877.[95] The way body parts are used to symbolize communicative exchange seems vague only at first glance. Apparently, only the latest media technology is ever good enough to model the brain. New terms and paradigms change along with the relevant apparatuses.[96] Corresponding metaphors and metonymies are both symptoms and effects of historically specific epistemic constellations and competitions. One good example is offered by switching problems, which are a central reference point for information theory and cybernetics. This was particularly true of research on neural networks in addition to Claude Elwood Shannon's modeling of information theory on telegraph and telephone switches.[97] Warren McCulloch and Walter Pitt's logical calculus of nerve activity from 1943—one of the foundational texts of cybernetic thought—transformed the behavioristic view of the network, in which the connections of individual telephone lines had served as a model for brain processes.[98]

In his theory of reflex, Ivan Pavlov cited the telephone exchange as a model analogy for the mechanism of "temporary communication" of the nerves. In a public speech in 1909 titled "Natural Science and the Brain," Pavlov explained the mechanisms of the conditioned reflex as follows:

> Suppose that instead of the present system of telephonic communication effected through the central telephone exchange, that is, temporary communication, all the subscribers were permanently connected with one another. How expensive, inconvenient and indeed impracticable it would be! All that is lost in this case by the conditional nature of the connection (one cannot get connected every moment with every subscriber) is largely compensated by the wide range of possible connections.[99]

Pavlov's analogy, which he introduces rhetorically with "I shall make a comparison," is based on a summary rejection of psychological approaches to understanding brain activity. Instead, he favors a strictly physiological view of higher nerve functions and wants to "record them objectively" according to the rules of their relationships.[100] The telephone exchange suggests itself as an example of the objective, understandable, temporary production of connections. The logical conclusion, from Pavlov's perspective, is of the exchange as the site where conditioned reflexes are produced. Their temporary twitching makes telephone switches one of the highest sectors of the central nervous system (of society and the media). As a laboratory for producing complex reflexes, the telephone exchange as Pavlov understands it reacts to environmental stimuli that it modifies, amplifies, and diminishes. This concerns not only the relationship of system and environment but also the very actions of the operators based on conditioned reflexes.

THE TELEPHONE EXCHANGE AND ITS PHYSIOLOGICAL DOUBLINGS: NEURONS AND ASSOCIATIONS

In the case of early telephone exchanges, parallel analog models of physiological networks abound. They range from the associationist psychology of William James and the study of neuronal structures to the debate over brain physiology between Camillo Golgi and Santiago Ramón y Cajal in the 1890s.[101] James's remarks on association and train of thought in particular read like a continuing description of telephonic exchanges. In his *Principles of Psychology* (1890), however, he distinguishes his view from the association of ideas in the tradition of John Locke. James insists that not ideas but "objects thought of" are the foundation of all mental connections.[102] This form of object reference does not just make things (and hence an object reference rather than abstract ideas) the focus of interest. It is itself part of a connectionistic formation of cultural techniques and knowledge in the late nineteenth century. Much like Pavlov, who spoke of it even more explicitly, James writes in *Principles of Psychology* of "calling up" the association or "switching" the nerve stream.[103] But far more remarkable is James's introduction to a kind of if–then transmission technology for the train of thought in a diagrammatic representation (figure 5.11).

Using the example of the memory of a dinner party, he prosaically explains the form in which traces of memory are reactivated. While the nerve areas a, b, c, d, and e are all stimulated at the end of the party ("act A"), the areas l, m, n, o, and p represent the stored return home through the cold night ("act B"). The thought of A must awaken the association with B. In the same way, a renewed discharge of neurons connects the end

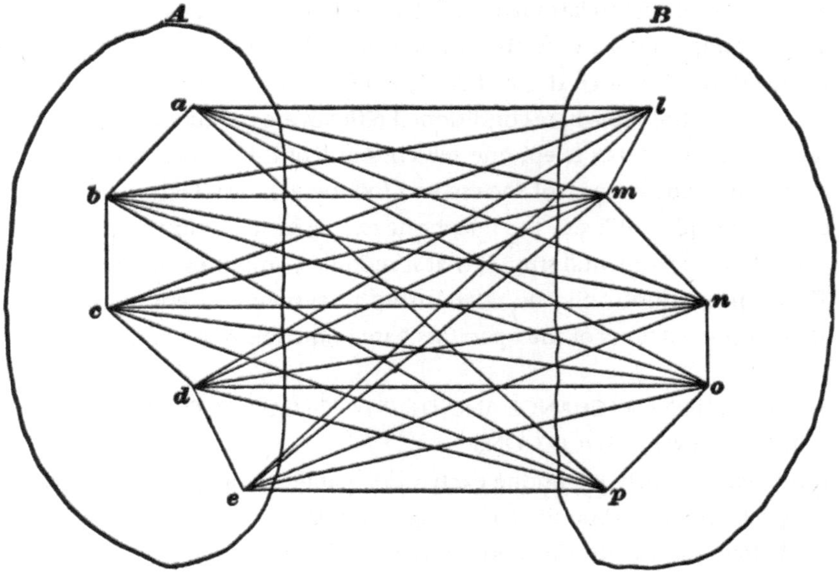

FIGURE 5.11
William James's diagram of the train of thought. Drawing, 1890.

of the party to the long walk afterwards.[104] This sort of ideal automatism of connections also drives the telephone network: A and B would no longer be connected by manual switching but, as in William James's model of the train of thought, proceed entirely automatically.

5.4 DISAPPEARING VOICES, GROWING NETWORKS

> Maintaining and joining, the telephone line holds together what it separates. It creates a space of asignifying breaks.
> —Avital Ronell[105]

Spiritualism, religion, and ritual—the flip sides of the new electrical and photochemical media in the nineteenth century—are also their social and cultural-technical foundation. Ancient rhetoric had already sought to communicate the voice as voice. Telephony as a cultural technology for speaking is indebted to medieval and early modern magic, which was maintained in the invention of apparatuses and media.[106] The conjuring of spirits—which Bell's spiritualist assistant, Watson, was not alone among his contemporaries in pursuing[107]—joined forces with the capitalization of ideas at the patent office. Jonathan Sterne has demonstrated how closely the "voices of the dead" were related to the embalming techniques of Victorian burial culture, the consumption of chemically preserved canned food, and the rise of sound recording and broadcasting. An ethos of preservation established the frame for the sound recording: "Recording was the product of a culture that had learned to can and to embalm, to preserve the bodies of the dead so that they could continue to perform their social function after life."[108]

But if the telephone has always been the "telephone of the Beyond," what role is assigned to networked connections with this nonplace?[109] In "The Resurrection Co.," a short story that Walter Rathenau wrote in 1898, the dead from Necropolis, Dakota, start to phone one another.[110] They repurpose to their own ends telephones installed in the cemetery to ensure that no one would be buried alive. But rather than wanting to speak to the living—to whom the mortician Strowger dedicated his invention—the dead wish to communicate directly with one another—coffin to coffin. The telephones are quickly removed, but the bells have to keep ringing as long as the Dakota and Central Resurrection Telephone and Bell Co. remains in business.

Voices without living bodies and voices from the beyond turn out to be unwanted but unavoidable guests.[111] The knowledge of communication was composed of the telephonically uncanny shadow of death. Whereas the dead could dispense with wires and apparatuses, as in Rathenau's "Resurrection Co.," the living had to reassure themselves again and again of their language and hence of their bodies, as electrically transmitted voices were always disappearing. The technological innovations to expand the territory of networked machines, including the inductive charging ("loading") of lines by means of loading coils that was common from 1900 onward and the establishment of long-distance telephony using amplifier tubes in 1912, worked to counter such loss.

"'What did you say? Are you still there?' signal the fact that translational activities govern the milieu of all telephonic utterances, doubling the translation of sound waves back into voice simulation."[112] The absent "presence" of the voice must be produced anew at every moment, simultaneously as a speech act and a technological process. Communication via network is fragile—a fact that remains apparent to all users of a manual exchange but was supposed to fade away with every advance in the mechanization and automatization of telephone networks. The embodiments, materializations, and performances of all the networks that form the telephone system "by and through" wires thus became that much more important. The telephone and the telephone network are at once a physiological and a physical matter, in the nineteenth century and after.[113]

Just as Bell's invention long remained an open, ambiguous object, Strowger's apparatuses enjoyed success after a considerable delay. Only years after the undertaker's death did Strowger switches begin to be installed on a large scale—accompanied by intense discussions among international experts. In Germany, they played a groundbreaking role before World War I and to some degree also thereafter.[114] North America nevertheless remained dominant, above all thanks to the massive commercialization of the telephone research at Bell Labs, which also led to the digital encoding of long-distance telephone communications.[115]

The organizational forms of telephone companies were at least as important as the material foundations of switching. Their effect on smaller networks, which were at first geographically and economically separate and then joined into one large network or system, deserves a history of its own. In the first decade of the twentieth century, independent telephone companies in the United States, which were betting on automatic exchanges, found themselves in serious competition with AT&T, the parent company of the Bell System. AT&T, headed by Theodore Vail, then undercut their prices and, in 1913, even agreed to increased government regulation in exchange for a quasi-monopoly. In the process, the United States set out on a government-supported path to establishing a comprehensive national telephone network, which, in this case, was a hybrid of public infrastructure (common carrier) and an enduring private monopoly.[116]

The story of linking individual telephony networks to form a heterarchy and what German telephone engineer Hans-Dieter Schmidt has called "the largest machine in the world" played out as an improbable formation of a global network. It began to emerge in many localities and often retained its local character as it expanded. Manual exchange technologies never disappeared entirely, just as mechanical connections exist alongside digital technology. Manual telephone exchanges and switchboards with ten thousand connections led to endless series of Strowger switches. Telephone books listing names gave way to country codes. Inexpensive local calls set the stage for extremely expensive long-distance calls, which were replaced by the cheap international calling card. Unfortunately, chronological genealogies tracing leaps in technological progress have to leave out this enormous diversity of nonsimultaneities and improbabilities.

By starting out with Watson, Strowger, and Rathenau, perhaps we can tell another nonteleological history of telephone networks and their connections. Given the disappearing voices of the dead, it would be a tale of uncertainty about the end of human bodies and the beginning of sociotechnical networks that the telephone incarnates like no other networked and networking apparatus of our everyday culture.[117] In that story, the practices of producing connections described so well by the word "switching" would stand for the historically, materially, and situationally transforming operational styles of network cultures.

One aspect of its self-reflection would be the recoding of its forms of connections. Just as William James followed the train of thought in the diagram, diagrams of networks produce what they show. But what do the graphs of graph theory really tell us? And how could they become so popular that they contributed to the universalizing tendencies of formal network theory? These questions lead us into the history of knowledge in the natural sciences, sociology, and anthropology. What we understand today as social networking is also indebted to a long process of scientification that will be discussed in the following chapter.

6 A VISUAL HISTORY OF THE NETWORK DIAGRAM (I): FROM THE MODELS OF THE NATURAL SCIENCES TO THE CALCULATION OF SOCIAL NETWORKS

Simple geometric figures have long been objects of fascination and models for knowledge. The net joins circles, lines, polygons, and trees in the long history of diagrammatic forms for ordering space. Yet, the net is distinct from the abstract Euclidian figures on which geometric operations are mainly based because it has a flexible relational structure. Like the tree of knowledge, it is a material symbol that can serve as a model. But it goes beyond the order signified by the tree by alternatively referring to two objects: a fishing net and a spider's web. "A net is not a tree," Umberto Eco once declared, and went on to observe that tree-shaped diagrams can themselves become encyclopedic labyrinths: "The main feature of a net is that every point can be connected with every other point, and, where the connections are not yet designed, they are, however, conceivable and designable. A net is an unlimited territory."[1] The abstract model of a net, according to Eco, has neither center nor outside. He refers to Gilles Deleuze and Félix Guattari's botanical metaphor of the rhizome as the most suitable image.[2]

The ideal properties attributed to the net and the rhizome in post-structuralist philosophy are scarcely matched by real objects so far known to us. On the contrary, the possible connection of every node with every other one is increasingly difficult to visualize, the greater the set of actors or data is. Likewise, even with the best virtualization technologies, unlimited territories are distinguished by the geometric invisibility of the infinite in a two-dimensional space. All the same, the popularity of network visualizations persists. One must ask why. After all, graph theory has other methods for rendering form—for example, relationships represented by matrices and in databases. Moreover, the deficits of the net diagram, which include a lack of dynamism, the relative difficulty of representing transformation in time, inadequate explanation of emergent developments, and lack of clarity about the difference between system and environment, are well known. Evidently, there are pragmatic reasons why diagrammatic nets are enlisted to hold other nets.[3]

That is why the three chapters of visual history in this book are dedicated to semiotic practices and operations whereby the net diagram plays a special role in the formation of networks: in the sciences, the production of goods and logistics, and the counterdiscourse of critical conspiracy theory. The role of the diagram, above all, is to form models. It is, "as a medium, also the conveyor of something that it transfers with application to something."[4] As models, network diagrams produce what they know. They are simultaneously models *of* and models *for* something.[5] They allow scientists to order, mark, and explore a given network. Meanwhile, formal network analysis relies on a combination of automated data processing and human interpretation of the graphs generated: particular ways of seeing to draw comparisons.

6.1 MARCELLO MALPIGHI'S THINGS AND THE MIRACULOUS NET

> Let us come back to yourself considered as network, and that network's formation.
> —Théophile Bordeu to Jeanne Julie de l'Espinasse in Diderot's "Le rêve de d'Alembert"[6]

When Marcello Malpighi described his microscopic look at the capillaries of frog lungs in 1661, he did not hesitate to describe the structure he discerned as a *rete mirabile* (miraculous net). The Italian physician and microscope virtuoso (1628–1694) communicated this discovery in two letters under the title *De pulmonibus* (On the Lungs). They were addressed, first, to his teacher Giovanni Alfonso Borelli, who offered inspiration and criticism in a number of letters of his own. Borelli insisted that illustrations be included with the publication. Otherwise, Malpighi's insights would not be understood.[7] The correspondence, which appeared in print, addresses a learned audience across Europe that was already familiar with the description of the circulation of blood that William Harvey published in 1628.[8]

Malpighi, whose career would take him from his studies and chair in Bologna to a position as personal physician to Pope Innocent XII and a professorship at the Sapienza in Rome, described his microscopic observations of pulmonary alveoli as follows: "in the external part of the lungs, when exposed to light, a certain wonderful network seems to spread, by which you would say that emerging protuberant single bladders are bound together. The same is observed, obscurely, in the section of the lung inside."[9] The most fascinating unexplained feature is that the pulmonary capillaries allow fluids to enter the artery and exit again on several sides. Malpighi's frogs were subjected to the anatomist's torture: he dried them out, boiled them, introduced substances into their windpipes and blood vessels, drained their blood, sucked out their air, and more.[10] Magnified 180 times under a microscope, the transition of arteries via capillaries to the vein that Harvey had explained as tissue porosity could be demonstrated experimentally.[11] The exact processes of exchange in the area of the capillaries remained hidden to the eye, however, despite the best mechanical efforts. It was with good reason, then, that Malpighi described the filigreed structure as a "miraculous net" in its functionality. This choice of terms was traditional. Similar descriptions were not unusual in ancient medicine.

A mediating function was attributed to net-like tissues, even before Malpighi—for example, in René Descartes's *Traité de l'homme* (Treatise on Man), written in the 1630s and published posthumously in 1662. Here, the French philosopher identified the *réseau* (network) as the central part of the brain, from which individual *filets* (nets) branch out as transportation routes for the animating *esprits animaux* (animal spirits—from the legacy of the Galenic pathology of humors).[12]

The history of early modern anatomy is too complex to admit treatment here. The crucial point is that the use of the microscope to make things visible that was cultivated in the seventeenth century as part of the epistemic culture of *anatomia artificiosa et subtilis* led to a visual modeling of bodies and their organs.[13] By drawing and identifying individual parts, the net made the object of knowledge visible in the image and captured it as a diagram. Its object reference was subject in Malpighi (who followed the Galilean school of drawing in this respect) and Descartes to the conditions of iatromechanics, which translated objects into representations drawn from book culture. Frog capillaries became "epistemic things," to borrow a concept from Hans-Jörg Rheinberger, and small machines: not only static structures but also active in the distribution of fluids. Malpighi himself made the comparison with the *opus reticulatum* of Roman architecture to describe what he sketched as the capillary net (figure II in figure 6.1).[14]

Malpighi's descriptions of anatomical nets—like his descriptions of the structures of plants[15]—are among the best starting points for a visual history of the network diagram. They mark the translation of a material object into a diagrammatic drawing, without which further modeling of relational connections in networks would scarcely be conceivable. Malpighi drew a scale model of a physical net. At the same time, for him and Borelli, the drafts of the publications of treatises were themselves a *modello* of the knowledge they contained.[16] Material nets can be made visible; they have a material presence. The pictorial status of networks as relational processes and hybrid quasi-objects, in contrast, is a precarious matter in which analog modeling risks contravening theoretical assumptions.[17]

In early modern medical and natural history, a unique aesthetic of the net combined with anatomical objects, especially the circulation of blood. Network models were invested with a reticular semantics that would be invoked, in altered form, in the modern era, usually in connection with urban circulation.[18] In the eighteenth century, physicians of the Montpellier school and, following them, Diderot developed a view of human anatomy in which net structures could scarcely be separated from biosocial networking energies.[19] Finally, in the descriptive tableaus of natural history, the net itself became a self-reflexive figure of knowledge that recorded the order of living creatures.[20]

How did the diagrammatic aesthetic of the net continue to evolve in the nineteenth and twentieth centuries? The history of knowledge in this chapter is dedicated to this question. Above all, it addresses mathematics, chemistry, and sociology. First, however, some notes on pictorial and media theory concerning the visuality and operativity of the network diagram are in order.

6.2 NETWORK VISUALIZATIONS BETWEEN THE IMAGE ACT AND THE ACT OF WRITING

> The diagram is *between* writing and image.
> —Sybille Krämer[21]

Network diagrams are systemic images because they continuously strike the limits of paper surfaces and computer user interfaces.[22] Epistemologically, they might more accurately be called "net diagrams." In this form, calculated

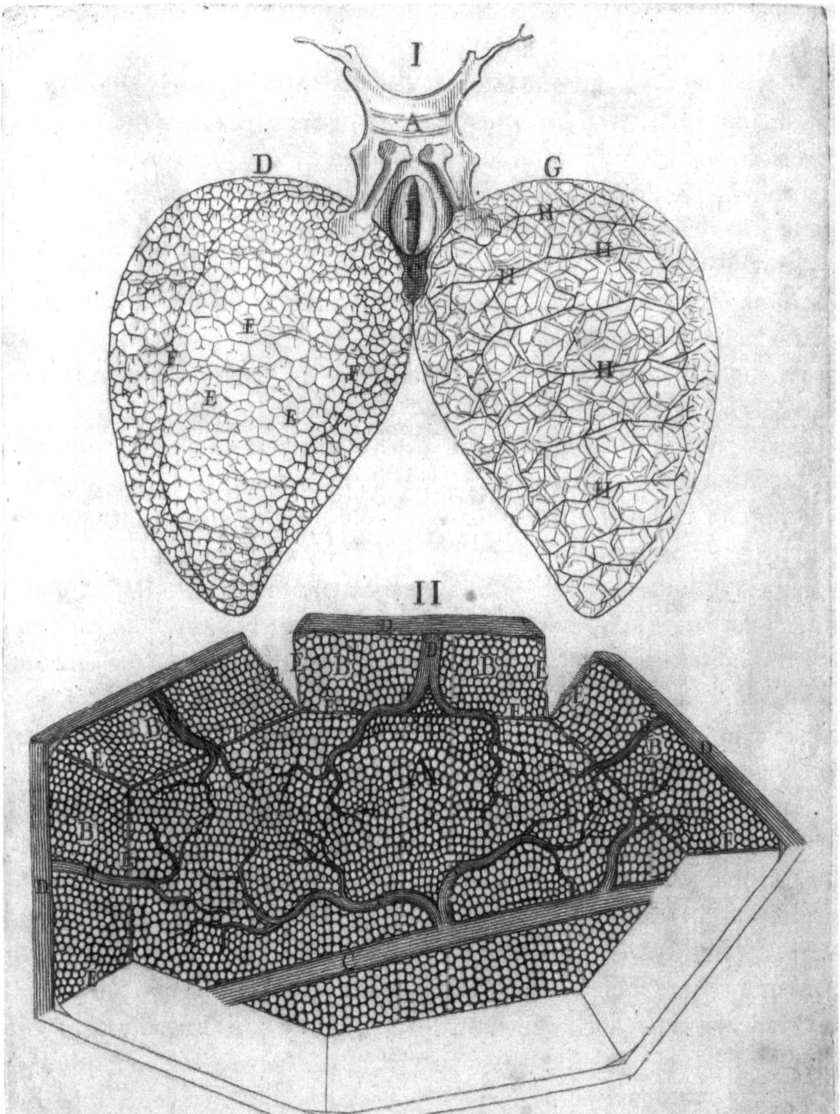

FIGURE 6.1
Marcello Malpighi, Anatomy of a frog lung with *rete mirabile*. Copperplate engraving, 1663 [first drawing, 1661]. 15.5 cm × 9 cm.

relationships are expressed between homogeneous elements of one and the same component, the so-called node. The size of a dot, the length and form of lines, and the expanse and form of planes have no meaning, if one accepts the definition in Jacques Bertin's classic *Sémiologie graphique* (translated as *Semiology of Graphics*).[23] Meaning is produced by the topology of the relationships becoming visible, for which we can disregard conformality, even though the viewer's understanding is guided by geometric conventions.

Strictly speaking, a drawn net only becomes a network diagram when its nodes are heterogenous and form an interconnective and heterarchical network of agents.[24] The continual expandability, aggregation, or disentangling—that is to say, the fluctuation—of networks cannot be rendered in pictures and can be rendered only poorly in tabular matrix form. Accordingly, the images in which social connections are supposed to be recorded push toward dynamization by way of media techniques of animation and simulation. The "drawing things together" of scientific representation, as described by Bruno Latour, is joined by "simulating things together."[25] But its current forms, which extend into swarm research, also work with the scaling possibilities of network calculations.[26] In its topological and relational forms, the network diagram has become the standard medium for the network society.[27] For social structures, it establishes a scale relationship with which the contours of social macrodimensions can be obtained from the communal microdimensions of groups.

An emerging, productive, interdisciplinary research field on the theme of diagrams and diagrammatology provides inspiration for visual studies, philosophy, and cultural studies.[28] It was already nascent in French post-structuralism, most clearly in the work of Gilles Deleuze, Michel Foucault, and Michel Serres.[29] Serres, in the very first pages of *Hermès*, his philosophy of communication published in 1968, made the network diagram a topological model for a new historiography.[30] Serres entitled his introduction "The Communication Network: Penelope," recalling the weaving patiently performed by Odysseus's faithful wife. The book begins with an epigraph from the Orphic tradition. Before Persephone received Dionysus from Zeus, she was working on a tapestry that was supposed to depict the entire universe. This mythological grounding introduces a mental experiment typical of Serres's thinking about networks: "Let's imagine, drawn in a space of representation, a network diagram. It is formed, for a given instant (because we will see well enough that it represents one state of a situation in flux), by a plurality of points (vertices) connected by a plurality of branches (paths)."[31]

In this diagram, no point is privileged. All of them have their own, changeable energy, which in turn can influence the other points. The paths establish relationships between the vertices. They transmit stable connections that Serres calls "determination flows." "It is therefore a network whose internal differentiation is maximized at will, as irregular a diagram as possible."[32] If it were regular, it would take the form of a grid.

Because nearly everything in such a figure can become a "vertex" or "node," the network is first an abstract philosophical structure that serves as the epistemological foundation for the formation of models. If used abstractly, "it can be a mathematics, a graph theory, a combinatorial topology, or a theory of schemata."[33] At the same time, for Serres, the network permits a different form of empirical historiography: nonlinear, with reversible causalities, occasional feedback loops, and complex interruptions—"anyone who wishes to influence is suddenly influenced by the result of his influence."[34]

In this way, Serres outlines a way of thinking about history that orders the historical imagination and narrativity in the form of a network graph. The course of past events becomes flows in a diagram, and events become nodes: history is rendered as a flux of circumstances whose complexity can be modeled and grasped in a network diagram.[35] To this end, Serres employs a vocabulary in which elements of twentieth-century mathematics—especially graph theory—meet up with ideas from cybernetics and information theory. Beyond these contexts in the history of science, his historical network modeling is indebted to an aesthetic reading of myths that testifies to a fondness for games. Chess, for example, becomes the paradigm for the struggle between two different networks that permeate, determine, and transform each other over time.[36]

Serres does not, however, leave things at the myth-inspired mathematics of the first pages of *Hermès*. He also makes the diagrammatic foundations of his epistemology explicit in a small book on Vittore Carpaccio's painting. Here, he succinctly identifies the concern of the descriptions of paintings, offering "assertions on the third [field]"—not philosophy of communication or mathematical proof but rather aesthetics.[37] Serres develops a unique view of the image, reading every painting by Carpaccio in terms of aspects of its narrative geometry. He translates the action and structures of the painting into points, coordinates, vectors, and flows. He decodes the diagrams on which the pictorial is based by translating them into a narrative. This approach results in an image in which the geometric figure opens pictorial space in the first place. It is not about a fixed form and structure but rather about a permanent, network-like transformation that operates geometrically and discursively at once. Serres sketches this procedure, translated into a rhetorical question: "Geometry as the art of transformations—how could one define it more precisely?"[38]

Such a procedure calls to mind Erwin Panofsky's efforts to understand constructions of perspective from antiquity to Impressionism as symbolic form.[39] But unlike Panofsky, Serres does not draw lines on the painting retrospectively in order to discover them having been there all along. He presumes a floating performability of the diagrammatic that cannot be pinned down.[40] Every aesthetic is relational, constantly in motion, even when the image might appear to be something fixed. The dynamization—and Serres is almost always concerned about a form of *dynamis*—is indebted to narrative, especially in a historicizing form.[41] Moreover, Carpaccio's paintings are tied for him to the topography of Venice, with its bridges and canals.

One can say that Serres subverts something that Gottfried Boehm has called "iconic difference" for the sake of a narrative form.[42] In Serres's interpretation, Carpaccio's paintings no longer appear to be something from which viewers can separate themselves. Here, the diagrammatic aspect of the image is a form of immersive visual magic that works with the vocabulary of the natural sciences. Network diagrams always seek to point beyond themselves and capture dynamic network relationships in the form of an "immutable mobile" (Bruno Latour).[43] To adopt the categories proposed by Horst Bredekamp, on the one hand, they are intrinsic image acts in which the form comes into its own as form; on the other hand, they belong to the tradition of the schematic image, which brings to life and corporealizes what is modeled.[44] Both contribute, in placing before one's eyes, to the production of scientific evidence. Finally, network diagrams do not just make statements about relationships. With the narrative reading in Serres's sense, they become net*work* diagrams.[45]

6.3 MATHEMATICS, CHEMISTRY, GRAPH THEORY

> If we correctly model the network assembly, our final network should closely match the reality.
>
> —Albert-László Barabási[46]

Are diagrams simple? Do their iconic qualities reduce complexity? Charles Sanders Peirce addresses these questions in a piece from 1895 he called *Short*

Logic: "A diagram is a kind of icon particularly useful, because it suppresses a quantity of details, and so allows the mind more easily to think of the important features. The figures of geometry are, if accurately drawn, such close likenesses of their objects that they are almost instances of them." Diagrams, however, can also be dissimilar to their models: "Many diagrams resemble their objects not all in their looks; it is only in respect to the relations of their parts that their likeness consists."[47]

This double form of relationality produced by drawing also characterizes the pictorial status of formal and formalized nets and networks. A geometric figure of nodes, lines/edges, and spaces in between, which at first appears simple but is capable of extreme variation, serves as our point of reference. Peirce's "relations of . . . parts" play a crucial role. Explicit modeling with the aid of the topological net diagram, which has become a fundamental component of graph theory today, goes back to the eighteenth century. Some problems of the "geometry of position" have been known for longer still.[48] It is all the more astonishing, then, that prior to the twentieth century, nets and networks were by no means a privileged subject of mathematics. Only later—and, even then, in very limited way—did a field of knowledge develop that understood "Les réseaux (ou graphes)," to quote the 1926 essay by French mathematician André Saint-Laguë, as an autonomous subsection of both theoretical and applied mathematics.[49]

My thesis is that over a period of about two hundred years, mathematical network knowledge was in a state of latency. Formally, graph theory (i.e., network theory on the basis of the network diagram) could have been applied much earlier to phenomena in nature, technology, and society. The fact that this first occurred in the twentieth century and was then, in a sense belatedly, interwoven into semiotic practices and the *Lebenswelt* provokes other questions. Accordingly, my second thesis is that this turn, which began around 1930, cannot be explained without the visuality and visualization of connections in graph theory. Notation of calculations alone did not achieve broad cultural currency and popularization. Only as writing increasingly switched to visualization and as computability was tied to older visual traditions did network knowledge become a formalized part of the modern epistemic order.

The history of science involving graph theory before and after 1936 hosts a gallery of imposing ancestors. Albert-László Barabási, in his network manifesto *Linked: The New Science of Networks*, likes to tell stories about heroes: "Graph theory boomed after Euler with contributions made by mathematical giants such as Cauchy, Hamilton, Cayley, Kirchhoff, and

Pólya."[50] Such a perspective omits the fact that, first, objects and problems seek out their theories and, second, theories generate their own objects and problems. The following scenes that played out before 1900 are dedicated to two subfields of what would later become known as graph theory. First, there is the problem of a knight's path across a chessboard and the optimal route between twenty points—which Saint-Laguë still dismissed as late as the 1920s as "popular mathematics." Second, I will briefly address the notation of chemical formulas in which the word "graph" and the associated visual form take on a new meaning. Whereas the curves drawn in coordinate systems—which continue to be called "graphs"—express measured and calculated structures in written form, "chemicographs" function much more relationally and topologically.[51]

ALEXANDRE THÉOPHILE VANDERMONDE, WILLIAM ROWAN HAMILTON, AND GAMES WITH GRAPHS

Leonhard Euler's approach to the paths across the seven bridges of Königsberg in 1736 already employed a form of mathematical modeling that simulated spatial connections. Crossing the bridges mentally instead of visiting them, Euler figured out whether one can cross all the bridges over the Pregel by traversing each just once. He did so by expressing the routes as combinations of letters. His process and his answer—that no such itinerary is possible—are still considered the foundation of graph theory.[52] Because Euler's solution is so popular (in canonical histories of mathematics and computer science and as a legend circulating on the Internet), the fact is often overlooked that his model does not feature any topological graphs (figure 6.2). As more recent research has shown, the mathematician restricted himself to reading and writing in order to produce a model of "seeing without seeing."[53]

Euler's polyhedron formula initially held richer consequences for the history of mathematics than his pastime with the "geometry of position" at Königsberg, which he never visited. When working on the problem, he recalled his feverish dreams as a young scientist concerning the Königsberg problem. A short work on the knight's path across a chessboard that he wrote in 1759 for the yearbook of the Akademie der Wissenschaften in Berlin (although not published until 1766) is important.[54] Euler solved the problem in part by using an elegantly abstracted game board (figure 6.3) that lacks the lines between fields found in some of the sketches in his diary.[55] The French mathematician Alexandre-Théophile Vandermonde

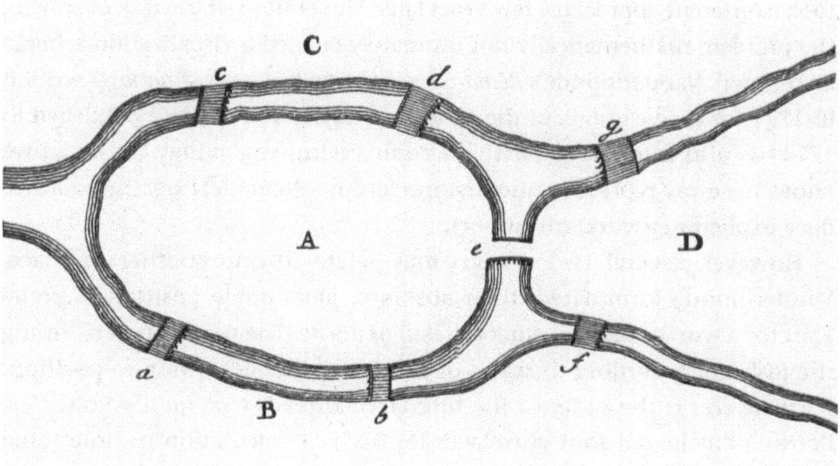

FIGURE 6.2
The Königsberg Bridge Problem. Copperplate engraving after a sketch by Leonhard Euler, 1736.

42	59	44	9	40	21	46	7
61	10	41	58	45	8	39	20
12	45	60	55	22	57	6	47
53	62	11	30	25	28	19	38
32	13	54	27	56	23	48	5
63	52	31	24	29	26	37	18
14	33	2	51	16	35	4	49
1	64	15	34	3	50	17	36

FIGURE 6.3
Leonhard Euler's knight's tour chessboard. Print based on a drawing, 1766 [1759].

took a different approach a few years later. He set himself the task of solving the problem mathematically but ultimately created a visualization scheme all his own. Vandermonde's *Rémarques sur les problèmes de situation*—written in 1771 for the yearbook of the Académie Royale in Paris and published in 1774—is also interesting for the accompanying engravings. As far as we know, his essay represents the first modern mathematical treatise to introduce explicit net(work) metaphors.

However, several tied threads may relate to one another in space, Vandermonde formulated, their absolute, measurable position is irrelevant for a worker making a net (*réseau*) of knots. The worker is performing the task not according to scales of size but rather according to position: what he *sees* is the order of the interlaced threads ("ce qu'il y voit, c'est l'ordre dans lequel sont entrelacés les fils"). A calculation or notational form should be based on what we would today call "the worker's tacit knowledge."[56] Vandermonde implemented this precept by assigning both two- and three-dimensional coordinates (figures 1 and 2 in figure 6.4). In so doing, he not only formed arithmetical rows as paths but also modeled a visual mathematics of knots.[57] In the overlapping, there is no material knot. The inexact counting notation creates the knot in the first place (figure 6.5).

In this framework, the knight's tour problem can be solved in a new way. Vandermonde enters the coordinates for all sixty-four squares of the chessboard in a two-line series and limits himself to the problem of symmetrical moves. For the movement of the knight, the arithmetical solution is based on either:

$$\begin{matrix} 2 \\ 1 \end{matrix}$$

or:

$$\begin{matrix} 1 \\ 2 \end{matrix}$$

being added or subtracted from the coordinates—for example, as follows:[58]

5 4 2 1 3 2 1 3 1 2 4 3 1 2 4 3
5 3 4 2 1 3 1 2 3 1 2 4 5 7 8 6

As an iconic materialization of four subgraphs of sixteen moves each, this symmetrically constructed path represents what would later be known as

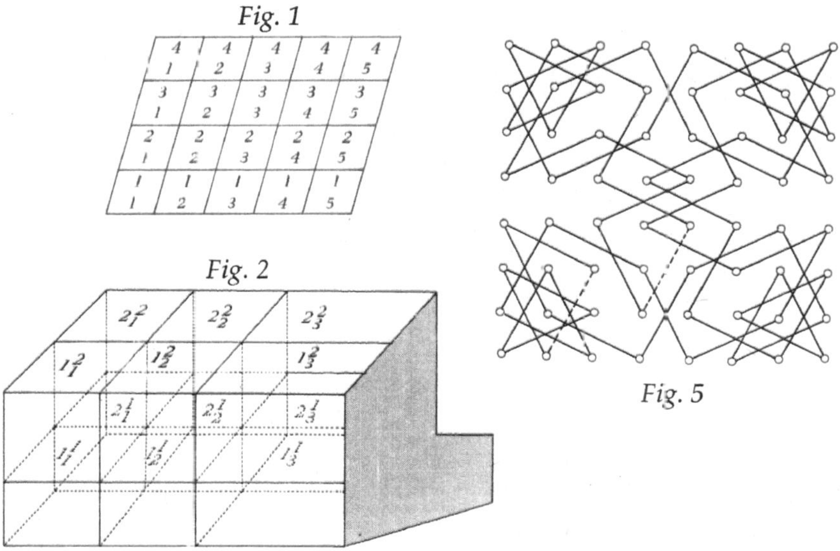

FIGURE 6.4
Alexandre-Théophile Vandermonde's topology, network graph of chess moves, 1774 [1771].

either a "finite graph" or a "Hamiltonian path" (figure 5 in figure 6.4) in the specialized language of graph theory. Vandermonde's illustration was first in the long series of solutions to this mathematical problem since the Middle Ages to replace the space of the material chessboard with a graph.[59]

This epistemic circle is a model *of* something whose pictorial content seems to be realized by the notational system in the first place, which then produces a net figure that has lost the chessboard as its base. On the one hand, the spatial analysis starting out from the material object is fundamental; on the other hand, one underestimates the graphic representation, which deliberately aims for an aesthetic effect, if one considers it merely a side effect of the arithmetical notation.

Practices of graph theory up to the present have been characterized by tension between picture, path, calculation, and visual model. Against this backdrop, it pays to consider the Icosian Game, which otherwise tends to be treated as rather marginal in the history of science. The game was developed by the Irish mathematician Sir William Rowan Hamilton between

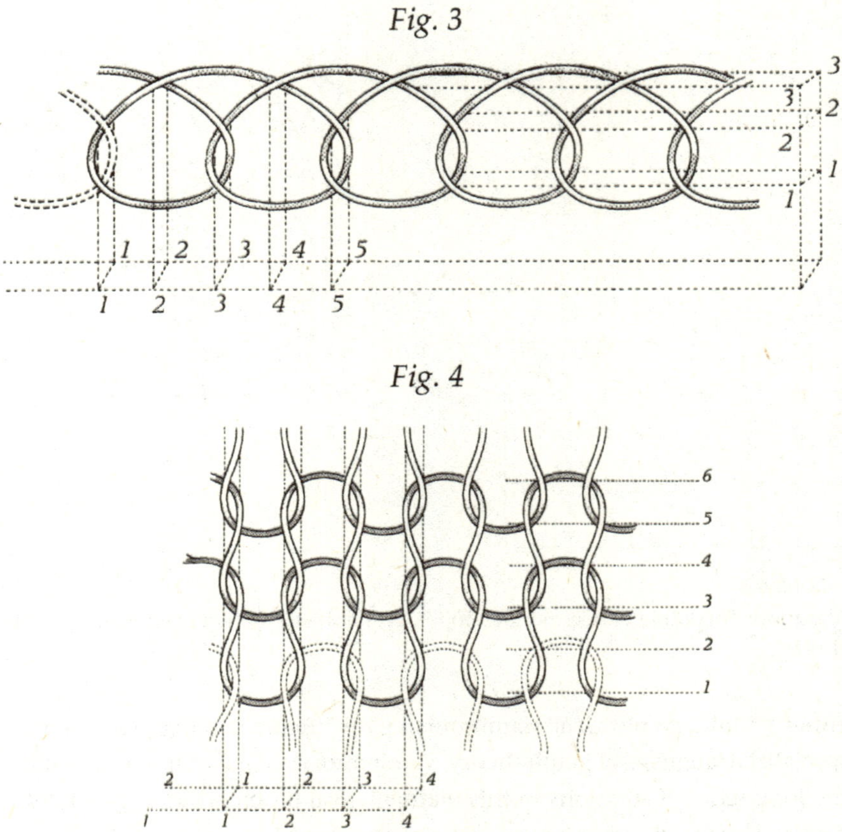

FIGURE 6.5
Visual mathematics of knots in Alexandre-Théophile Vandermonde, 1774 [1771].

1856 and 1859. Its name alludes to the twenty vertices of a wooden dodecahedron. Although the project turned out to be a financial failure for the Victorian game industry, it occupies an important place in the history of graph theory. A look at one of the four surviving examples suggests a comparatively harmless barricade game (figure 6.6), whose rules, written by Hamilton himself, represent an invitation to rack one's brains (which presumably did little to boost sales).[60]

Some or all of the twenty pieces of the Icosian Game are placed on the board. The goal is to find a circular or noncircular connection along the line and thus actuate a flexible switching. For example, player 1 could take

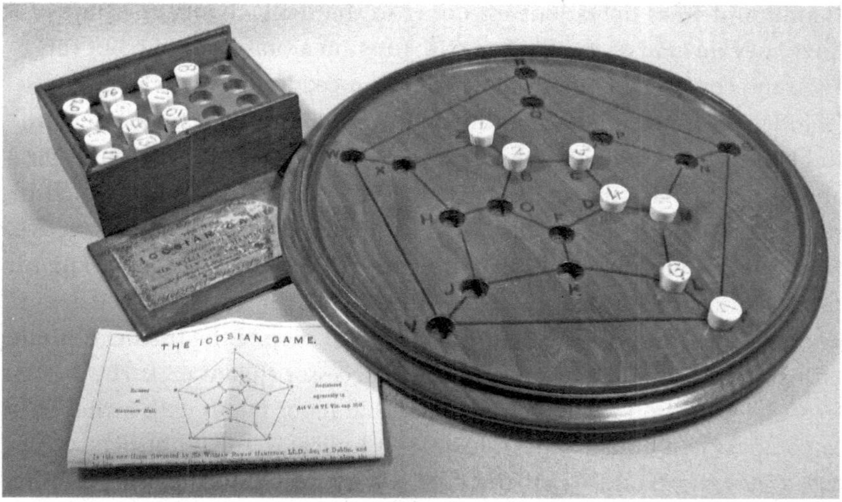

FIGURE 6.6
William Rowan Hamilton, Board, pieces, and rules of the Icosian Game, published by John Jaques and Son, 1856–1859. © 2021 Hordern-Dalgety Collection, http://www.puzzlemuseum.org.

five positions, and then player 2 would have to place the remaining fifteen stones so that they form a closed circle. The letters can be cycled through alphabetically or not. Alternatively, they can be specified by the other player as the endpoint.

Assuming this is a visual model, what are we dealing with here? Is it a purely theoretical and heuristic model to try out an "as if"?[61] If so, *for what* would it stand? One could say that we are dealing with a model for a mathematical diversion. Hamilton was extraordinarily proud of his game. He felt it represented his ideas on noncommutative algebra. Noncommutative means x times y is not equal to y times x—that is, the substitution (*commutare*) of variables that is normally possible with multiplication does not lead to the expected result. In noncommutative algebra, multiplication with the same variables can result in a different product, which is particularly important for understanding geometric ring forms outside of Euclidian space. This is also explained in the "Hints on the Icosian Calculus" on the lithographed directions for the game. The algorithmic rules for manipulating a line offer a hint about whether the Icosian Game is programmable.

Hamilton defines operations for doing so that help change the state of a line. They include reversal (BC to CB), rotation around a final point (BC to DC), and considerably more complicated procedures.

Johan Huizinga understood the riddle or riddle contest to be one of the fundamental culture-building forms that define relations between playing and knowing.[62] If one considers Hamilton's idea for the game, which he sold to John Jaques & Son for twenty-five pounds, from this perspective, it reveals nothing about a situation of scientific competition. The English mathematician Thomas Penyngton Kirkman had already published a more general solution to the problem in 1856.[63] But despite the financial failure of his first game, Hamilton produced a successor (figure 6.7). The Traveller's Dodecahedron no longer has a variable set of pieces with a challenge to compete in a riddle contest. Instead, it indicates in its subtitle, "A Voyage round the World," that the goal is a contest with oneself. In all its variations, the colonial journey goes from Brussels to Zanzibar more than a decade before Jules Verne had Phileas Fogg travel around the globe in

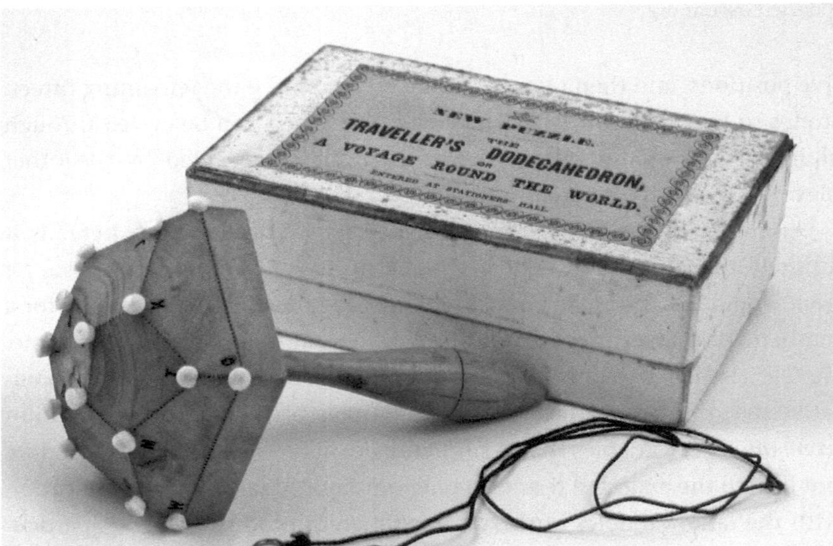

FIGURE 6.7
William Rowan Hamilton, The Traveller's Dodecahedron, circa 1860. © 2021 Hordern-Dalgety Collection, http://www.puzzlemuseum.org.

eighty days in 1873. The object has a maximum of twenty nodes, connected by a ribbon representing the route. The rules locate the twenty nodes geographically as follows: B = Brussels; C = Canton; D = Delhi; F = Frankfurt; G = Geneva; H = Hamburg; J = Jeddah; K = Kashmir; L = London; M = Moscow; N = Naples; P = Paris; Q = Québec; R = Rome; S = Stockholm; T = Tobolsk; V = Vienna; W = Washington; X = Xeres/Jerez; Z = Zanzibar.

Here, space is not *conceived primarily topologically*, as in the Icosian Game. Geographical and geopolitical references reappear in the context of play. Connections between the cities make sense, for example, when one travels from Moscow via Tobolsk to Kashmir. The preponderance of European metropolises is hardly surprising. Nor are stations in the heart of Britain's colonial dominions unexpected. The alleged margins of this network of world transportation are, however, interesting. Jerez can perhaps be justified economically as the capital of Kashmir. Jeddah on the Red Sea is fascinating because of its role as an Arabian financial metropolis. Tobolsk draws on its fame as a former capital of western Siberia.

The second edition of the Icosian Game begins with two travelers at the same time in the same city. The first four cities are visited jointly. After that, the second player has to find a way of reaching all of the remaining fifteen cities exactly once before returning to the starting point at the end. If efforts are successful, a circle has been completed. The goal of the game is to get back to where one started. The paths to a mathematical solution were—and are—not trivial. Even today, there is no efficient method for solving the problem of the Hamiltonian path.

Seen together, both games are, above all, models *for* something. They are early and—in contrast to timetables—not yet marketable route planners without a GPS function. Anyone using today's navigation systems or booking train connections on the Internet is encountering a modified version of Hamilton's game in which an opponent has determined where the blocking pieces stand. It is remarkable how long the knowledge made explicit by Kirkman and Hamilton was considered relevant only to popular mathematics. Graph theory remained latent in the history of knowledge.[64]

GRAPHS CAN DO ANYTHING: JAMES JOSEPH SYLVESTER'S DREAM AND CHARLES SANDERS PEIRCE'S VISION

The media properties of the game make possible, in the words of Natascha Adamowsky, "the organization of abstraction." Inventing a game is almost creating a model. Conversely, every representation of an abstraction can be understood as a playful design.[65] The emergent "serious" sciences, however, evidently feared too close an association with games or play. The cautious introduction of three-dimensional models of atoms in chemistry in the 1860s was met with the mockery of professionals. Not entirely without reason, chemists practicing in Britain feared associations with children's construction sets and model kits. For the first public lectures, held by the German chemist August Wilhelm Hofmann 1865 in London, table croquet sets were the preferred material from which to assemble three-dimensional sculptures of atoms.[66]

As a teaching aid, these "glyptic [i.e., sculptured] formulas" were intended not to imitate the structure of actual atomic bonds but rather to model all their relationships visually (figure 6.8).[67] Such plastic models remained controversial, however, and so-called chemicographs gradually became the standard for two-dimensional chemical notation. This pictorial turn in chemistry includes founding scenes as prominent as Euler's Königsberg study and the feverish dreams of the polyhedron formula: August Kekulé's famous visions of chemical structures. Kekulé's tales still sound like the origin myth of atomic notation and are, in a sense, part of a rhetorical figure of chemical bonds in general.[68] His most famous dream, in which the benzene ring and its alternating single and double bonds evoked the

FIG. 19.

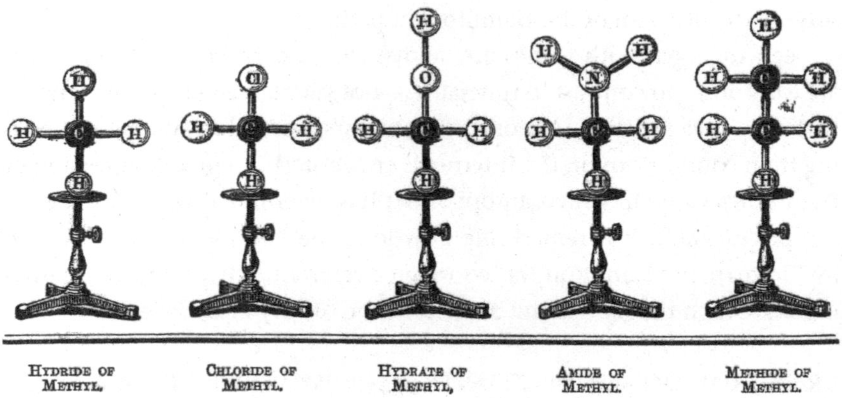

FIGURE 6.8
August Wilhelm Hofmann, models of atoms of bonds with methyl/–CH^3. Based on the models presented to the Royal Society in London on April 7, 1865.

symbol of the world serpent Ouroboros, has received a literary monument in Thomas Pynchon's *Gravity's Rainbow* (1973):

> Young ex-architect Kekulé went looking among the molecules of the time for the hidden shapes he knew were there, shapes he did not think of as real physical structures, but as "rational formulas," showing the relationships that went on in "metamorphoses," his quaint 19th century way of saying "chemical reactions." But he could visualize. He *saw* the four bonds of carbon, lying in a tetrahedron—he *showed* how carbon atoms could link up, one to another, into long chains . . . But he was stumped when he got to benzene. He knew there were six carbon atoms with a hydrogen attached to each one—but he could not see the shape. Not until the dream. . . . Kekulé dreams the Great Serpent holding its own tail in its mouth, the dreaming Serpent which surrounds the world.[69]

The American mathematician James Joseph Sylvester (1814–1897) also used a dream scene to introduce the idea he published in 1878 of a unity of chemical and mathematical notation systems. As he pursued ideas for how a mathematician could bring together chemistry, physics, and biology, Sylvester found that the model occurred almost all on its own: "I was agreeably surprised to find, of a sudden, distinctly pictured on my mental retina a chemico-graphical image serving to embody and illustrate the relations of these derived algebraical forms to their primitives and to each other."[70]

Writing down, once again, seems to have preceded visualization.[71] One could say that nineteenth-century "images of objectivity," left "as natural as possible," whose importance Lorraine Daston and Peter Galison have demonstrated so persuasively, were produced the other way around here.[72] When Sylvester applied mathematical substitutions to free radicals—the parts of molecules that will bond with all other substances—he found mathematical replacements for chemical elements. The result of this process was a catalog of chemical bonds (figure 6.9).[73] With Sylvester, nature did not write itself. It is not a model *of* something. Dreaming mathematical modeling produces the diagrammatic structures in the first place.

Sylvester's mathematical approximations testified to ongoing uncertainty about the invisibility of chemical bonds. The algebraic notation introduced by Jöns Jacob Berzelius beginning in 1813 was supplemented from the late 1850s onward with a more visual form of recording.[74] To depict valences—open opportunities for bonding—Scottish and English chemists such as Alexander Crum Brown and Edward Frankland worked between 1864 and 1866 to standardize the system of so-called graphic notation. Brown

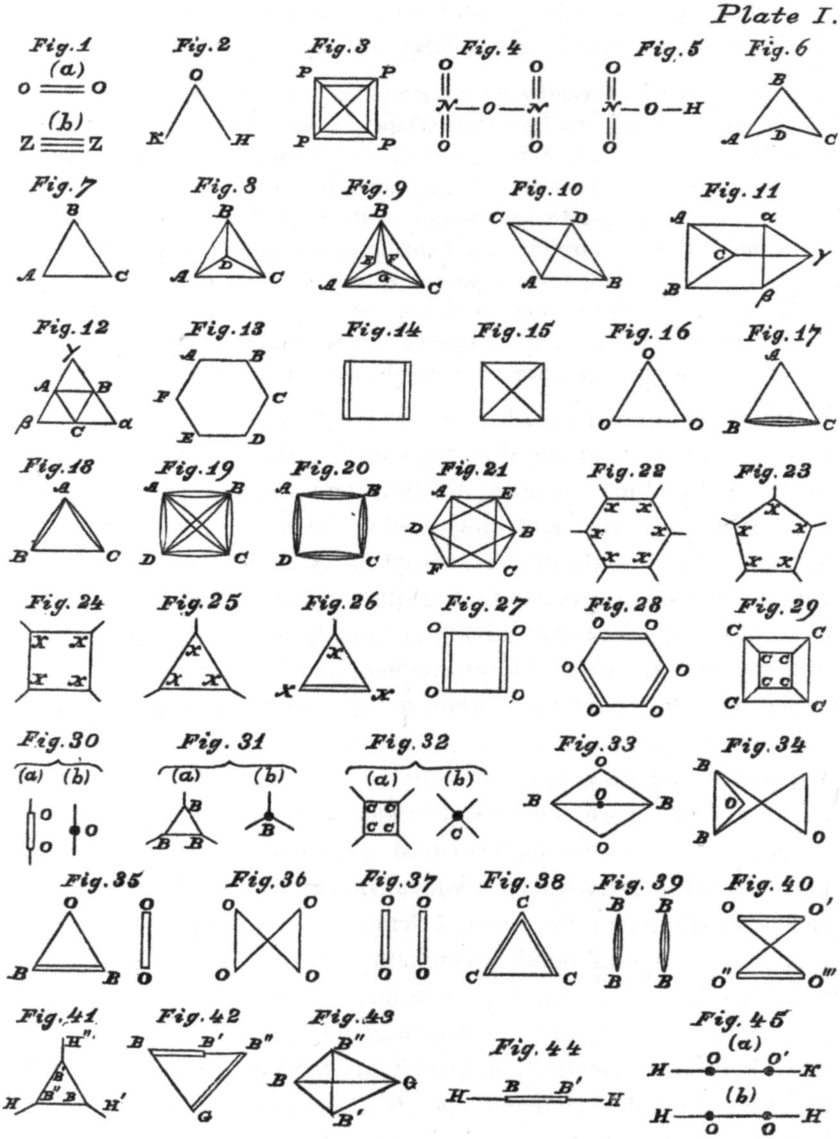

FIGURE 6.9
James Joseph Sylvester, Catalog of chemical elements, 1878.

provided visual information about isometry in 1864. The various materials can have the same letter notation but require a different structural formula to represent them. In his *Lecture Notes for Chemical Students*, written in 1865 and 1866, Frankland stressed that the illustrations represented neither the form of the molecule nor their relative position. They were strictly about bonds.[75] A theoretical model thus interacted with experimental practice in trying to do justice to the relationship of molecular forces.

One might think that Sylvester could simply fall back on practical experience in chemistry. His shortening of the word "chemicograph" to simply "graph" in an article in *Nature* in February 1878 did indeed become standard, even though his combination of algebra and chemistry did not gain acceptance over the long term.[76] The illustration he included once again drew on the various possibilities of notation in a catalog form that promised objectivity, beginning with the established way of writing the different possibilities of multiple bonds (see figure 1 in figure 6.9). Sylvester eventually omits the "concrete" level of atomic bonds in favor of a geometric rendering, which not only followed the established pictorial forms for Platonic solids but also continued traditions of Hermetic magic and contained the core of his mathematical theory of atoms. Thinking about covariance and invariance is crucial to the generation of the image. Whereas covariance describes the change by transformation, invariance defines what remains the same, even when the formula or the substance changes. The question of (im)mutability concerns chemical modeling on three levels: the purely mathematical, the chemical-algebraic, and the visual.

The written and pictorial dimensions stand in conflict. Transforming a formula leads to a new formula that retains its invariant parts. The metamorphosis of a diagram of chemical bonds necessitates not only the renotation of the mutual part but also a completely new composition of the graph.[77] At the same time, however, the greater effort can generate epistemic surplus value, particularly in the case of a graphic proof of "true representation."[78] Whereas figure 17 in figure 6.9 is an incorrect graph of the bond of three nitrogen atoms, figure 18 in figure 6.9 depicts it appropriately, according to Sylvester. Like all chemists of the day, he stressed that chemicographs did not in any way show the material. They were useful because they afforded "graphical proof" of the relational structure of molecules.[79] A good example of this is the juxtaposition of detailed and condensed geometric representation of possible bonds in figures 30–32 in figure 6.9, in which one can sense the leap to three-dimensional models of atoms. Despite all their limitations, chemical graphs and three-dimensional models of atoms evolved into a diagrammatic universal model. Charles Sanders Peirce—geographer, chemist, mathematician, logician, semiotician, and, incidentally, one of the first subscribers to the *American Mathematical Journal*—played a pioneering role in the transition. As topological graphs became more common, they were adapted to other areas of science, including linguistic logic and electrical switches (figure 6.10).

Wolfgang Schäffner has shown that diagrammatic representation and mechanizability merge in Peirce's existential graphs. More than combinations of signs and image acts, they "operate with statements that also actually exist in the corresponding discursive universe."[80] Peirce, who was as skilled as his chemist colleagues at describing the dreamlike act of semiosis, anticipated a movement that would have a lasting influence on the order of knowledge in the twentieth and twenty-first centuries.[81] Topological graphs can model everything inasmuch as they model connections: circuits, transportation networks, political activity, social contexts, biological food networks, the infrastructure of the Internet, computer programs (Petri nets), molecular structures, and firing synapses. The network diagram's popularity in the sciences is due to the continuous crossover between calculability and iconicity—the writing act and the image act come together to materialize the "work" of the network to be recorded in the diagram.

Especially from an epistemological perspective, graphs and network diagrams may seem to be relatively simple, pragmatic methods of recording without a specific context. But they have quietly developed a pathos of their own. Based on the older, in part magical code of binding and dissolving, they have become network generators that are simultaneously a model *of* something and a model *for* something. Network diagrams become

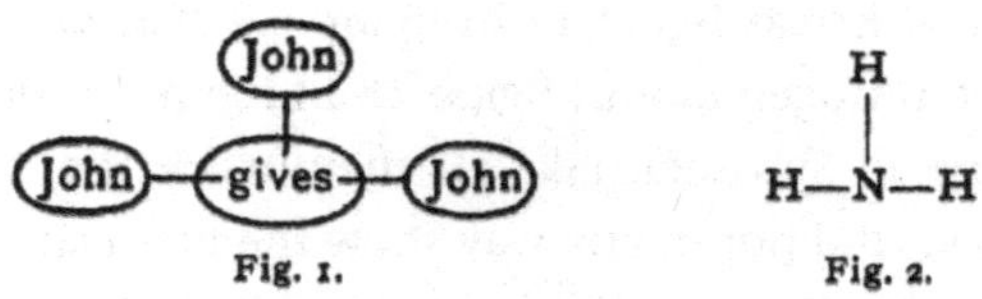

FIGURE 6.10
Charles Sanders Peirce, Analogy between logical and chemical notation, 1897.

a model *of* something as a result of cultural techniques of measuring and observing—for example, flows of data and traffic, the exchange between ecosystems, or migration movements. As a model *for* something, they lend network society not only opportunities to control the "space of flows" but also a diagrammatic form of self-description and self-reflection.[82]

6.4 JACOB LEVY MORENO'S ELECTIVE AFFINITIES: SOCIOMETRY, SOCIAL ATOMISM, AND THE BINDING FORCES OF SOCIAL NETWORKS

> God was a super sociometrist.
> —Jacob Levy Moreno[83]

The semiotic practices of chemistry and its diagrams left significant traces in another field of knowledge. Anyone exploring the sociological and anthropological literature of the mid-twentieth century will find concepts and visual arguments that refer to atomic models and their physical and chemical properties. Some American sociologists imported these forms of representation in scientific garb in a kind of happy parasitism while Claude Lévi-Strauss constructed elaborate structuralist models around the atom of kinship.[84] In any case, the "social atom" (figure 6.11) became an element in the visual depiction of human relationships,[85] although Lévi-Strauss did not employ a comic-book style but rather a sober diagrammatic formula for rendering relationships. Such epistemic theft marked the beginning of a sociological love affair with the visual codes of chemistry and molecular physics that would, in turn, become the foundation of data processing in the social sciences.

Such forms of exchange between the science of substances and the world of ideas about human relationship look back on a long history that includes hermetic and alchemical traditions, Lavoisier's revolution in chemistry, and Goethe's *Wahlverwandtschaften* (Elective Affinities) leading up to the twentieth century.[86] The sociometric drawings I will discuss represent a kaleidoscope of American-European elective affinities. Not for the first time, the social aspect of chemistry produced a chemistry of society in which emotional bonds drove the exchange.

On April 3, 1933, the *New York Times* published a brief article with the headline "Emotions Mapped by New Geography." The piece began with a sensational lead: "Charts Seek to Portray the Psychological Currents of

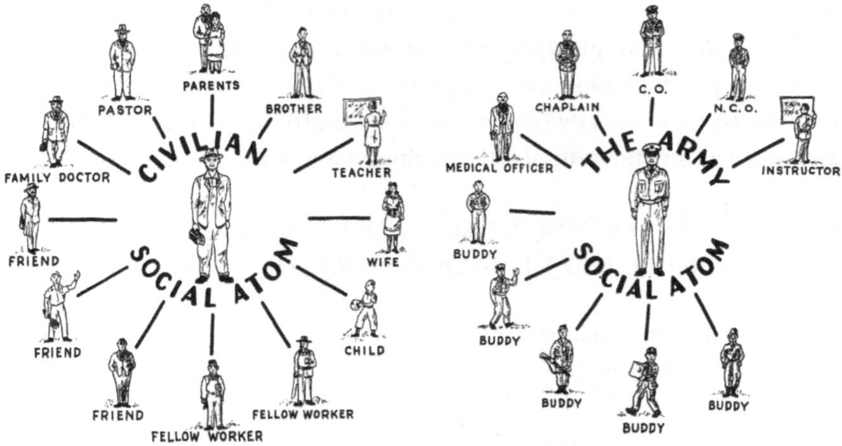

FIGURE 6.11
Ernest Fantel, *The Civilian and Army Social Atom, Before and After*, 1951. The "social atom" shows the transformation of personal bonds between the environments of family and military. It not only presents the relationships of attraction, repulsion, and neutrality around an individual, but also represents, according to Moreno, the culture of roles around a person as cultural atom.

Human Relationships. FIRST STUDIES EXHIBITED Colored Lines Show Likes and Dislikes of Individuals and of Groups. MANY MISFITS REVEALED Dr. J. L. Moreno calculates there are 10 to 15 Million Isolated Individuals In Nation."[87] Readers learned of the invention of a new science called psychological geography. It was presented, the article continues, at the annual meeting of the New York Medical Society at the Waldorf Astoria hotel with graphic charts. The color charts of the social scientist Jacob Levy Moreno (1889–1974) had been handed out to the press ahead of time by Dwight Anderson, the publicist of the Medical Society.

Moreno had first published the results of sociogeographical research conducted with social psychologist Helen Hall Jennings in a book titled *Who Shall Survive? A New Approach to the Problem of Human Interrelations* in 1934.[88] The second edition, from 1953, which incorporated the "social atom" and was dedicated to Moreno's third wife, Zerka, had a different subtitle: *Foundations of Sociometry, Group Psychotherapy and Sociodrama*. That same year, Grete Leutz translated the work into German as *Die Grundlagen der Soziometrie* (The Fundamentals of Sociometry). The subtitle, *Wege zur Neuordnung der Gesellschaft* (Paths to a Reordering of Society), suggests the social-utopian and ultimately religious quality of Moreno's thinking.[89] Moreno was, among other things, an anti-Marxist, anti-psychoanalytic (or anti-Freudian), and semi-messianic thinker with an unclear relationship to his Sephardic Jewish origins, always searching for inspiration and support in his work from new female "muses."[90] He was also the head of an improvisational theater and, as such, a psychodramatic group therapist.[91] After his European years in Vienna and Berlin, he immigrated to the United States, and there, from the early 1930s on, he headed a reform school for girls in Hudson, New York.

The spatial layout of the New York Training School for Girls resulted in the forms of sociogeographical observation that Moreno applied on a larger scale in his presentation in New York. The sociometric concept of a network resulted from a mix of ethnographic and experimental components in which the sociometric test developed by Jennings was the basis for recording the personal relationships and "psychological currents" of a community of Black and white girls. Their status as outsiders and the questions of social differences according to ethnic membership (race) play a constant role in Moreno and Jennings's reflections. The researchers' case studies, which were made possible with the cooperation of the superintendent Fannie French Morse, revolved around the everyday activities of the closed educational space.

Moreno's wish to identify the pattern behind repeated escapes of fourteen girls on fourteen consecutive days in the autumn of 1932 led him to undertake mapping on a massive scale. Five charts transform the camp with 435 residents in sixteen cottages into a topological structure. The girls' rectangular buildings are redrawn into a relational atom characterized by circles and connection lines (figures 6.12 and 6.13). Attraction (red), repulsion (black), attraction/repulsion (red/black), and indifference (blue) are represented visually. The course of these different "psychological currents," measured by questions from housemothers, generates a model based on energy.[92] What is first called "repulsion," borrowing from electromagnetism, is quickly renamed social "rejection." The attractions and rejections of individual currents are classified on various questionable levels: sexual, "racial," social, industrial, and cultural.[93]

But all of this is not enough for Moreno to explain the fundamental principle of social connections, which he traces back to a particular form of grafting.

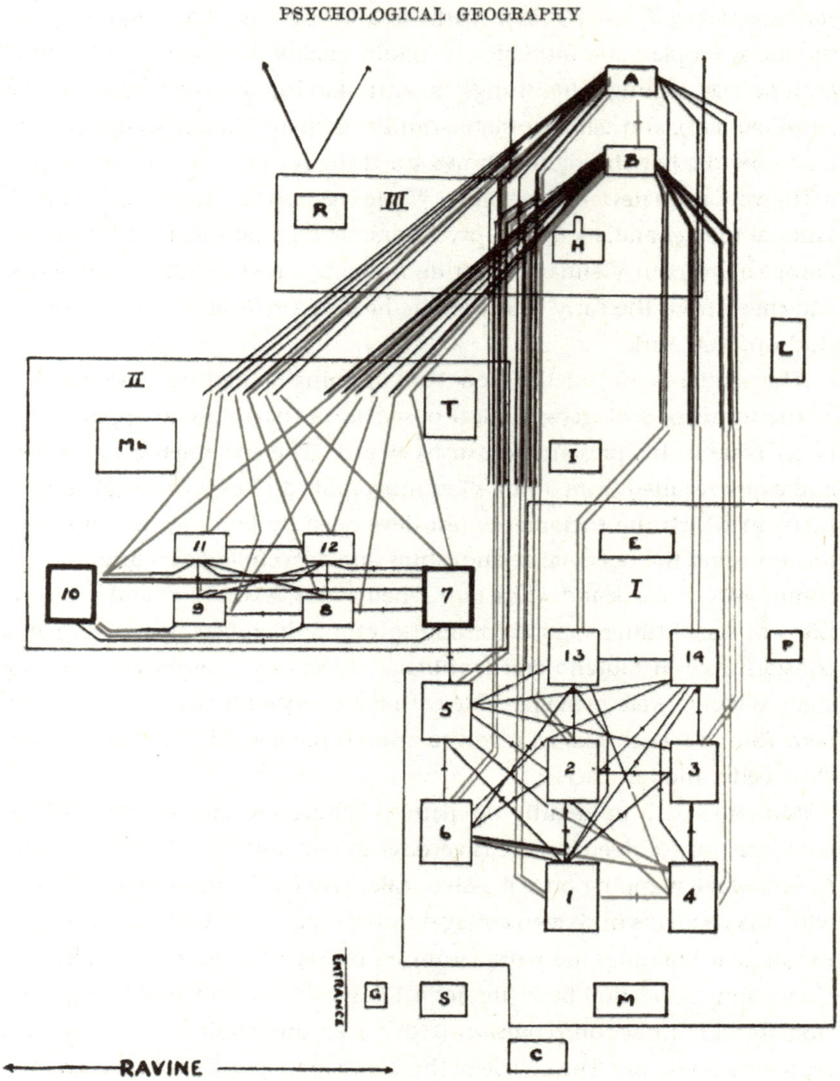

FIGURE 6.12

Jacob Levy Moreno (?), *Psychological Geography Map II*, 1934. This map shows the topographic space of the camp in Hudson. The "psychological currents" between the individual cottages are indicated by red (attraction) and black (rejection) lines. Red and black connections represent clashing emotions: half attraction and half repulsion.

"There are still deeper layers. We had suspected that beneath the ever flowing and ever changing currents there must be a permanent structure, a container, a bed which carries and mingles its currents, however different their goals may be."[94] He concludes that the fourteen runaways were part of a "hidden" network. This observation follows a method of diagrammatic notation that Moreno had already employed to record his psychodramatic improvisations. Without the analog modeling of connection lines, no network vision is possible. Net-like graphs work here as a performative social bonding agent that guarantees the freedom and independence of the individual—one can detect a very American spirit here—and, at the same time, becomes the controlling superorganization as the society matures. But Moreno's enthusiasm for mapping previously hidden networks got the better of him. He began pronouncing universal theses informed by Gestalt psychology:[95]

> The networks represent the oldest form of social communication. Traces of them are already in subhuman societies. They are collective formations, the individual participants are unconscious of all the networks in which they partake, although they may be aware of one or another link between some of the individuals, or realize that such networks exist. An individual cannot move out of networks, just as he cannot move outside of his skin. Networks pre-exist him and pre-exist the official groups of which he is a part.[96]

As scientifically dubious as the surveillance regime of Moreno's housemothers may have been, it was supposed to measure social exclusion precisely. Sixteen percent of the girls under examination were not considered worthy of contact by others. With a brief reference to comparable studies, Moreno extrapolates from the Hudson data to the whole population of New York. Of seven million individuals, 1.12 million must find themselves in an isolated position. In a social imperative, this implies, "Seek out closeness to your fellow citizens!" As an explicit call, the appeal had its roots in Moreno's work with group therapy and the improvisational theater of spatial positioning.

Hand-drawn charts III and IV (figure 6.13)—the first of which is said to have been around twenty by fifteen feet—represented the topological response to the question of which girls would most like to live with which others. Based on these social preferences, Jennings and Moreno identified five networks in Hudson. They were determined based on intense drawing of maps that were tabular matrices of rejection, attraction, and indifference (figure 6.14). In the mathematical language of graph theory, Moreno and Jennings followed every path to where it broke off and then turned to

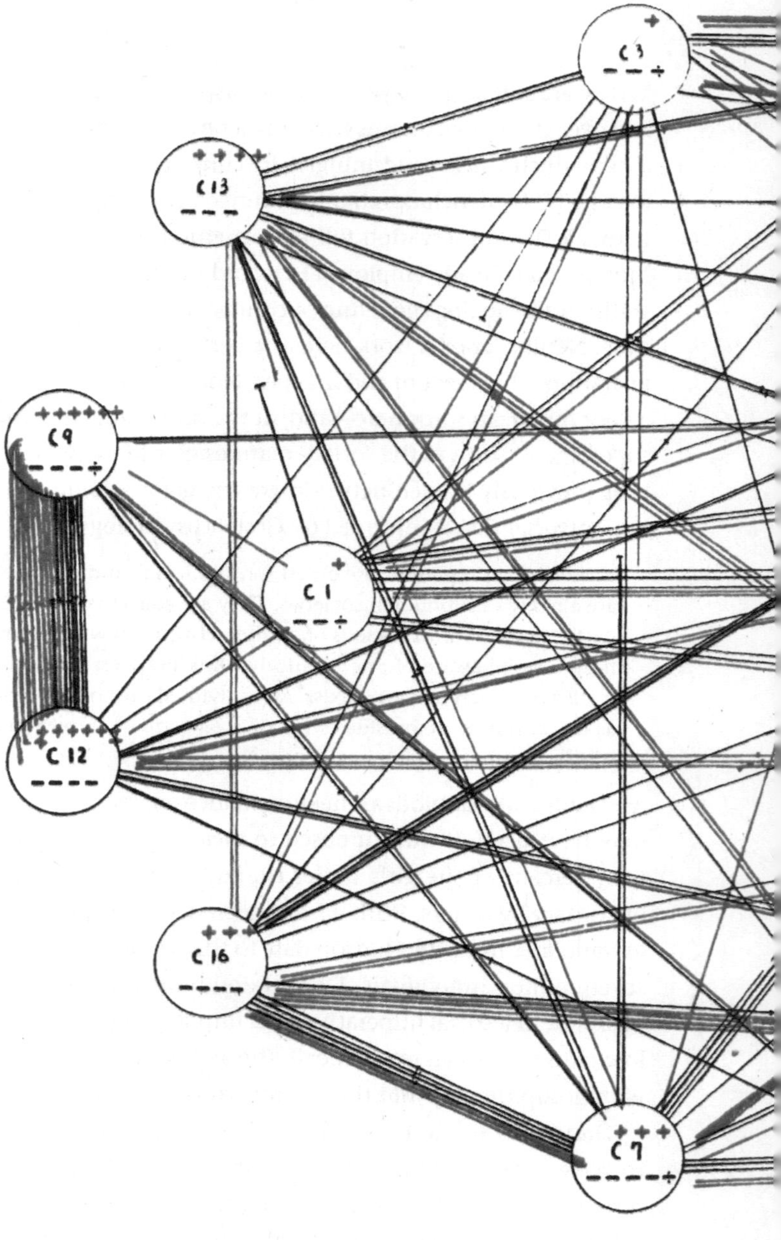

FIGURE 6.13

Jacob Levy Moreno (?), *Psychological Geography Map IV. A Reduction Sociogram*, 1934. A reduction of the overall map on which plus and minus operators provide the scale. Every plus codes ten attractions and every minus ten rejections in one cottage. Double points on a sign halve the points. Every red and black line between the cottages stands for five attractions or rejections. The total can be calculated by multiplying by five. This technique reduces the quantity of lines so that "large populations" can also be mapped.

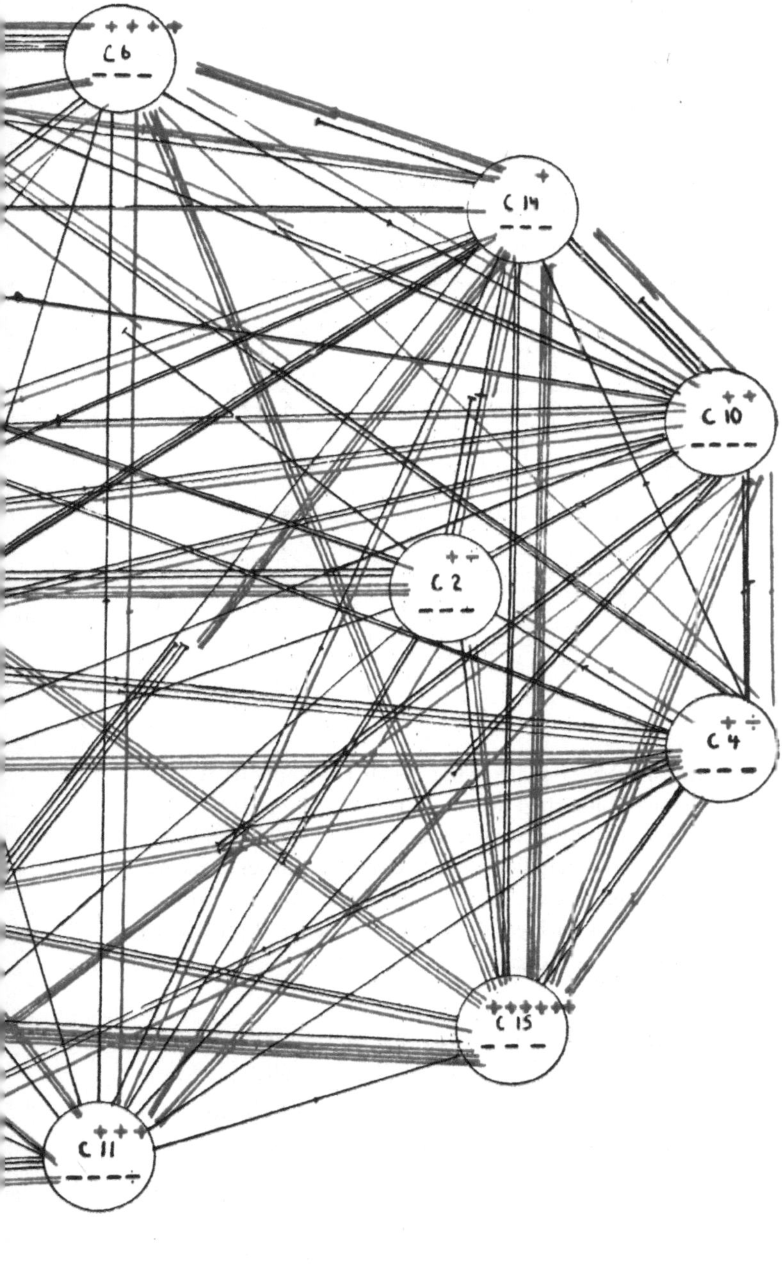
C6
C14
C10
C2
C4
C15
C11

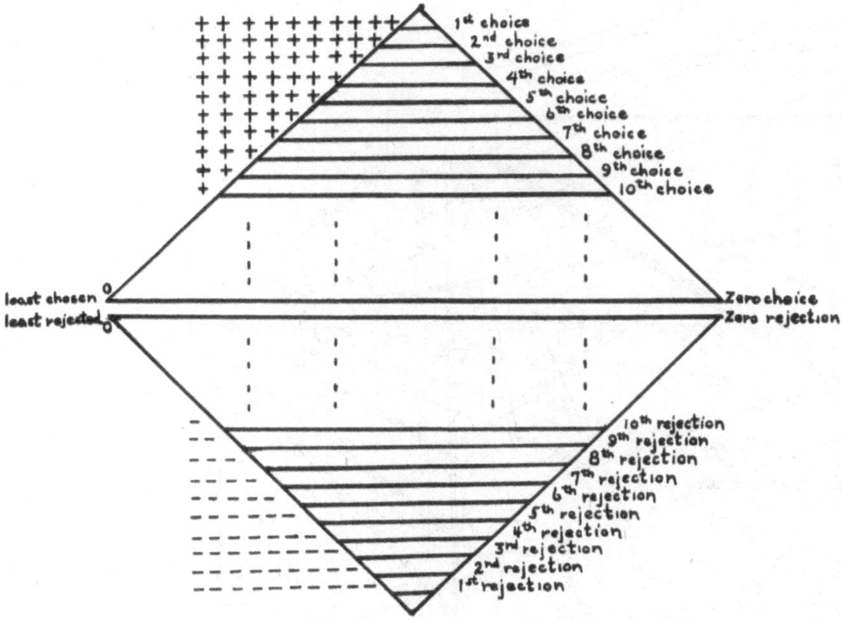

FIGURE 6.14
Jacob Levy Moreno (?), *Hierarchic Sociogram*. Drawing, 1934. The upper triangle represents the "field of positive forces" and the lower one the "field of rejection." In the case of the question of who would most like to live together with whom, both show the person preferred (plus, first choice, etc.) or biggest rejections (minus, first rejection etc.). Moreno illustrated by means of such gradations—for example, the number of isolated individuals who were not selected. As a general schema, it pins society down to hierarchies of rejection and attraction.

the next chain. The structure that resulted from this mapping of yes/no or rejection/acceptance excludes relationships of indifference, which Moreno regarded as just as important for the social dynamic. He thus already came very close to the functionality and aesthetic of digital networks.

Without the possibilities offered by digital computers, sociometric matrices are best suited for managing data in tables. The relative degrees of social connection are determined by entries in charts. The relationships are coded first by various operators such as plus, minus, and zero.[97] In quantitative social science today, the digital distinction of one and zero, connection and nonconnection, has become standard.[98] But all ethnographic fieldwork begins locally. Groups and communities in which one attempts a complete sociometric survey became the field for discovering social networks.

As for Jacob Levy Moreno and Helen Hall Jennings's sociometry, for Alfred Radcliffe-Brown and the Manchester school of research into social networks, the "network" served to structure and explain data from field research.[99] Here, too, the concept helped to abolish false distinctions between the "primitive" and the "modern." Social anthropologists noticed and analyzed interpersonal networks in small towns in North America and in Norwegian fishing villages as well as in African cities.[100]

The turn from macrotechnological network infrastructure to microsocial networks during the second half of the twentieth century began at the scientific and geographical margins.[101] Moreno's stories of the reform school, the sociometric arts of drawing, and Jenning's precision in empirical data collection established the parameters for the articles they collected in the journal *Sociometry*. For example, in their reports "Social Attraction Patterns in a Rural Village," published in 1937 and 1938, the sociologists George A. Lundberg and Mary Steele profess their faith in an objective analysis of social activities as "atomic behavior."[102] Their site of investigation was an unnamed community in Vermont with a population of around a thousand, whose social structures were to be determined by surveying around two hundred families (94 percent of the population). Lundberg and Steele were primarily interested in analyzing the energies behind or preceding recognizable processes. People were understood as so many elementary particles. Imagining and depicting them as social atoms raised the question of structural rules, including the social relationships revealed.

In their first attempts at mapping, which used data from questionnaires, direct sociometry soon reached its limits, which stemmed from the density of the network to be recorded. The lines condensed into an "inextricable maze which tended to obscure the significant clusterings which we wished to exhibit."[103] Steele and Lundberg, however, confronted the challenge: "If we choose to regard any social group as a system of energy flowing through a more or less intricate pattern of intersecting channels, then the charting of these channels becomes a primary problem in the explanation of the behavior of the group."[104] Employing several scales used by sociology of the 1930s to indicate socioeconomic and cultural status, the field researchers collected spontaneously offered information from housewives about their closest friends in the community[105] and pursued an energetic schema of "attraction" and "repulsion."[106] The result of a mixture of chemistry, economics, behaviorism, veiled psychoanalysis, and undisguised structuralism was the cyberneticization of an entire small town in eight elegant diagrams.

These diagrams represent the existing networks or "constellations" in which people with a large number of incoming and outgoing connections occupy the center.

The most prominent example—the "Lady Bountiful," who is rendered in the style of a cabbalistic sefirot tree—was drawn by the student Jane Pitts, who participated in the project (figure 6.15). Grouped around the endearing and generous old widow of sixty, with the geographic localization M31 and the socioeconomic score 258, are many people who, according to the direction of the arrow, tend to be in a one-sided relation to this matriarch of the dominant family. They, in turn, have a large number of connections to the seven other networks. Lady M31 feels a connection to M16—physician, politician, and, in sociometric nomenclature, a "satellite." The latter, in turn, is appreciated by another satellite, the banker's wife B7. The other six diagrams also operate in this geometric, narrative mode à la Serres. The fifth map, for example, shows how the centrally placed wife of a factory foreman serves as a social hub. One exception to the continual game of the debit and credit of the social capital of connections is shown in the final diagram, centering the lonely and excluded inhabitants of Vermont. Having no connections signifies unimportance and social exclusion (figure 6.16).

Such paper arrangements might be judged a plaything from the nursery of sociological network studies. Related methods, such as Mary Louise Northway's "target sociograms" of the 1940s, served as a form of ludic, dynamic modeling of social relationships (figure 6.17). That, however, would mean underestimating Lady M31, who even after Lundberg and Steele's report haunts the history of English-language sociology as a charitable undead who is very much alive. The data from the surveys were not only entered into a machine in 1950 but also, around 1980, used as the basis for the first three-dimensional real-time computer visualization of social networks, of which only a few images survive today (figure 6.18). Once again, sociologists borrowed the semiotic practices of chemistry—with the difference that entire software packages, such as the crystallography application ORTEP, were now repurposed to new ends.[107] There was initially great enthusiasm for doing so, although the software implementations of the relationships of chemical forces—one need only think of multiple bonds and valences—still do not easily reveal sociological insights.

The early years of the twenty-first century brought a rise of quickly manipulable network simulations. In addition to solutions that have been

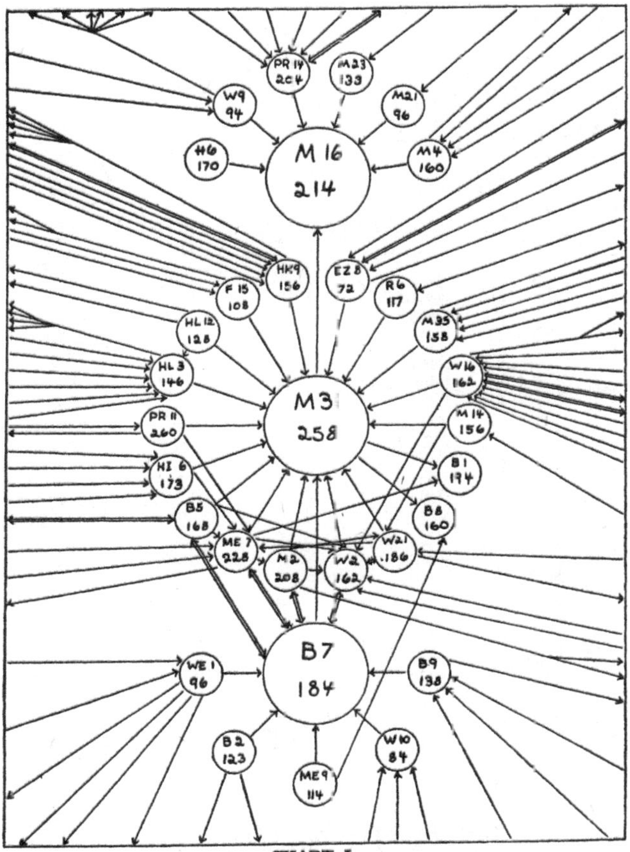

CHART I

The most important "friendship" constellation in the village, centering on the matriarch of the leading family with her principal satellites, a physician-politician (M 16) and a banker's wife (B 7). Each person is represented by a circle. The letter and the first number in the circle is the code symbol of the person. The second number is that person's score of socio-economic status on the Chapin Scale. Each arrow represents a choice made or received according to the direction of the arrow. Mutual choices are represented by double-headed arrows. (On account of charting difficulties the following lines are omitted from this chart: Me 7 and M 2 to Pr 14; We 1 and W 16 to Hl 3.)

FIGURE 6.15

Jane Pitts, *Chart I: Friendship Constellation* ("Lady Bountiful"). Drawing, 1938. Instead of color notation, this diagram uses arrows to code mutual attitudes. On the evaluations of the Chapin scale used, see note 105.

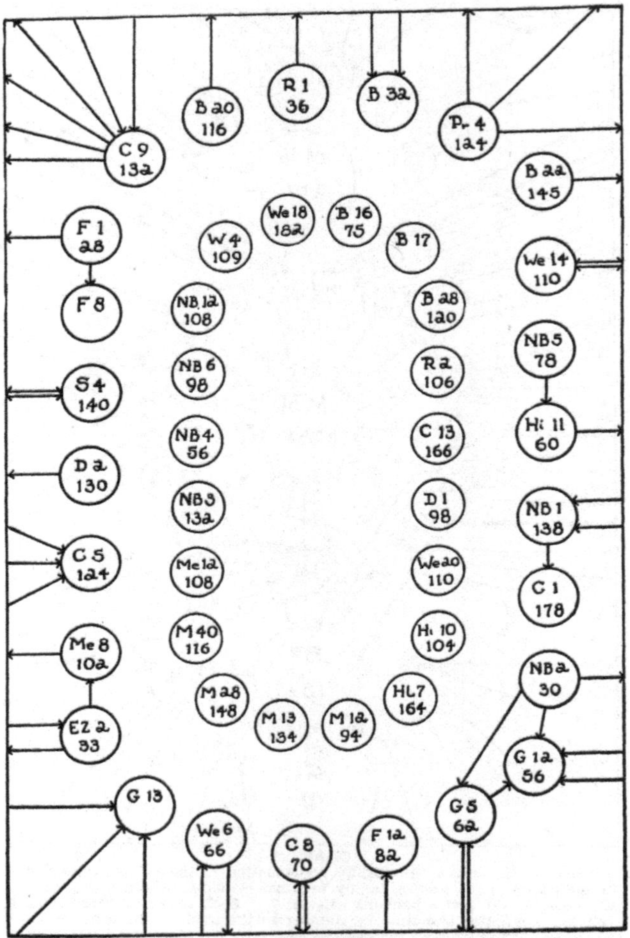

CHART VIII

Isolated and semi-isolated individuals. This chart consists of all persons (a) (inner circle) not making or receiving any choices in the village and (b) persons not appearing in any of the other charts or on their margins. All except three of the inner circle are recent migrants, persons with normal social contacts in adjoining areas or neighboring villages, and some summer residents. Mean socio-economic score, inner circle, 117; outer circle, 93, by far the lowest of any group in the village. The group makes 21 outward choices toward 18 individuals, a per capita of 1.1. 21 choices are received from outside by 14 individuals, a per capita of 1.5. Centripetal index, 1.3.

FIGURE 6.16

Pauline Lee: *Chart VIII. Isolated and Semi-Isolated Individuals.* Drawing, 1938. In the connectionism of sociometry, social isolation means no connections drawn.

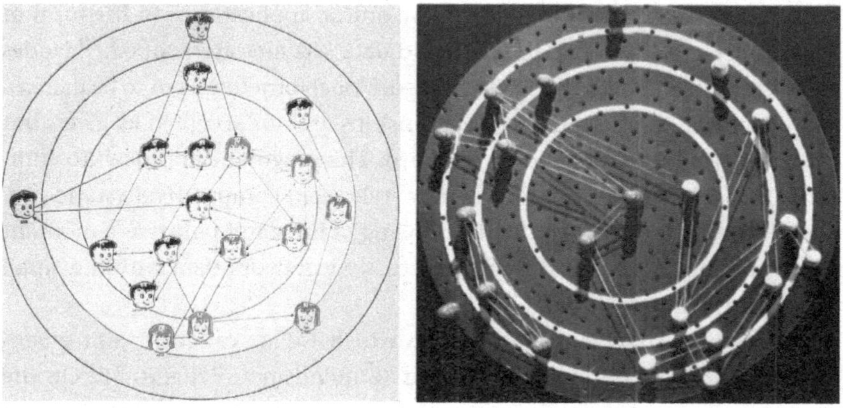

FIGURE 6.17

Left: Pat Grant, *Target Sociogram of a First Grade Class*. Graphic, 1952. Right: Dorothy McKenzie, *Target Sociogram Board*. Model, 1952. Both Grant and McKenzie were students of Mary Louise Northway, who taught at the University of Toronto. Her sociometric field research was dedicated primarily to children and young people. As in the work of Lundberg and Steele, the focus on the center plays an important role in the representation of social status.

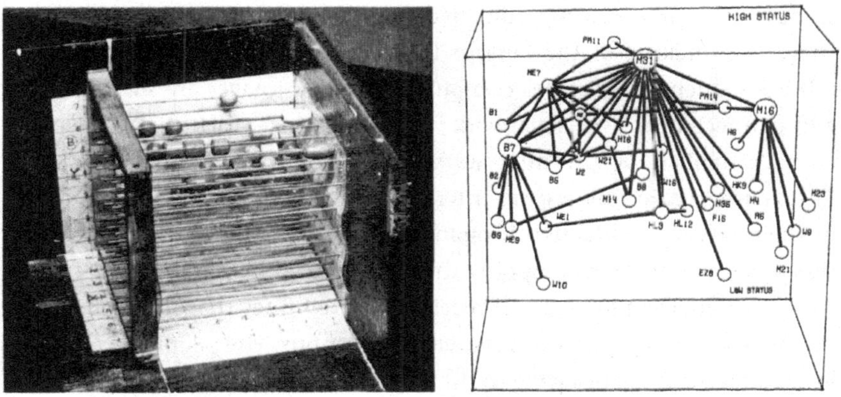

FIGURE 6.18

Left: Francis Stuart Chapin, *Tabulating Machine*, 1950. The stepfather of George Lundberg and inventor of the Chapin Scale wanted to illustrate with this machine the social status of the small town in Vermont, which he did not see embodied in earlier diagrams. Lady M31 is there positioned at the top in his machine in accordance with her social status. Right: Alden S. Klovdahl/ORTEP, "A Three-Dimensional Computer-Drawn, Visual Representation of the 'Lady Bountiful,' Network Rotated to Present a 'Side View.'" Computer graphic, 1981.

standard for many years, now there are online applications, in the form of Java applets, for example, to which old data sets are often added.[108] Nodes can be moved and lines or edged removed. Visual form turns out to be increasingly variable. More than ever, relationships are not accepted as given but rather actively constructed and modeled. The pragmatic approach to computer data can make the diagrammatic style seem comparatively static. Yet, the movement-image includes the drawing of lines and points as a moment of calculation,[109] and it attempts to represent network data within a small space of registration.[110]

"We cannot draw closed the net in which we are caught," Walter Benjamin observes pithily in the fragment "Capitalism as Religion."[111] On the contrary, pressure toward continual interconnective expansion is what makes up networks. Unfolding in time is just as possible via narrative and discourse as via image acts. The cultural techniques of speaking, writing, drawing, and calculating meet up in the network diagram, which for the most part supports Serres's epistemology: "Iconic communication transports *all abstractions and all histories*, theorems and narratives."[112]

One cinematic example of this continual crossing of writing and the visual dramatizes a large-scale historical sociological project as a multiple scene of networking. *Kinsey*, directed by Bill Condon and released in 2004, shows the making of the famous first Kinsey report in a complex four-minute sequence. Faces and computer-animated images are superimposed and connected as if in a network.[113] The sequence's discourse on gender is made explicit by the interviewee repeatedly asking, "Am I normal?" The viewer's gaze follows the interviewers around Liam Neeson's Kinsey in their surveys conducted throughout the United States. Condon directs this process as an act of mapping that is expanded nationally (figure 6.19). As in sociometric practice, red marks animated lines that follow both the course of a journey and interpersonal connections. The people speaking are located on the map, resulting in the impression of a national actor-network that is as large as can be imagined.

The film's informational aesthetic takes the form of a situated, connected, collective narrative. At first, individual voices and faces can be distinguished, but the network gradually develops into a sort of audible composite photograph. In edited images that follow one another rapidly, individual heads almost seem to fuse, and the voices overlap. The cinematic action lends a narrative and energetic charge to the network diagram. In

FIGURE 6.19
Bill Condon et al., Sequence from *Kinsey*, 2004.

contrast to many other diagrams—think of Vandermonde's chessboard or Moreno's topology of the camp—here the map is preserved as a frame of reference. The figure is not separated from its ground.

Kinsey uses complex montage and computer-generated images to tell a story of a constantly spreading and continuing national "grand narrative." Condon's subtle presentation counters American puritanism with gentle optimism. The mapping sequence is therefore also an argument for the liberalism of narrative and sexual connections. The emotional circulation expands on a national level in a way that Michel Foucault described as "centrifugal" in his history of governmentality. While the disciplining of the individual by a surveilling network of institutions concentrates "centripetally," enclosing and isolating spaces, a liberal apparatus of security aims at continual expansion of circulation. It leaves things be only so that social and economic exchange can continue to grow: *laissez-faire* as the optimal foundation for action.[114] One notes in *Kinsey* the desire for a sexual liberalism freed of compulsions and rigidities without any explicit reference to the capitalist economy. The network form of the narrative, however, points to a fundamental split: nets can support, but they also entangle and ensnare. Those who pull close the socioeconomic net in which they are standing will trap themselves.[115]

The theory of social networks, especially in the school of sociometry influenced by the immigrant Moreno, is closely connected with a specifically American capitalist constitution of the subject. The network diagram becomes the scale of the social against the backdrop of a conception of individuals acting autonomously, whose social atomism satisfies the demands of economic liberalism. Within this framework, the macrodimension of society and the microlevel of the actors can fuse in the discourse of the image. The ability to connect and to associate thus becomes the measurable part of the collective status. Collectivity results from connectivity. The diagrammatic strategies of social networks suggest a fluctuating utility of loose connections that are supposed to tighten into a network in accordance with actors' desires.

Traces of the disposition described here are still present in the popular discourses on Web 2.0 and the rise of social media platforms. By contrast, research on social networks based on graph theory and forms of visualization has become increasingly differentiated. There is more awareness of the limitations and problems of formal analysis of social relationships by scientists who employ the methods critically.[116] Nevertheless, one cannot avoid the problem of the effective validity of the production of knowledge. Just as models of atoms could be employed independently of chemical theory only to vouch pragmatically for the reality of relationships of bonding, network diagrams in the sciences are models *for* something. As a medium of performative self-description, they open up a specific space of presentation in which existing and possible actions are negotiated. In network diagrams of society, scientific logics inform the imagination of human interaction without being able to do justice to their inherent meaning.

Few network diagrams get by without a form that points to the spatial distribution and distributing qualities of infrastructural networks. The form has already wormed its way into the aesthetic of the network, as it were. Whereas in the beginning organic references, such as the circulation of blood and the capillary net, are necessary, views change with the phases of innovation in the industrial world. Transformation processes often occur in the interplay of urban architectures (which grew largely unplanned) and their visual exploration. Few networks reveal that more clearly than the formation of the London transportation systems, whose synchronizations in time and space are the subject of the next chapter.

7 TRANSPORTATION: MAP, NETWORK, AND SYNCHRONIZATION AROUND 1930

Frank Pick must be pictured as a happy, if also overworked, person. When he was hired in 1908 as a publicist for the Underground Electric Railways Company of London, no one suspected that the young solicitor and statistician would change the public face of the city forever. As an ambitious publicity manager, Pick (1878–1941) made history in transportation, architecture, design, and art with his skillful commissioning policy. He saw to it that stations were redesigned with clearly visible signage and more room for the company's posters. Thanks to his initiative, Edward Johnston's sans serif Underground type was introduced as a uniform typeface in 1916, and the red-and-white roundel with a blue crossbar was standardized. Above all, Pick relied on the poster as a promotional medium—a decision that created a prominent place for avant-garde graphic art in Britain. Cubism, Futurism, Vorticism, and Bauhaus became shared cultural property on the postered walls of the Underground. Pick's ambitions, which were sometimes perceived as benevolently dictatorial, did not shy away from architecture, which he integrated into the corporate design of London transportation. Charles Holden's functional and elegant interwar buildings became part of an overall concept that paid attention to detail and self-confidently aspired to modernity and efficiency.[1]

Henry Charles Beck should not be imagined as an unhappy person, even though he earned his living with irregular positions. From 1925 onward, Beck (1902–1974), whom his friends called Harry, worked as a draftsman in the Signal Engineer's Office of the Underground Electric Railways Company. He was employed as a commercial designer of images ranging from illustrations for the in-house magazine to advertisements, hired for short-term assignments. He sketched the design for the now-famous Tube map on a double-page spread in his exercise book while out of work in 1931 (figure 7.4). After he was rehired, a fellow draftsman encouraged him to submit his design to the publicity department. The publicity department, directed by Frank Pick, rejected his work as "too revolutionary," by Beck's account.[2]

One year later, Beck reworked his drawing and met with a better response. Pick remained skeptical about the Tube map, but he took a risk on it by having 750,000 copies printed in January 1933. In February, another 100,000 were printed. Harry Beck was paid five guineas (£5.25), the equivalent of around US$700 (in 2021), for a project whose basic form he had produced while unemployed. His freelance work became less steady. He was never hired for a permanent position. In exchange for handing over his copyright, Henry Charles Beck insisted that he be allowed to integrate all the changes to the network into his design himself.[3] London Transport's repeated refusal to grant him that right embittered him. Until the 1990s, his name fell into oblivion.

What was it like to ride the London Underground in Pick and Beck's day? And what changes did their innovations bring about? This chapter attempts to grasp the rhythm and nature of urban transportation, which, in the 1920s, became a model of perception based on its material media culture.[4] The protagonists of this tale are not primarily Frank Pick and Harry Beck, although both played major roles. Elements of "anonymous history"[5] stand side by side: daily passengers, large crowds of people, urban activities, tickets, subway maps, station architecture, time clocks, reworked timetables, diagrams of delays, interlocking towers, state reorganization of infrastructure, fragments of films, and literary digests. In short, what mobilizes a network that allows one to move through space? Beck's map shows how the perception of urban space was geared to nodes and their connections—an element in the increasingly temporalized networks in twentieth-century network history, in which fluctuations of traffic were paired with cultural techniques of synchronization.

7.1 SPACE: HENRY CHARLES BECK DRAWS THE METROPOLIS

> If you're going underground, why do you need to bother about geography? It's not so important. Connections are the thing.
>
> —Henry Charles Beck[6]

In London in 1891, the German engineer Ludwig Troske wrote about the first steam-powered underground railway, in operation since 1863:

> For the foreigner, traveling on the subterranean lines is not especially comfortable at first. Here, too, it takes some practice and adaptation before easily and quickly getting the hang of it. Having left the bright air of day and arrived in the dimly lit station, one's eyes first try to get used to their new surroundings. . . . Probably these different trains have certain distinguishing features up front on the locomotives . . . except that in the murky light of the station and the relatively rapid arrival of the trains, such signs help only those in the know. Soon, however, even the foreigner gets used to this unusual operation and becomes familiar with and can use the various aids for finding the right train. With the help of a map from the London railway network on the train, he will then find his destination station as well.

Both boarding and exiting require concentration "because the prescribed announcement of the station names is often unintelligible. Otherwise, the name of the station is only applied in tiny letters to a few lampposts." Information on the walls gets lost in the busy confusion of advertising signs.[7]

In London in January 1933, London Transport—a conglomerate of six railway lines operated by five railway companies listed on the stock exchange—had a new map printed of the Tube (figure 7.1). On the back was a list of station names with the polite request, in the style typical of Frank Pick's publicity work: "A new design for an old map. We should welcome your comments."

Whereas the commentary of London subway riders has largely been lost, the transformation of a topographical map (figure 7.2) into a topological diagram has repeatedly attracted attention from researchers. Art and design historians have situated the modified Tube map in the milieu of Constructivism and, following Roland Barthes, understood it as one of the mythologies of everyday life.[8] Scholars often regard this map as an "immutable mobile," in Bruno Latour's use of the term.[9] But practices of subway riding and the organization of urban transportation in the 1920s and 1930s are usually overlooked. Instead, emphasis falls on the modernist imagination of the machine traversing an ordered space-time that seems to have been condensed into the Tube map.[10] The perspective goes hand in hand with the discovery of a long-forgotten designer. The cultural memory of British railway enthusiasts has led, after an intentional suppression of Henry Beck, to his popularization, if not canonization, in the past twenty-five years.[11] Many scholarly assessments, including this one, have been part of this memory boom.

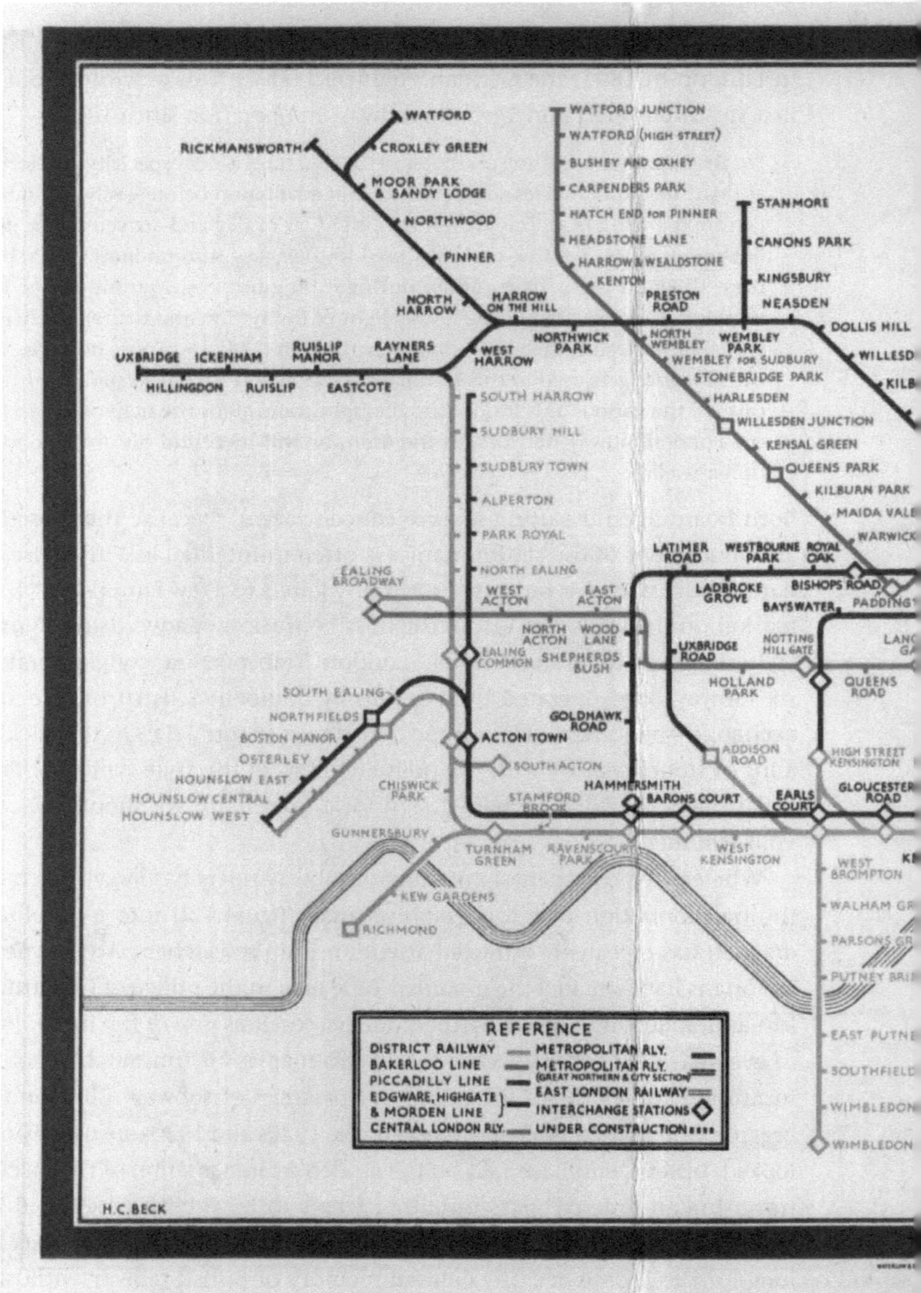

FIGURE 7.1

Henry Charles Beck, *Tube Map*. Printed by Waterlow & Sons Ltd., London, Dunstable, Watford. Color lithograph, 1st ed., January 1933. 22.8 cm × 16 cm. © TfL from the London Transport Museum collection.

COCKFOSTERS
OPEN MIDSUMMER 1933
ENFIELD WEST
SOUTHGATE
ARNOS GROVE
BOUNDS GREEN
WOOD GREEN
TURNPIKE LANE
MANOR HOUSE
FINSBURY PARK
ARSENAL (HIGHBURY HILL)
DRAYTON PARK
HIGHBURY & ISLINGTON
CANONBURY & ESSEX ROAD
HOLLOWAY ROAD
CALEDONIAN ROAD
KINGS CROSS ST. PANCRAS
OLD STREET
ANGEL
MOORGATE
HIGHGATE
TUFNELL PARK
KENTISH TOWN
CAMDEN TOWN
MORNINGTON CRESCENT
EUSTON
EUSTON SQUARE
WARREN STREET
GOODGE STREET
HENDON CENTRAL
BRENT
GOLDERS GREEN
HAMPSTEAD
BELSIZE PARK
CHALK FARM
ST. JOHNS WOOD
BAKER STREET
GREAT PORTLAND ST
REGENTS PARK
FARRINGDON
ALDERSGATE
RUSSELL SQUARE
CHANCERY LANE
POST OFFICE
BANK
LIVERPOOL STREET
HOLBORN
TOTTENHAM COURT ROAD
BRITISH MUSEUM
OXFORD CIRCUS
ALDWYCH
COVENT GARDEN
MANSION HOUSE
ALDGATE
SHOREDITCH
TO BOW ROAD BROMLEY WEST HAM PLAISTOW UPTON PARK EAST HAM BARKING UPNEY BECONTREE HEATHWAY DAGENHAM HORNCHURCH UPMINSTER & SOUTHEND
MONUMENT
ST. MARYS
STEPNEY GREEN
LEICESTER SQUARE
STRAND
BLACKFRIARS
CANNON STREET
MARK LANE
ALDGATE EAST
WHITECHAPEL
MILE END
TRAFALGAR SQUARE
TEMPLE
CHARING CROSS
SHADWELL
WAPPING
WESTMINSTER
VICTORIA
WATERLOO
LONDON BRIDGE
LAMBETH NORTH
BOROUGH
ST. JAMES PARK
ELEPHANT & CASTLE
ROTHERHITHE
SURREY DOCKS
KENNINGTON
NEW CROSS GATE
NEW CROSS
OVAL
STOCKWELL
CLAPHAM NORTH
CLAPHAM COMMON
CLAPHAM SOUTH
BALHAM
TRINITY ROAD (TOOTING BEC)
TOOTING BROADWAY
COLLIERS WOOD
SOUTH WIMBLEDON (MERTON)
MORDEN
UNDERGROUND

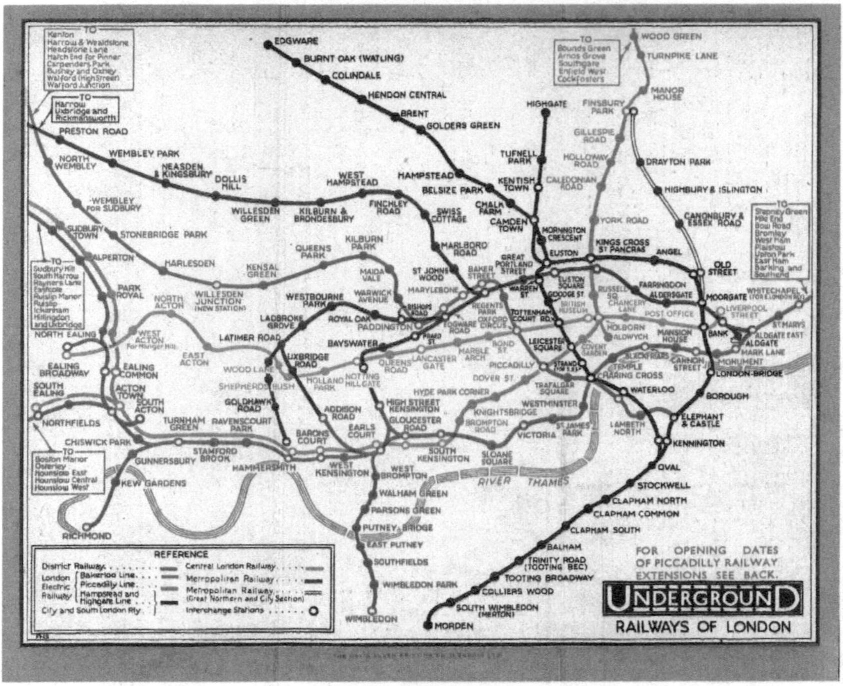

FIGURE 7.2
Frederick H. Stingemore, *Underground Railways of London*. Printed by David Allen Printing Co., London. Color lithography, 1932. 14 cm × 16.6 cm. © TfL from the London Transport Museum collection.

If one follows Mr. Beck to his later triumphs, the view shifts from the geometric clarity of the map to historical contingencies and contradictions. Beck's colleagues in the graphic design department of London Transport are known to have likened the Tube map to electrical circuit diagrams. Beck himself made fun of this similarity with a caricature published in March 1933 in the company's *Train, Omnibus and Tram Staff Magazine* (figure 7.3).

While the River Thames functions as a logarithmic curve, the Bank of England stands under high tension. The British Museum becomes the "Wonder plug," and because, as on circuit diagrams for radios, an antenna is integrated rather than Piccadilly Line, there has to be a ground ("Earth") in the south as well. For all its humor, this circuit-diagram thesis can be taken further. The Tube map adopts practices from the directing of trains

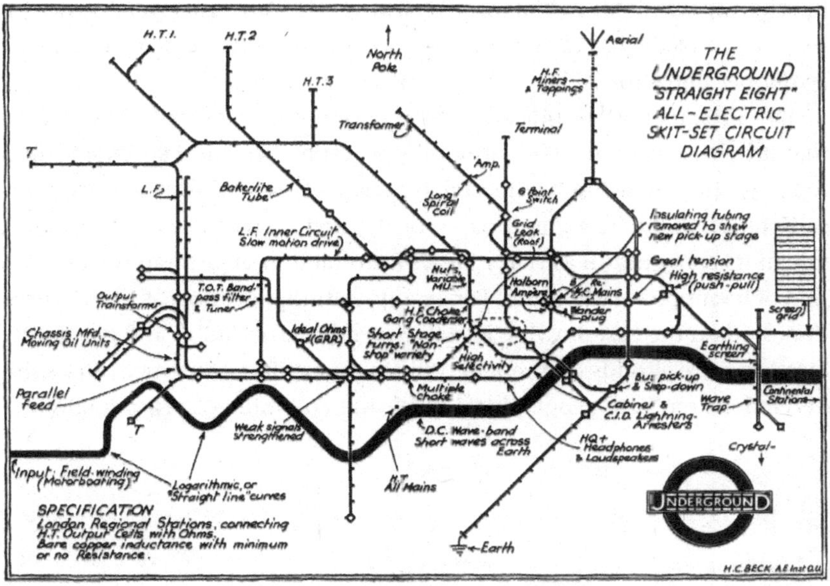

FIGURE 7.3
Henry Charles Beck, *The Underground "Straight Eight" All-Electric Skit-Set Circuit Diagram*. Drawing in *Train, Omnibus and Tram Staff Magazine*, March 1933.

and the use of switch towers and therefore integrates specific techniques of synchronization.

The choice to dispense with topographic precision of scale is also evident from the final surface detail: the River Thames is integrated, in highly abstract form, as a natural marker. The consistent use of 45° angles lends the network a systematic regularity previously unknown. The Tube map fulfills one mathematical criterion for a net, only to undermine it again for the sake of providing a successful infographic: the angles do not have to correspond to the actual geography, nor do the lengths of the lines.[12] Then comes the graphic high point: the stations with no opportunities for transfer are represented not by circles but by small ticks. Beck also integrated an already familiar technique for bringing the periphery closer to the center: distances in Central London are expanded, while distances between stations in the suburbs—so-called Metroland—are compressed. The distances in the graphic depiction do not correspond to actual density.[13] Bringing the surrounded area close is in keeping with the change of social distribution in urban space. From the late nineteenth century onward, social structures were rearranged by cheap mass transit because housing outside of London's center became affordable for the poorer classes as well.[14]

The Tube map is an indicator of the boom in topological views of space that it helped promote. For London, topography became a specific form of topology in which the fleeting relationality of traffic flows was geometrically rationalized. The graph paper of the first sketch plays just as large a role in its constitution as Harry Beck's introduction of geography under the sign of its supposed exclusion. For example, the application of white correction fluid on the first drawing reveals that geometricization and geographic location were at odds (figure 7.4). Within this shift, a constructive positing of relations replaced geographical surveying and mimetic reproduction.

FIGURE 7.4
Henry Charles Beck, First sketch of the Tube map. Pencil and ink on graph paper, 1931. 19 cm × 24.1 cm. Victoria & Albert Museum London, no. E.814–1979. © TfL from the Victoria and Albert Museum collection.

The Tube map imported the aesthetic of the Eastern and Central European avant-gardes, from Constructivism to the Bauhaus, to England. In turn, Frank Pick's strategy for popularizing modern art for public spaces appealed to artists such as Man Ray and László Moholy-Nagy, both of whom proposed spectacular poster works for London Transport in the 1930s.[15] During that period, Moholy-Nagy also designed a world network map for Imperial Airways that applied Beck's topological aesthetic of connectivity to air travel and aligned it in parallel with the topographic map.

Another aspect of enthusiasm for Beck's design—which has even spread to a current guidebook: "We'd be lost without Harry Beck"[16]—is the repeated emphasis that the evolution of the map to the diagram stands at issue. This is an overgeneralization. Following Nelson Goodman, one can say that all maps, as geometric figures, are already diagrams, but not all diagrams are maps.[17] In its diagrammatic form, Beck's invention is still a map for everyday navigation, although "the DIAGRAM" is often capitalized in descriptions of the Tube map. This does not mean, however, that diagrammatic techniques pervaded the practices of the railway business at the time. For the history of networking, it raises the question how spatial forms and times of transportation networks are constituted by diagrams between image, writing, number, path, and time. Standardized clock times and synchronized traffic flows are central conditions for successful switching and transfer of passengers. The abstract aesthetic of the Tube map, which reorganized the network of the London Underground along urban lines, presents a perfectly timed network that is supposed to connect to and integrate other forms of transportation. This was associated with a form of temporal and socioeconomic synchronization that was always the goal, however impossible in actual fact.

Beck's maps addressed a range of phenomena. First, they belong in the context of the merger of five independent railway companies that operated the six inner-city lines, each of which was listed on the stock exchange. These lines were notorious for feuding. From July 1, 1933, they formed an integrated transportation system under the name London Passenger Transport Board (LPTB). The chaos of different tracks and prices is evident in the diverse tickets from the first third of the twentieth century.[18] Desire for a "penny transport" analogous to Rowland Hill's standard postage in the 1840s runs through the cultural history of the London railways.[19]

Second, the Tube map continues an iconic design that began with the introduction of the disk logo in 1908 and the house typeface, Underground or Johnston's Railway Type, by Edward Johnston. This typeface was used consistently from 1918 onwards and promoted by Frank Pick, the director of London Transport.[20] Third, the map enabled passengers to move, especially when changing lines at stations. The graphic representation and control of moving human bodies interested Beck more than anything else until the end of his life.[21] Fourth, it involved the attempt, already made in the field of time measurement and regulation technologies, to synchronize space and time beyond the demands of the individual rail line.

7.2 TIME: LONDON TRANSPORTATION; SYNCHRONIZATION ALONG THE LINE

> La forme d'une ville / Change plus vite, hélas, que le cœur d'un mortel.
> —Charles Baudelaire, *Tableaux Parisiens*

> No human heart / changes half so fast as a city's face.
> —translated by Richard Howard[22]

SCENARIO 1: PAPER, HAND, TELEPHONE

In nineteenth-century Britain, traffic of the British steam-powered overland railways was coordinated by a combination of optical control signals, timetables, and Charles Wheatstone's five-needle telegraph. Operating an urban railway posed entirely different challenges. Overland traffic demanded the introduction of a standard time based on schedules almost as a matter of course.[23] In an urban environment, the most crucial issue is achieving a high frequency of trains. Electromechanically networked control was originally necessitated by the railways' long distances, but it also proved vital in the accelerated rhythms of urban space. There are three technical scenarios of synchronization for following the course of a train, some of which require storage on paper, some blinking lights on an illuminated diagram. After initial skepticism, the telephone increasingly came to join the alliance of telegraph and train.

The British trade journal of the time, *Railway Gazette*, offers diverse examples of the connectivity of centralized, automated telephone communication, switching towers, and movement coordination, including the control rooms of long-distance trains. A photograph from 1932 showing the control room of the London & North Eastern Railway in Darlington,

FIGURE 7.5
Control room of the London & North Eastern Railway, with graphs recording the course of trains on the left. Darlington, 1932

between Durham and York, offers insight into the practices of how movement was charted (figure 7.5).[24]

In this case, train traffic is controlled by a hand-drawn line, the written form of which was simply called "graph" in the trade. With every report, an entry "in relation to time and place" was made on a sheet of paper divided vertically into hours and minutes, ultimately resulting in a continuous line.[25] The topological form of Beck's map related to the office work of the daily practice of controlling trains. Both were indebted to communication by means of an adapted coordinate system. The Tube map not only followed typographical, aesthetic considerations but also produced standardized notation systems that made it possible to operate trains.

SCENARIO 2: PNEUMATICS, LEVER, ELECTRICAL IMPULSE

A ten-minute educational film from 1921, which survives only in fragments, shows the complexity of controlling urban train tracks. In the rhythm of its intertitles and images, *Signalling on the Ealing and Shepherd's Bush Railway* dramatizes the exacting efficiency of electromechanical controls.[26] Despite the interruptions in the second half—the edits are abrupt, and the intertitles announce sequences that are never shown—the film presents the details of manual switching in sequence. The moving train works as a kind of automatic writing system, which triggers signals when it moves a lever at switching points. When the train passes, not only is the signal arm reset but the train's current position within a section of track becomes visible on an illuminated diagram (figure 7.6).

The corresponding intertitle of the film is "Illuminated diagram informs signalman position of all trains." This simple statement conceals several unspoken contradictions. As it continues, the film shows only how the switching is communicated to the signal room. Even if it were not just electropneumatic switches that provided feedback on the vehicle's current state, it signals movement of a train that has just happened at an undefined speed.[27] This uncertainty about time—in *Cybernetics*, his programmatic text on regulating technology, Norbert Wiener tellingly chooses train signals and switches as his example for mechanical feedback[28]—was countered by the efforts of railway employees but never overcome entirely. The slogan "The fewer the

FIGURE 7.6
Illuminated diagram for locating trains. Digital film still from *Signalling on the Ealing and Shepherd's Bush Railway*, 1921 [2007]. © TfL from the London Transport Museum collection.

signals, the fewer the trains" led to the development of signal systems that would permit trains to run more frequently by reducing distances between the switches.[29] The speed of the trains was subject to the greatest contingency because the moving object increased the necessity to synchronize time. In this context, it is unsurprising that a massive effort began after World War I to measure statistically the frequency of trains and their delays, taking their stops into account.[30]

British railway equipment of the 1920s, most of which was supplied by Westinghouse Brake & Saxby Signal Company, distinguished between two or three states, with "clear" and "danger" or "on" and "off" being most common. The optical signals with precisely three angles—0°, 45°, and 90°—suggest a possible source for the specific geometricizing of Harry Beck's map. But in the tunnels of the London Underground, optical signals were of little use. So, signal lights with just two colors were employed. The code "go"/"stop" was most applicable where trains traveled with the same speed and comparable acceleration and deceleration. In the history of railroads, this binary distinction presumably originated in the warning systems that were bitterly necessary in London, so-called fog repeaters, which had replaced older acoustic warnings by a "fog man."[31] There was a material distinction between "on" and "off": visibility or no visibility.

The presence and absence of trains transmitted by the medium played a crucial role in their remote control. The sophisticated storage methods of subway clocks, which constituted a distinct form of network time in the close sequence of underground trains, mirrored these enduring fleeting moments.

SCENARIO 3: PAPER, CLOCKS, NETWORK TIME

What happened when a subway train was late? How did the system deal with the higher number of passengers on football Sundays in London around 1930? The processes for measuring, controlling, and correcting generated diagrams of delays, revised timetables, and signal report failure forms. An example of a revised timetable from the handbook *Handling London's Underground Traffic* of 1928 shows how fluctuation, asynchronicity, and resynchronizing were dealt with (figure 7.7). The only color illustration in the volume demonstrates a minor change to train planning on the District Railway (the District Line, indicated in green, is how West Ham and Chelsea football fans still reach their stadiums today). In this case, the early morning times reveal the complexity of the adjustments: trains had to stop

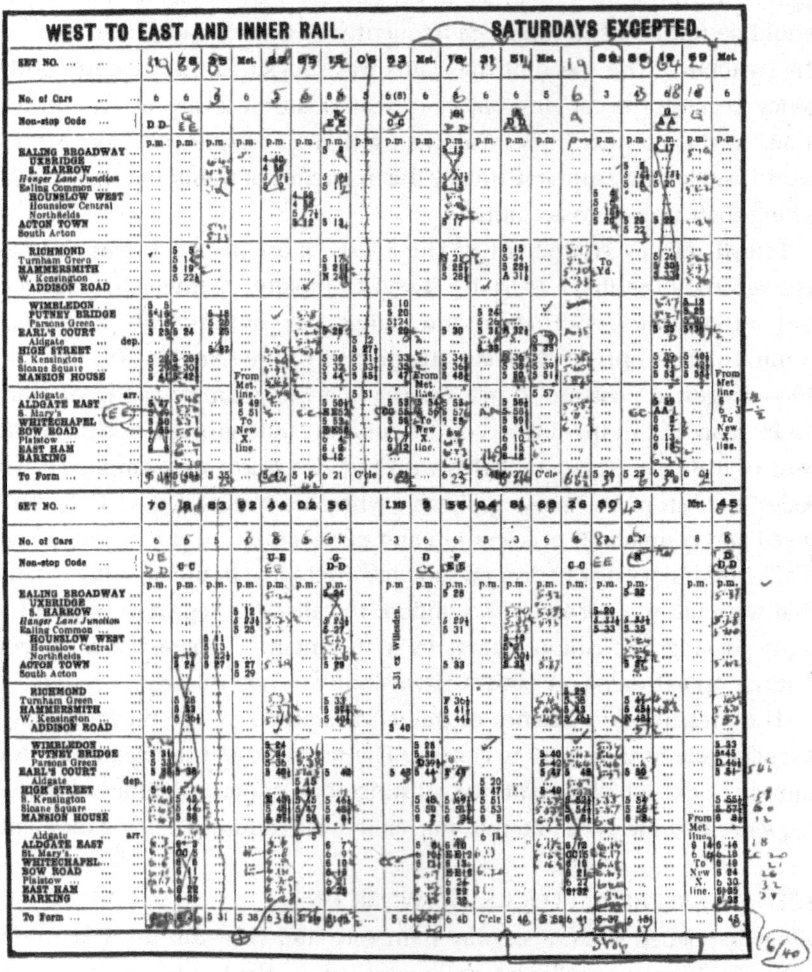

FIGURE 7.7
Timetable updated to reflect changes to the frequency of trains on the District Line, circa 1928.

at stations not otherwise served. Moreover, the London lines often were not linear but instead integrated several starting points. The trains of the District Line started both from Ealing Broadway station and from Richmond and Wimbledon. That made the junctions beyond Gunnersbury and West Kensington potential bottlenecks. Without general knowledge of the era and awareness that the individual lines were competing, this timetable is almost illegible. John Pattinson Thomas, operating manager and author of *Handling London's Underground Traffic*, described the scenario as follows:

> Throughout the various routes covered by the trains of the District Railway there are ten branches. A train may be delayed at Wimbledon, for instance, owing either to unsuitable fog-signalling arrangements or overcrowding, and by the time it reaches Earl's Court it has missed its path and has to be sandwiched in at the wrong place. This may mean that it fouls the working of a Circle train at South Kensington, causing trains to block back on the bottle neck on a tight 1½-minute headway. Signal to signal checking then occurs, and from an initial 2-minute delay on the branch line an accumulated delay of 10 or 12 minutes may be set up.[32]

The railroad employee's sense of honor called for one qualification with British understatement: "Delays, however, are happily of rare occurrence."[33]

Apart from the timetable, however, the intervals of trains also had to be constantly timed and stored along the stretch of track. The maximum capacity of a line or a loop was forty-eight trains per hour, which had to be turned around within seventy-five seconds at the terminus. If the signaling systems were ideally equipped, the minimum distance between two trains was fifty-four seconds, with a target of thirty seconds for the train stop.[34] The train conductors performed a crucial function in the coordination of time, despite the possibility of remote control for the acceleration and braking of the train via relays. Stations with heavy traffic had an automatic clock for sequencing the trains. This was known as a headway clock (figure 7.8).

This clock was reset to zero as soon as the train left the station. The next engineer could therefore see how far ahead the previous train was, so that he would not get too close without noticing—"unconsciously."[35] The headway clock also made it possible to record the sequence of trains on paper. The resulting headway chart represented and created a very different kind of "Underground time." Each mark made by an inking ribbon stood for one departed train (figure 7.9).

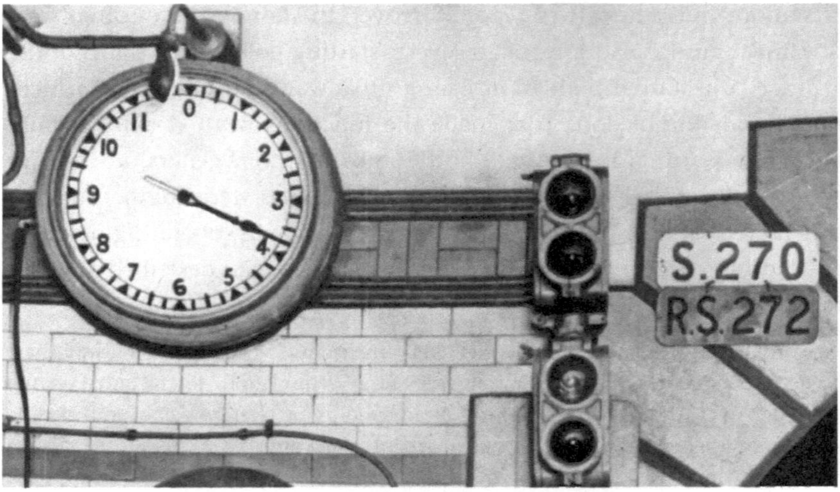

FIGURE 7.8
Headway clock indicating the distance from the train ahead, circa 1928.

The tighter the sequence, the more difficult it was to distinguish the marks on the paper. The day of headway charts was about nineteen traffic hours. If one combines this form of storage for each of the six lines, a genuine form of networked time results. It follows the telegraphic synchronization of clocks in the nineteenth century but nevertheless has its own specificity.[36]

7.3 TIME: LONDON TRANSPORTATION—NETWORK SYNCHRONIZATION

> The metro is not a site of synchronism despite the regularity of many people's schedules.
>
> —Marc Augé[37]

All forms of synchronization along the line, which included boarding and exiting, said very little about the specific operations of a railway network. By means of electrical transmission, the Taylorist railway timeclocks could be combined into a train recording diagram framed like a painting (figure 7.10). Here, too, the disk rotated at the speed of a clock, while the

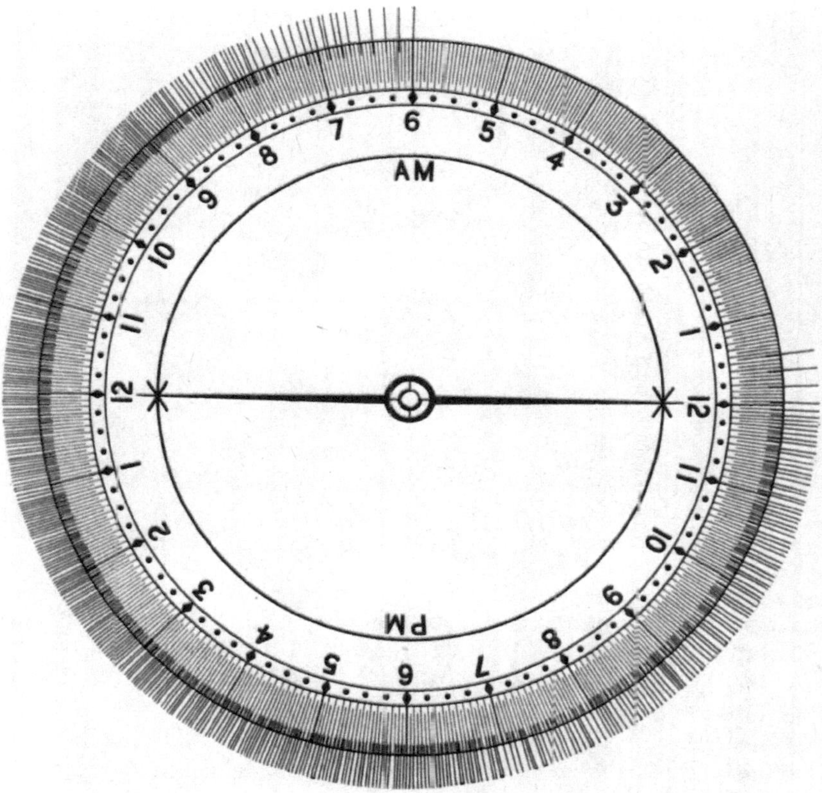

FIGURE 7.9
Headway chart recording the frequency of trains on one day of operation, circa 1928.

train departures indicated the simultaneity of network operation. The train recording diagram thus represented a timekeeping and synchronizing equivalent of the topological Tube map.

Its installation in the control office on Leicester Square in 1928 was supposed to serve as a model for the future equipment at all train depots "so that the engineering and operating staff will be advised at once, and simultaneously, of the occurrence."[38] Not only was this sort of will to synchronization important for operating transportation, it also made it possible to turn off the electricity on schedule. A simultaneous "traction off" as soon as all trains have come to a stop was the model of economy.

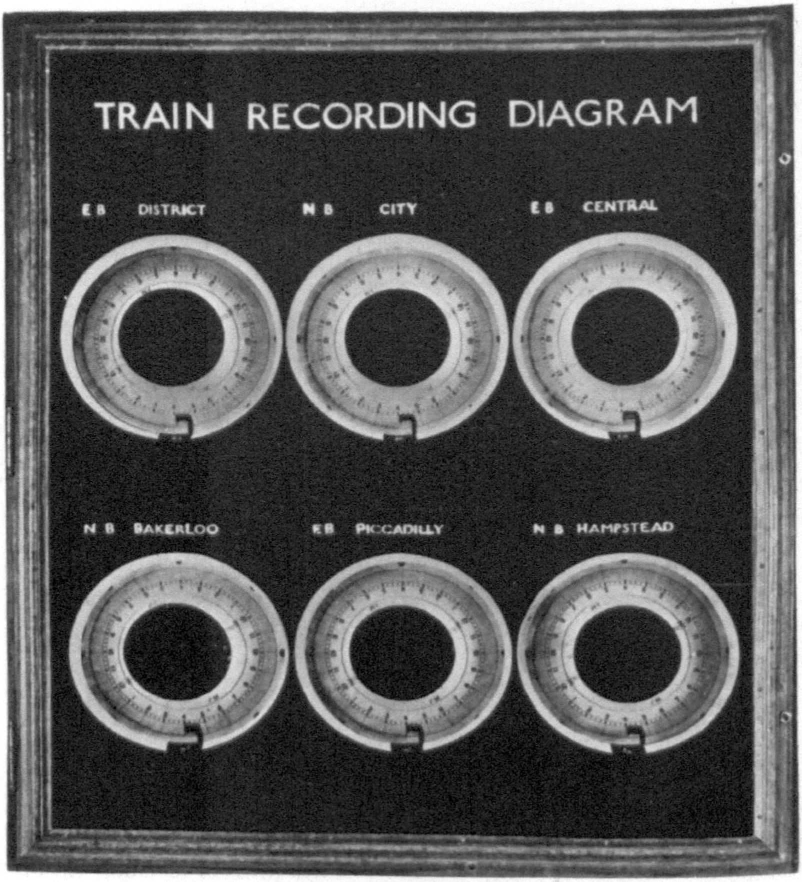

FIGURE 7.10
Automated *Train Recording Diagram* for the entire London network, with electrical transmission to headquarters and control centers, circa 1928.

TACTICS: BOARDING, TRANSFERRING, EXITING—OR "CONNECTIONS ARE THE THING"

There is at least one other story of the always desired (but never attained) synchronization of the London Underground network. That would be the story of boarding, transferring, and exiting and its acoustic clocking and visual navigation. In his unfinished novel *Der Verschollene* (translated as *America*) of 1927, Franz Kafka shared the elements of such a story in an artfully compressed, elliptical form. The protagonist, Karl Rossmann, who has emigrated to America, experiences riding the subway in the fictional metropolis of Ramses:

> They would rush to the nearest underground station almost at a run, Karl carrying the basket; the journey flashed past in a second, as if the train were being pulled through a vacuum, and they were already getting out and clattering up the stairs at the other end without waiting for the lift, which was too slow for them; then the great squares appeared, from which the streets rayed out star-fashion, bringing a tumult of steadily streaming traffic from every side; but Karl and Therese stuck close together and hurried to the different offices, laundries, warehouses and shops to do the errands which could not easily be attended to by telephone, mostly purchases of a minor nature or trifling complaints.[39]

Kafka always presents the subway journeys of Karl, who knows constant travel in enclosed spaces very well from his job as an elevator boy, as elliptically compressed in an Expressionist gesture. They are merely sketched. The passage itself "flashes past in a second" or occurs in the space between two sentences: "But he soon came to a decision, counted out the money he needed for the fare, and rushed to the underground station. When he got out at Clayton he heard at once the noise of many trumpets."[40]

Whereas in Kafka's Ramses the confused blaring of out-of-tune trumpets does not occur until the end of the train trip, everyday rail travel in London was and is characterized by announcements and noises.[41] Anyone taking the escalators to the Hampstead Line—now the Northern Line—in 1921 was admonished loudly by a device with the apt name Stentorphone: "Please keep moving: if you *must* stand, stand on the right: some are in hurry, don't impede them!"[42] The contemporaneous use of gramophones with amplifiers, as was common at Oxford Circus for example, wore out roughly one vinyl record per week. Even today, in the elevators at Covent Garden station, one has to listen to the distracting words of a prominent radio moderator to pass the time while riding. Educational measures to counter panic when boarding and exiting were recorded on film in the 1920s. The introduction of trains with doors that opened automatically, which was intended to speed up train dispatch, led to the production of amusing titles such as *The Right and Wrong Way to Board an Electric Train.*[43]

Thomas's handbook of 1928 also identifies transferring, the connectivity typical of a transportation network,[44] as one of the most problematic aspects of managing the Tube: "A different type of connection occurs where the

Underground line crosses under or over, but does not physically connect with, another Underground or foreign railway line. This may necessitate (as it does at Charing Cross, the interchange point of District, Hampstead and Bakerloo trains) the synchronization of the times of the first and last trains each day, and this is always carefully carried out."[45] The regulation of passenger flows between the lines was a priority, not only when redesigning stations—for example, Piccadilly Circus station, which reopened in 1928 after being renovated (figure 7.11). It intensely preoccupied Harry Beck and

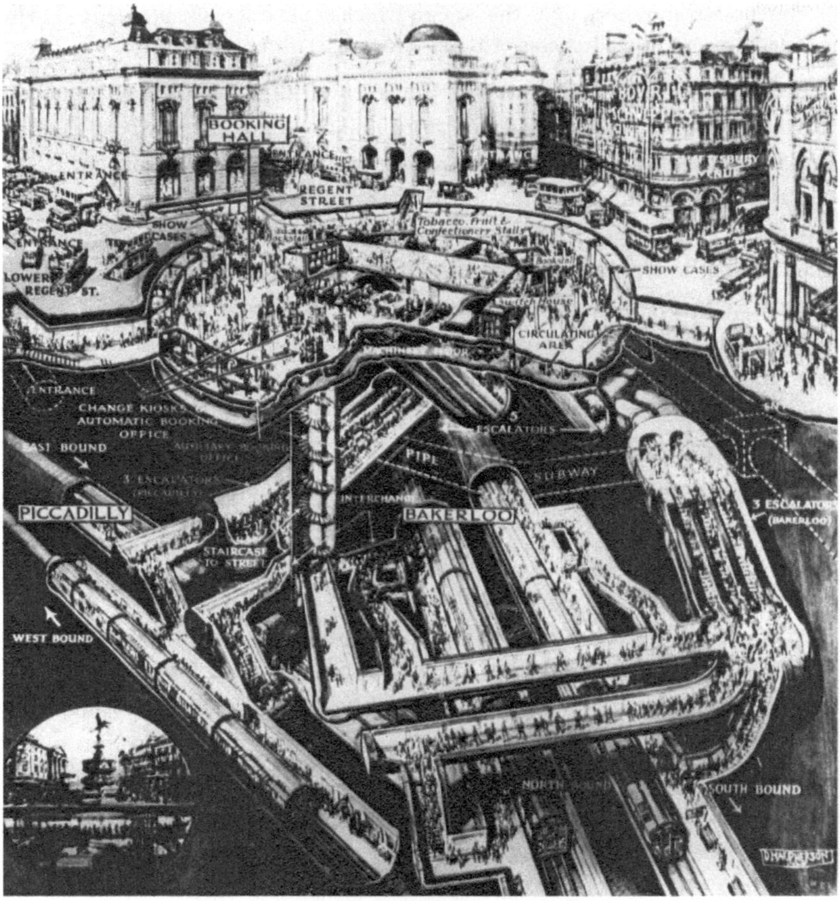

FIGURE 7.11
[D. Macpherson?], *Piccadilly Circus Station*. Print, 1928.

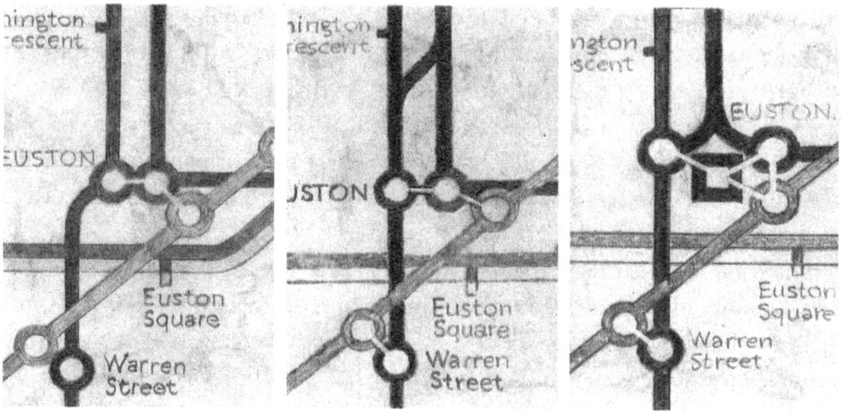

FIGURE 7.12
Henry Charles Beck, Pencil sketches on the topology of Euston and King's Cross stations, 1961. Courtesy of Capital Transport Publishing.

other illustrators. It was not just a matter of reducing the maze of tubes to clear lines. Over the course of his life, Beck incessantly tinkered with the graphic representation of transfer (figure 7.12). In old age, he described his practical theory of space best: "If you're going underground, why do you need to bother about geography? It's not so important. Connections are the thing."[46] Charles Holden's architecture for stations and Frank Pick's guidelines for design also aimed at a practical aesthetic of connectivity. Outdoors, stations were illuminated as entry points in urban space. Indoors, Pick sought to aid navigation through the passageways to the tracks with clear, indirect lighting on the ceilings.[47]

The tactics of passengers, including clever transfer, occurred on the fly and are difficult to reconstruct, despite drawings and architecture.[48] The crowd of anonymous individuals, the transitory mass ornament of transportation, always remains unknown.[49] Following Marc Augé, one can say that the experience of underground train travel is rarely a collective one, even though it takes place in a shared architectonic space. The Metro is not a place of synchronized mass transportation, despite the collective rules—no smoking, *passage interdit*, and so on. We all celebrate our parties and birthdays on our own here, sketch our own topography of love affairs, work, pleasures. . . . [50] The route is always chosen and experienced individually, but in the joint interweaving of our personal paths, our participation in a collective urban identity crystallizes: community without celebration and solitude without isolation, while the names of stations produce a ritual sacredness of the Metro as a place of memory.[51] One can also say that the metropolis of the 1920s *was* a sphere of transportation: "the active and passive reception of new transportation technologies permit[ted] previously unsuspected formulations of collective emotion and behavior."[52]

There are few historical documents that can do justice to the specific quotidian quality of descending into the underworld of the city. Silent witnesses, such as ticket readers, register the numbers of travelers. Statistics on the number of passengers have tried to master the phenomenon of the crowd. The efforts of the Union internationale des chemins de fer to introduce a universal railway semiotics from 1930 onward yielded an ensemble of strategies but not of individual practices of movement and orientation.[53] In an action film such as *Bulldog Jack* of 1935, the passage itself becomes part of the plot.[54]

The detective serial *Bulldog Jack* develops in its eight-minute high-speed finale a personalized form of synchronization along the line. Just as the villain Morell, with Indian jewels he has stolen from the British Museum, takes control of a train, the detective Bulldog Drummond tries to stop him.[55] While the hero gets into a scuffle with Morell, his partner follows the racing train, identifying the names of the stations on the fictional line and crossing them off the graphic representation with a pencil (figure 7.13): "Holden Brew—only four more stations! . . . White City—only three more! Ooooh!! . . . Two more, two more! Ooooh!! . . . Only one more! This is the end. Ooooh. We're gonna crash, we're gonna crash!" The greater the speed, the more the film emphasizes this form of indicating location, while the cuts showing the architecture of the stations and tubes succeed each other at breakneck pace.

The characters in *Bulldog Jack* typify a new kind of urban passenger who is constantly attempting reorientation. There was an enormous difference between the experience of traveling the long-distance railroads of the nineteenth century and the everyday route underneath metropolises.[56] Traveling by subway produced an active aesthetic of disappearance.[57] The space-time routes between the stations, usually overlooked, return as a structuring aspect in the action film. In moving pictures—as in the switching diagrams of the railroad companies—the need to position the actors continually intensifies. Following Bernhard Siegert and Michel de Certeau, one can say, "Traveling with or on maps means *constantly jumping back and*

FIGURE 7.13
Orientation along the line in *Bulldog Jack*, 1935.

forth between *carte* and *parcours*."[58] This leap creates a temporal distance that is countered by synchronizing time "along the line" and for the entire network. In the middle of the architectonic space, passengers' attention focuses, above all, on points of connection and intervals of transportation.

STRATEGIES: SYNCHRONIZATION AS INTEREST OF THE STATE—ELECTRICAL AND RAIL NETWORKS

Can a state of synchronization be stored and be made politically enduring? In addition to the synchronization of diagram and apparatus, one can also speak of a synchronization in governmental interest in the context of the Tube map.[59] Efforts were made to counter the notorious unreliability of Britain's electricity supply with state planning from 1926 onward and, in the case of the Tube, from 1931 onward. The Tube map was designed while both the subway network and the national electricity supply were undergoing centralization in parallel.

The British railway world has always been distinguished by many independently run lines. The same is true of British energy production and power plants. In 1913, London had more than sixty-five electrical power plants, thirty-two different voltage levels for transmitting electricity and another twenty-four for distribution, and around seventy methods for measuring use and billing.[60] In 1923, the trade journal *Electrical World* commented on the resulting problems with sober humility: "The largest city in the world [with a population of 4.7 million] offers an excellent example of what electric supply ought not to be."[61] Such problems were programmatic as of 1926. From 1930 to 1936, they were addressed by devising a national grid to provide electricity, leading the railway companies to build their own power plants, in contrast to other cities such as Berlin and Chicago.[62]

Developing the London Tube into a joint network that could be used smoothly turned out to be problematic and protracted.[63] On the initiative of the Labour Minister of Transport, Herbert Morrison (1888–1965), a bill was sent to the legislature in 1931 to establish the London Passenger Transport Board. Even after the fall of the Labour minority government, the legislation continued until Parliament passed the London Passenger Transport Act. Meanwhile, Lord Ashfield and Frank Pick were conducting nerve-racking meetings to win over the owners of the separate train, tram, and omnibus companies to the project. The tradition of exhausting exchanges between railway companies had begun in 1907 with the Underground Electric Railways Company of London, which combined several competing and feuding companies.[64] Reports on the complicated negotiations under the direction of Lord Ashfield were featured in the trade journals *Railway Gazette* and *Modern Transport* between 1930 and 1933.[65]

With the Tube map's large-format presence in stations and 750,000 copies in circulation after its first printing, political pressure could finally accomplish the long-overdue centralization of an outdated form of organization. Improved workability turned out to be an advantage during the course of its introduction.[66] Henry Charles Beck's intuitive grasp of the daily paths of Londoners combined with state interest in establishing the LPTB, which began operations on July 1, 1933.[67] At the same time, the Tube map made more rigorous use of the grid that was emerging to supply electricity on a national scale than the electrical engineers of the Central Electricity Board (CEB) did, who acted more locally.

Both infrastructures could only be made to function somewhat reliably with the help of state authority. The Tube map can therefore be understood, following Michel Foucault and Gilles Deleuze, as a machine-like diagram of control that installed a mental map of the city.[68] It generated a space of bright, airy circulation, the antitheses of which remained uncontrolled things and bodies, to be sure, but with the materiality of architecture surrounding all traffic in its flows, fits, and starts.[69]

The proximity of the form of socioeconomic organization, practices of urban space in the network, and an aesthetic of relations has rarely been captured better than by the young graphic artist Lilian Margery Dring (1908–1998), who was educated at the Kingston School of Art and the Royal College of Art. In 1938, she offered Frank Pick a poster triptych depicting the LPTB in the form of the ancient god Hermes as a "modern god of transport." Dring's hope for a career as a poster artist never came to fruition, a consequence of the Great Depression and World War II, although she did produce a series of posters for the British Youth Hostel Association. Consequently, she concentrated on designing everyday visual objects in textiles: embroidered pictures, placemats, cushion covers, and curtains, some of which are shown today in the Victoria and Albert Museum.[70] The remarkable poster design she submitted impressed Pick, but it was never printed because the projected costs were too high. The three-part image shows Hermes, the messenger of the gods of ancient Greece, with the company logo as his heart, Beck's lines for the Underground as his arteries and veins, and winding cars as a snake around his arm (figure 7.14).

Like Beck's own circuit-diagram version of his map, the image plays with the city's landmarks: the god's mouth is represented by the BBC building, the ears by the Royal Albert Hall, the eyes by the National Gallery, and the neck by Tower Bridge. With this form of urban self-description, the artist was evoking older traditions of urban arteries in the medium of the circulation of the blood and body.[71] And she combined them with the Constructivist scheme of relations that characterizes Beck's network aesthetic. These two historical transformations are held together by Hermes. The reference to antiquity incorporates the reticular aesthetic of the early modern period as well as the geometric rigor of the modern era.[72]

The combination of body politic, socioeconomic synchronization, and urban infrastructure was pursued even more rigorously in Russia. When constructing the magnificent metro stations in Moscow, which began to operate in 1935, the Stalin regime gave little account to topographic features and constructed the Beckian underground lines quite literally. And whereas today it is still possible to get around halfway reasonably with the old Russian technology, and both RFID tickets and CCTV cameras store considerably more than cards and clocks would have suggested, the Tube is dependent more than ever on manual synchronization. The *touching in / touching out* of the surveillance state with the RFID chip of the Oyster card contrasts even in the twenty-first century with old-fashioned, handwritten

FIGURE 7.14
Lilian Dring, *ELPETEBE/LPTB: Modern God of Transport.* Design for a poster triptych, circa 1938. 36.8 cm × 74.9 cm. © TfL from the London Transport Museum collection.

notice boards and human announcements that let harried Londoners know which lines are out of service again.

Under such circumstances, it was generally expected that the daily breakdowns would build up into veritable transportation chaos during the Olympic Games in 2012. In the end, however, London transportation proved to be well prepared: commuters were warned about critical transfer stations such as London Bridge and generally encouraged to avoid riding at certain times and

places. The strategy focused especially on ensuring that the Queen Elizabeth Olympic Park and its station, Stratford, could always be reached by as many routes as possible. According to the official numbers, the load transfer within the network was successful, despite minor interruptions: a 15 percent drop in passengers, who had clearly taken other routes, enabled a 35 percent increase in traffic in the center of London, with sixty-two million rides during the Olympics and Paralympics.[73] Under the circumstances, this was nothing less than a masterstroke of communications and logistics—an astonishing testimony to Londoners' relaxed approach to the asynchronicities of their Tube.

7.4 SPACE: HOW TO DO THINGS WITH MAPS

When Harry Beck gave the city a figure of orientation in the form of a geometric construction, he excluded the disorderly, dirty, and arcane qualities of metropolitan mass transit. The network of the Tube map is abstracted from the densely agglomerated mass of bodies and stress to which circulating passengers expose themselves day after day. It is not about the waves of crowds on streets and squares that many contemporaneous literary dramatizations of urban space associated with traffic.[74] It is not the pattern on the surface of the street that makes the map but the design of a graphic user interface for the belly of the city.

Beck's sense for encoding daily connections was able to reshape social spaces. The principle that "connections are the thing" has been enshrined internationally, despite cultural differences in perceptual dispositions, and not just in travel maps.[75] The modifications—angles of 60° rather than 45° are particularly common—suggest that cartographic approach of space above and below ground is specific to every city. This is particularly clear in the case of Paris, where the topological variation of the map of the Métro never really prevailed over orientation around the surface markers of arrondissements and quartiers.[76] With the globalization of the Tube map, however, despite the local situations, a network orientation has been adopted transculturally in which places are connected to one another by leaps. Between them lives an intermediary space in at least two senses: first, the passageway, which oscillates between practiced indifference and hectic attention and, second, the blank, white areas of the map, which as nonplaces influence the function of stations all the more. The jungle of cities disappears in the artfully reduced density of transparent depiction of its transportation network.

In a short piece with the apt title *Zwei Flächen* (Two Planes) from 1926, Siegfried Kracauer describes such forms of urban contrast between dense, organically grown spaces and a geometrical order of squares. He contrasts the old port of Marseille, in which the "fauna of humanity" teems, with the geometrically arranged quadrangle of a plaza as a "square without mercy."[77] Bay and quadrangle are above ground, planar contrasts that blur together beneath the surface. One can say that the leap between bay and quadrangle is constantly being made in the underground space of transportation. In the middle of the ceaseless bustle of the crowd, the abyss of the depths, and the unlocalizable travels of trains, the stabilization of perception is incumbent upon the geometric map as user interface. The connection space below ground, despite its fixed architecture, can no more be stable than the rhythms of trains can produce a perfect synchrony. In its network, there is a continuing fluctuation of relationships that are reoriented and negotiated when transferring. If walking represents an activity in the city that is at once subjective and collective, changing means of transportation is like switching a cultural practice in which the connective net becomes the interconnective network.[78]

The Tube map works here as an "immutable mobile" with which the connectivity of things can also be produced linguistically on a daily basis. It reveals a space of timing in which fluctuating urban rhythms reach their ideal state in the simultaneity of the map. When it was created, Beck's invention was part of a larger cultural shift from substance to relation. It does not so much follow clear geometries as occasional moments of fleeting orientation, a Euclidian representation of an "other space."[79] In that sense, the topological map represents the always-updatable visual model of a behavior in and with spaces that is constitutive of network societies.

The question at the beginning of this chapter—What mobilizes a network that allows one to move through space?—would have to be answered as follows. It activates both in its physical realization and in its name the space of transportation networks in which only Hermes himself could achieve a synchronous simultaneity of action. And one can add that the architecture of steps, stations, and tubes as well as the semiological design of navigation shape urban spaces in such a way that the materiality of the network can recede into the background in favor of the circulation taking place.

Work on the temporal aspects of infrastructural and economic networks increased in the twentieth century. Timing, rhythms, and narrations of exchange demanded more and more attention. By means of temporalization, comparatively fixed nets evolved into relational and fluctuating networks. This is true even if one employs explicit forms of organizational networks. Just how much these things are, once again, a question of the correct timing is evident in the changed forms of production of material culture that are the subject of the following chapter.

8 A VISUAL HISTORY OF THE NETWORK DIAGRAM (II): NETWORK PROJECTS AND THE MATERIAL CULTURE OF CAPITALISM

At least since Daniel Defoe's epoch-making *Essay upon Projects* of 1697, "projectors" have been the showmen of capitalism. "If a project can guide someone through the *terra incognita* of ignorance, like a map or nautical chart, the question of the pilot or projector remains."[1] This "pilot" initially made himself known in treatises, which conveyed big designs with literary elegance. In nineteenth-century France, the networking projects of the Saint-Simonians still relied on the public circulation of ideas via newspaper articles before anything could happen.[2] But what happens when it is not dubious individuals or sects of clever engineers propagating risky plans but rather institutions that have made the logic of the project part of how they define themselves? The creative mode shifts from literature to graphic exposition.[3] Representative forms of undertakings change with a design practice that makes the production of material culture a central part of organization in network plans.

The connectivity of things takes on a new form as a result of planning, producing, and distributing in an explicitly networked form. In this recursive process, objects of material culture themselves become "network-like" via distributed business practices and transactions, in their internal technical structure, through the logistical acquisition of component parts, and in the organizational structure of the companies involved.[4] The ability to organize switches between networks is what distinguishes companies operating within corporate networks. In competition to organize production dynamically, a fight occurs over market share and hegemony that is not safe against the negative effects of networking either.

8.1 WESTERN: OPERATIONS RESEARCH AND NETWORK OPERATIONS METHOD

> *Project* is the term for an undertaking to be planned and executed, a task, a problem, a sequence of events, etc.
> —from a manual of network operations[5]

It isn't easy to install a computer. The undertaking is time-consuming, expensive, and associated with many uncertainties. That is important to bear in mind when picturing the decision made by companies to order central corporate computers in the 1950s and 1960s. In most cases, these newly acquired devices were supposed to perform tasks such as calculating salaries and sales. A dummy planning diagram from the Unilever corporation that used the installation of a computer as an example estimated the project would take seventy-six weeks. The striking feature of this planning diagram is the central, almost linear black path whose arrows pointing right are repeatedly interrupted by circles divided in half (figure 8.1). The steps in the process appear above the line. Noted below it is the project's duration in weeks. At the point of transition to the next activity, two numbers appear in a circle, separated by a horizontal line. The upper number indicates the minimum number of weeks needed to fulfill the task. Below it stands the maximum number of weeks for the corresponding step.

If one were dealing with just *one* task—in this case, the construction of the machine—the flow of events would be relatively simple. But interdependencies are the rule, not the exception. After management approval (board approval, week 2), the upper area of the diagram, for timing the personnel planning, is drawn in thinner strokes and circles. The programmers who are recruited (which takes a minimum of twelve weeks and a maximum of twenty) and then trained for three months (36/44) write the necessary programs within twenty weeks. Operators and input staff have to be hired (20/50) and trained. The lower half of the diagram shows events when the planned location has been selected and furnished. The interdependencies in this network project are not all that numerous and can be quickly understood: a computer is useless without the corresponding software and trained personnel. Temporal coordination is necessary to align processes with one another after sixty-two, sixty-four, seventy-two, and seventy-six weeks.

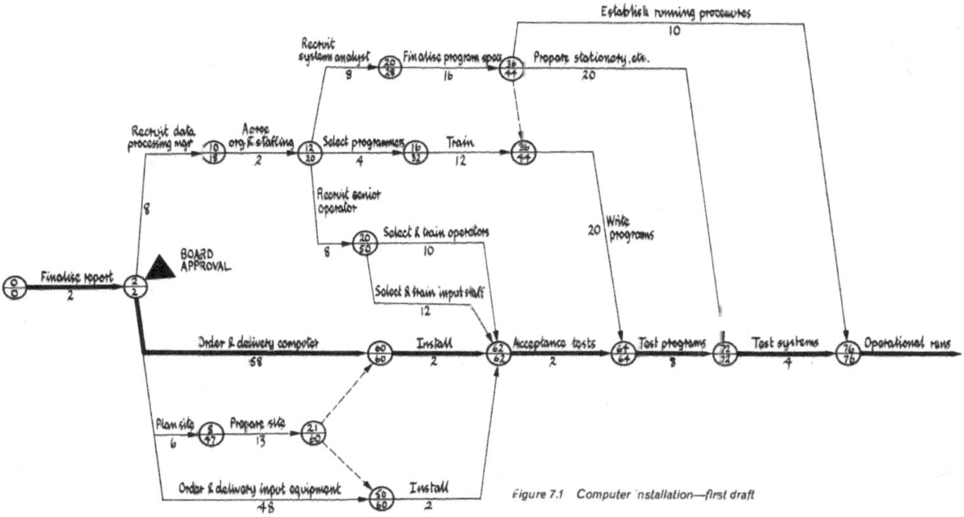

FIGURE 8.1
Kenneth Gordon McLaren and Eric Leonard Buesnel, Network planning diagram for the installation of a computer at Unilever. First draft, drawing, 1968–1969.

However, any project, no matter how precisely it has been planned, is subject to continuous fluctuation. The visual argumentation of the authors of the Unilever manual *Network Analysis in Project Management*, published in 1969, shows that thanks to deft coordination of the test phases, the duration of the project along the "critical path," emphasized graphically in bold, will run alongside a third and fourth design at sixty-four weeks.[6] But contingency increases with micromanagement. The critical path shifts to the human actors involved and in the process is split into three (figure 8.2). Further revisions of the network plan respond to (inevitable) delays, early arrivals, and shifts. In contingent cases, revising means resynchronizing by further optimizing, overwriting, cancellation, and, if necessary, redrawing and rewriting. At least in this example, however, the logical structure—the program of the diagram, in a sense—is preserved. The logic is only adjusted when necessary. Although the operational drawing is subject to continual updating, the "immutable mobile" of the diagram remains.

As a technical image, this network plan is characterized by a style between geometrical, technical drawing and carefully handwritten cursive writing.[7] In many respects, it exemplifies the historical interplay of institutionally

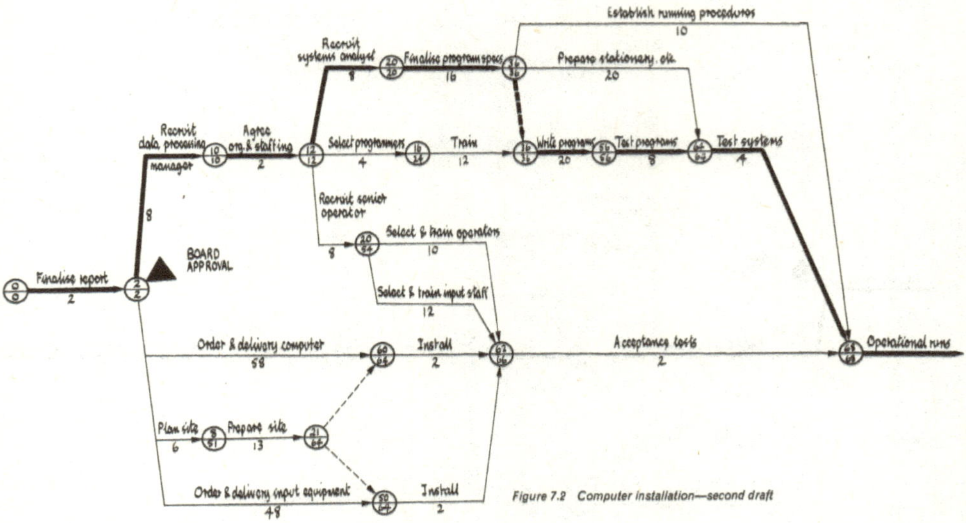

FIGURE 8.2
Kenneth Gordon McLaren and Eric Leonard Buesnel, Network planning diagram with critical path for the installation of a computer at Unilever. Second draft, drawing, 1968–1969.

run operations between computer control and human activities. The large derives from the small: microeconomic network planning produces globally networked companies that in turn depend on reports on their own activity. Network plans and their manufacturing styles express a changing relationship between knowledge, information, and the economy. Their visual calculations have provided Western capitalism with a material foundation for production. In the 1960s, they quickly became popular in companies that initially employed network-based methods internally before implementing them for the public as well.

Such efforts have not, however, always been given their due. For example, on the first pages of *The Rise of the Network Society*, the notable three-volume sociology of the network society, Manuel Castells focuses on the rise of the informational economy, whose beginnings he places in the 1970s. When the first edition of *The Rise of the Network Society* was published in 1996, Castells dated the genesis of a new formation of the global economy to the previous twenty years. It is not as if older forms of companies were unfamiliar with processes of distribution and decentering, but Castells was interested in a new quality whose historical origins went back at least to the network plans and network projects of the 1960s.[8]

Yet, explicitly networked forms of production based on the division of labor already gained traction in the second half of the eighteenth century. The necessary infrastructure, coordination, and logistical planning emerged in the lengthy process of industrialization.[9] The latter coincided with the rise of liberal techniques of governance, which in turn produced circulation aimed at connectivity and expansion, which would prove crucial for the nineteenth century.[10] New media technologies (telegraphs), means of transportation (railroads), forms of production (laboratory, factory), and economic structures (banking and stock-market networks) fed into and off of one another.[11] As the complexity of the production process increased, economic exchange formed new milieus. This meant the acquisition of raw materials by colonial exploitation as well as the organization and optimization of local trading chains. The issue of the best use of time and money and the best mobilization of bodies stands at the beginning of genuine network operations of capitalist enterprise. In what follows, therefore, I will leave aside, for the moment, the complex entanglement of companies and streams of finance and goods and examine material culture in the mode of the network.[12]

Logistics, with its practical, military origins, was established as an applied interdisciplinary field only in the twentieth century,[13] and operations research, which is closely related, likewise appeared on the scene quite late.[14] A definition of the discipline known by 1957 as operations research or operational analysis sums up its goal as follows: "Quantitative analysis of purposeful (i.e., military, industrial, or administrative) operations with intent to derive an integrated understanding of the factors controlling operational systems and for the purpose of supplying management with an objective basis on which to make decisions. Frequently involves representing the operation or the system by a mathematical MODEL."[15] The history that operations research wrote about itself as an enduring success points to a number of heterogeneous sources. They range from early modern game theory in the work of Girolamo Cardano[16] and the beginning of population statistics in the work of John Graunt[17] to the works of greats in the history of mathematics such as Gottfried Wilhelm Leibniz and Leonhard Euler, then to the treatises of telephone engineers such as Agner Krarup Erlang on probability theory,[18] the management guidelines of Frederick Winslow Taylor, and the motion studies of Frank and Lillian Gilbreth.[19] Another foundational act is usually attributed to the British government. Operations research is said to have begun in 1936 with the establishment of Bawdsey Manor Research in Suffolk, primarily to study radar technology for air defense.[20] At first, network analysis informed by graph theory did not play a determining role for studying the organization of paths of movement in time. Key figures such as Norbert Wiener were more interested in the use of electrical networks—that is, circuits that find application in the difficult problem of predicting the position of enemy objects in air and water.[21] Peter Galison aptly characterized this "ontology of the enemy" schooled on air defense and submarine warfare as a fundamental element of the "cybernetic vision."[22]

A friend–enemy distinction has remained in civilian applications of operations research, although concealed by the relative decorum of capitalist competition. As the first of the "system sciences" that emerged over the course of the 1940s, operations research translated the heated rationality of battle into the cool administrative logistics of the postwar era. The science of war increasingly transformed into logistics with which one could profit in the market from advantages in speed. It aimed to create the ideal combination of elements over the ideal performance of individual elements.[23]

The same applied to the organization of the production process itself, whose complexity produced a new mathematical, diagrammatic method of visualization in the late 1950s. As a medium of the internal communication of companies, analytical diagrams—organigrams, flow charts, calendric tables, and motion studies—had become indispensable during the first half of the twentieth century.[24] The development was always combined with planning based on calculating abstraction, more rarely, with mathematical formalization, and often with a new visual status for networks. Anyone opening up the journal *Operations Research* (founded in 1956) or a manual for project management will behold a number of illustrations called project networks, network plans, or net plans. Together with techniques known as the critical path method (CPM) and the program evaluation and review technique (PERT), it promised political and business leaders an efficient tool for optimization. As a wave of literature on project management flooded the market, two entities were consistently named as the points of origin. The DuPont company was credited with developing CPM,[25] and the Navy was said to be responsible for implementing PERT in the Polaris missile program.[26] Contemporary accounts often omit the French graph theory models of the Metra potential method (MPM).[27] They also leave out the new production management developed at Toyota from 1948 onward, which was long cut off from the West.[28]

National and Western changes in the culture of memory also determined how diagrams and visual forms were passed down. American engineers and managers derived the network plan from the Gantt chart, named after its inventor Henry Gantt (figure 8.5). The diagram's overview of activity moves directly into the network plan with the linear characteristics of an arrow from left to right. Instead of recording the dimension of time externally as one axis of the coordinate system, these diagrams integrate courses of time into the image itself as calculated flows. This is, however, preceded by the construction of logical relationships in order to record mutual dependencies within the project as a whole. Either activities are represented by arrows (activity-on-arrows) or in the nodes themselves (activity-on-node). The goal is to identify critical paths whose course corresponds to the longest possible itinerary through a network.[29] In a project network, time flows from the end to the tip of the arrow and is therefore irreversible, like shooting a projectile. The enjoinder to stick to the "master plan" is a crucial feature of these only somewhat flexible control networks.[30]

Other symbolic indicators are associated with typical forms of diagrams. Activities are divided and fragmented—in an intensification of the Taylorist principle—into subactivities.[31] Work steps that must be completed before the next can follow are indicated by arrows with a hatch halfway through. Dotted lines stand for a graphically necessary *nothing*—so-called dummies—that does not indicate any operation at all but supports the skeleton of the visual form. Purely graphic project assessment was conceived first and foremost for managers. Whereas the DuPont, Metra, and Navy methods were still designed to be used with mainframe computers, consultants reduced the construction to graph paper, ruler, compass, and eraser—with the eraser being the most important instrument.[32] The numbering of events refers solely to the calculability to be achieved by machine technology. More important is the ready-to-hand and personal entry into a surveyable representation, with which the project's continuous uncertainty can be addressed.[33]

The goal is to create a common language of planning and control for institutional communication.[34] Implementation is not a matter of course. According to the experience of all involved, it is the most difficult part.[35] For this reason, as a rule, hierarchically controlled network design is incumbent upon the boss or project leader, as this is the person who has the power to make decisions about the protocol for planning time and resources. Team diagrams are possible but, according to the handbook, should be written by no more than six people. In the keys of symbols in the manuals for project managers, moreover, several special network features emerge that align the logic of computer calculations with a modeled form. Thus, a mathematical model is sketched that reduces the complexity of processed data and condenses it into a visual representation.

This aspect of data processing turned out to be another source of uncertainty, not only because the use of the computer was always an additional step but also because the temporal overview proved indispensable in the network of interdependent actions. Additionally, the limitations of computer graphics did not permit any "neat" production of large-scale diagrams.[36] Instead, as a rule, graphic output of tables in ASCII were based on calendars and financial arrangements (figure 8.3).

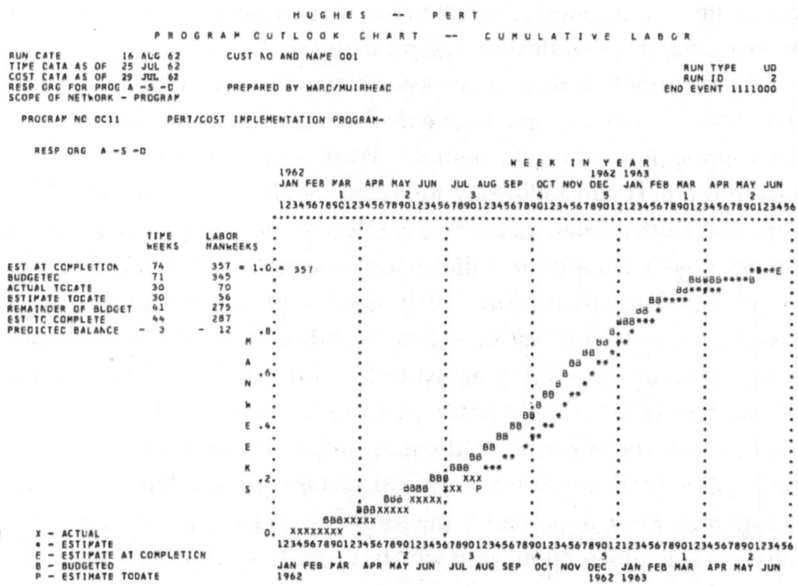

FIGURE 8.3

Hughes Aircraft Company, *Portions of Cumulative Labor Report Produced by the Hughes PERT Cost System*. Computer graphic, 1962–1963, published in 1964.

The limited variables of time and money were set apart, but the overall structure of the project became visible only in the breakdown into subtasks. The strategy created a network of linked projects whose individual tasks in turn became subnetworks.[37] The firm as a whole was transformed into layered network operations by means of which internal and external processes were integrated. For example, the Hughes Aircraft Company described itself graphically as a "fully integrated network" carrying out layered optimization processes in a continuous feedback loop (figure 8.4). Such methods of network operations mobilized and dynamized the programming of processes. The diagrams served as "immutable mobiles" of a time to market meant to be as small as possible. Feedback loops that slowed production and sales were to be avoided because they proved difficult to program and even more difficult to draw in complex diagrams.[38] The goal was a continuous flow of partial currents into one another—an organic functioning of the company in forward movement.[39]

Every network project runs in time—that is, it is deliberately linearized and channeled in a left-to-right schema, so that the sequences can be synchronized.[40] Once the logical relationships of the steps in the process have been established, attention turns to identifying the critical path. The

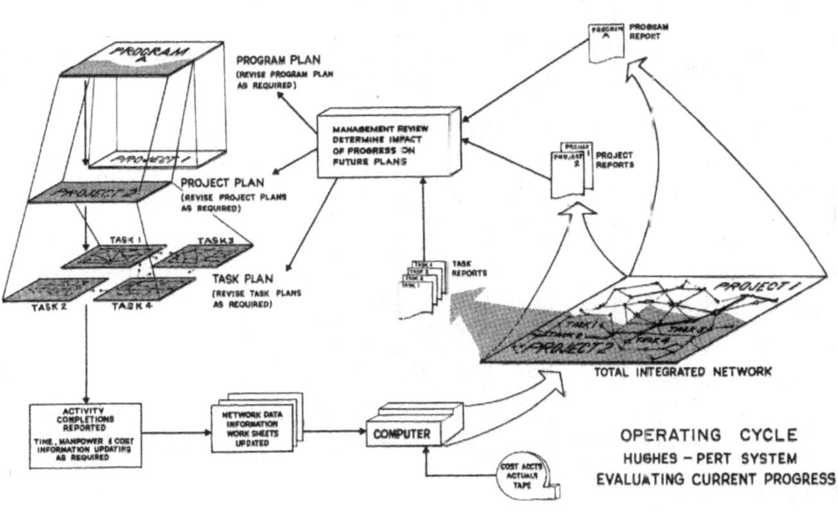

FIGURE 8.4
Hughes Aircraft Company, *Diagram of Hughes System in Operation*. Drawing, published in 1964.

topological network is now temporalized and can therefore be reconnected to the calendar (figure 8.5). The notation of crucial cross-connections reveals the options for notating the individual steps. The goal is a systemic closing and reduction of complexity that makes it possible to grasp visually the extent of the necessary changes and resynchronizations.[41]

Multiple revisions in the first sketches establish the architecture of the project. The corollary is often a geometrization that systematically removes the curved arrows and other ornamental elements in the revision of the initial design (figure 8.6). "Uncertainty" is addressed by a graphic form that is designated to discipline corporate action.[42] The image has to be clear, legible, and readily comprehensible.[43] It cannot follow aesthetic considerations because the participants would refrain from proposing and making changes. The size of the diagram should not exceed thirty by fifty inches.[44] This limitation points to the specific place of implementation: the walls of offices and

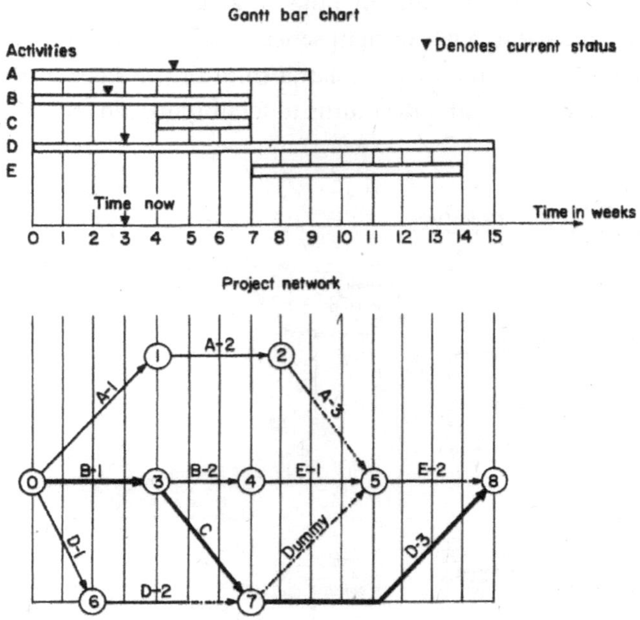

FIGURE 8.5
Joseph John Moder and Cecil R. Phillips, *Comparison of Gantt Bar Chart and Project Network.* Drawing, published in 1964.

conference rooms, preferably of management. Taylorism and information theory meet to optimize flows of communication and production.

But whereas the desired perfect order of factory and open-plan offices still emphasized the local site, network projects include and connect multiple sites in the diagram. The diagrams are scripts of production that are supposed to apply to all the complex organizational processes in time—whether surgical operations, assembling supersonic aircraft, organizing offices, or shooting a film.[45] Whether the matter concerns a missile engine such as the Polaris (figure 8.7, right) or seagoing vessels such as the SS *France*,[46] the network planning diagram conveys the message of the production of goods—that is, designing the form of material culture. This demand goes beyond the borders of political systems, since in the bipolar world order, operations research was also pursued in socialist countries.[47]

The network plan is also, because of its linearizing temporal character, a distinctive exception in the visual history of the network diagram. It is distinct from the schema of Claude Elwood Shannon's mathematical theory of communication inasmuch as critical paths in a contingent production environment are marked in time.[48] This "score" for space and time oriented around planning and action continues to be determined by the principle of sender-channel-receiver. Just as the message flows from A to B, the making of the draft plan is supposed to proceed directly to the sale. The network plan serves as a goal-oriented, calendrical assembly line in times of spatially distributed production based on the division of labor. Consequently, the time scale of the coordinate system becomes the measure of all things (figure 8.7, left), and the flow of information the guiding medium of manufacture.[49]

As a form that orders time, network operations diagrams raise the question of systemic instabilities and interruptions. Where is the noise? What is the parasite that intrudes? Like almost all technical networks, the network plan aims to mitigate the effects of unpredictable but unavoidable interruption, which is integrated into the planning from the outset. In the terminology of CPM and PERT, therefore, "float" is identified as an implicit countermeasure. It is the latitude between the minimum and maximum time of all process steps, allowing for calendric coordination and adjustment.[50] Flows in the network are anything but linear. They resemble the "percolating basin of a glacial river, unceasingly changing its bed and showing an admirable

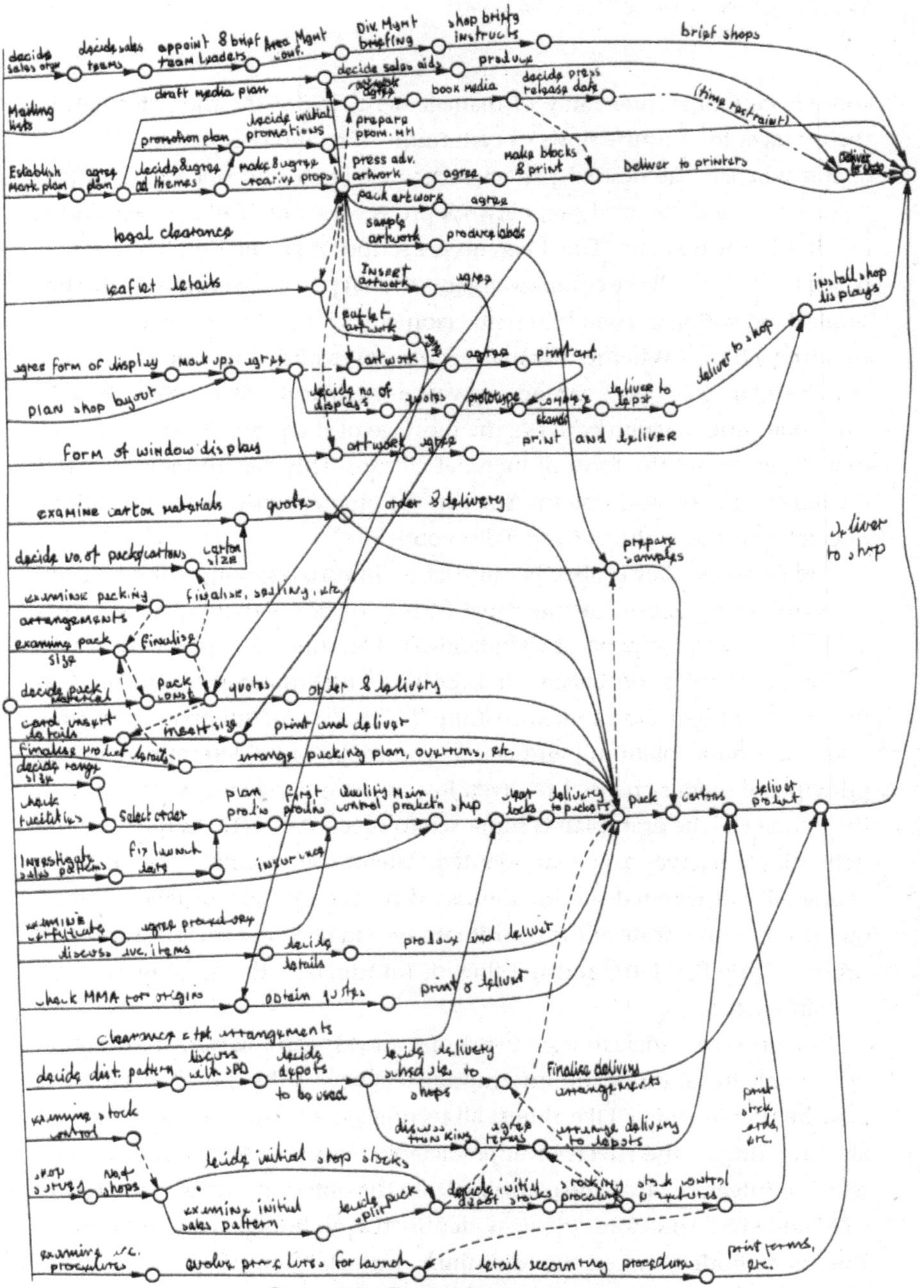

Figure 5.6 Typical first rough network

FIGURE 8.6

Kenneth Gordon McLaren and Eric Leonard Buesnel, Sketch and final drawing of a project network for the market launch of a product. Drawing, 1968–1969.

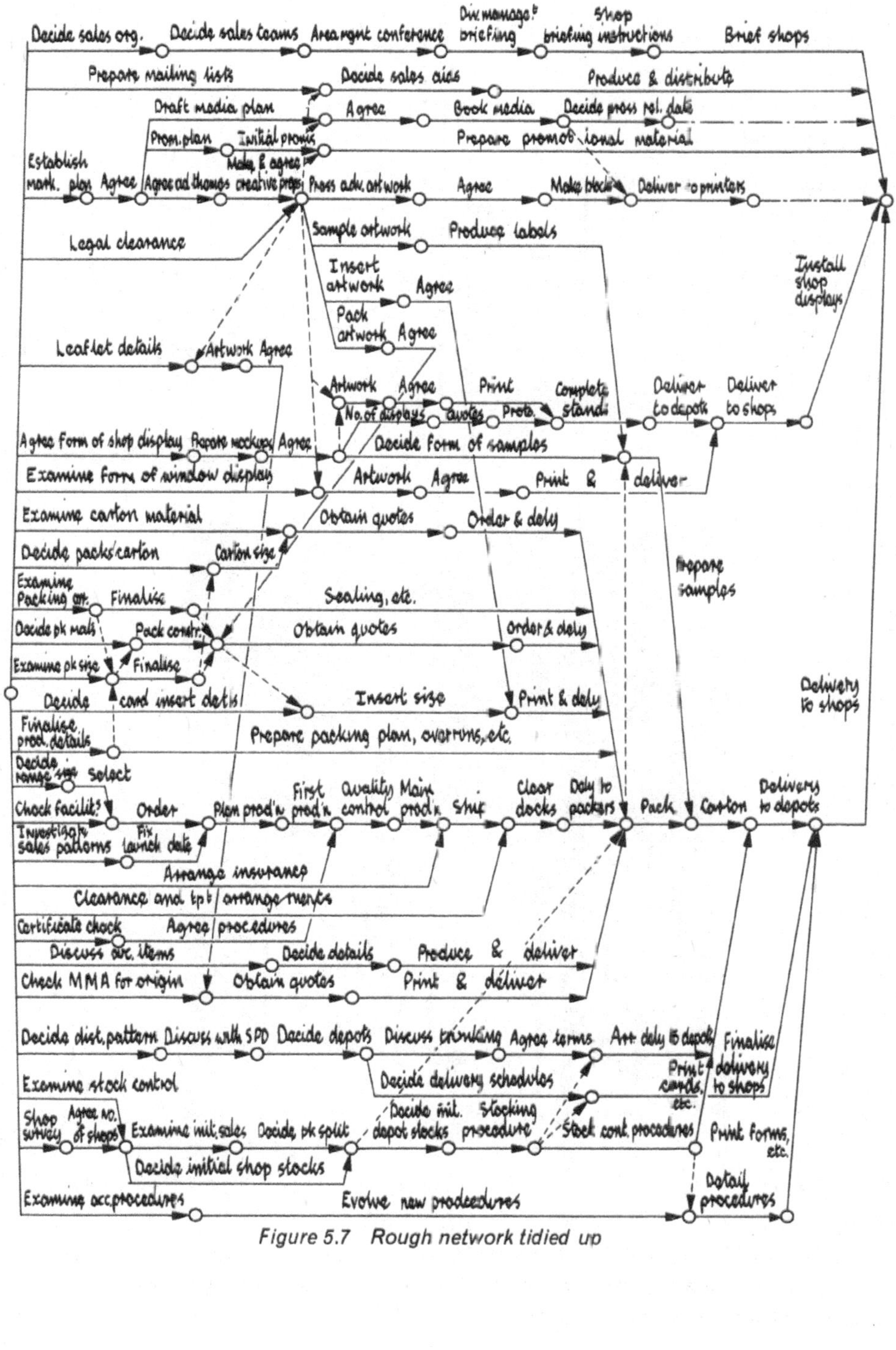

Figure 5.7 Rough network tidied up

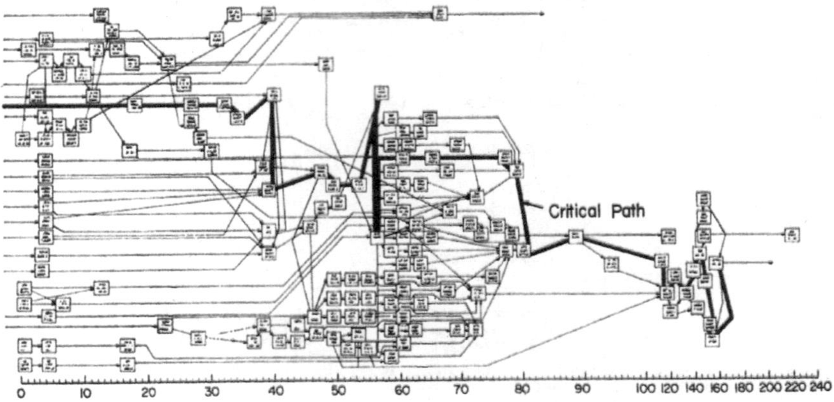

FIGURE 8.7
Top: Donald G. Malcolm et al., *System Flow Plan* for a missile engine for the United States Navy. Drawing, circa 1959. Below: Polaris A-1 missile in Cape Canaveral. Black-and-white photograph, circa 1960.

network of forks, some of which freeze or silt up, while others open up."[51] The failure of even one network operation along the critical path affects the project as a whole. No company can imagine itself safe from negative network effects. Distributed action increases contingency. But it would be even worse to neglect planning in the age of unsurveyable projects because the competition (both economic and ideological) is always suspected of having better forms of distributed production.

Manuel Castells, reflecting on his own Western perspective, has investigated this suspicion in the context of Asian business practices. Which production methods lead to which kind of network advantages in competition is situated culturally in keeping with the different social conditions in Japan, China, and South Korea.[52] Although the frameworks have shifted, Castells's conclusions remain largely accurate, even twenty-five years later. Virtual processes[53] have long since become an established part of corporate cultures because of digital technologies,[54] but the management of uncertainty remains key to corporate cooperation. Relationships between network companies are distinguished by lasting, reciprocal, situational dependencies. The logic of the network exceeds the capabilities of individual firms: anyone who does not operate in a networked way will vanish from the market.[55]

In the second edition of *The Rise of the Network Society*, published in 2000, Castells contrasted his reflections on the history of East Asian businesses with an American case study. Cisco Systems—which, as an infrastructure provider, was briefly the most valuable company in the world as a result of the dot-com bubble—stands as a corporate identity that is generated recursively from its own networking practices.[56] Using the network operations method, it can be shown, first, that constituting corporate cultures in this way was not just a phenomenon of the Internet age and, second, that their emergence is closely connected to the materialization of their own operational chains.

The pioneers of communication research in the 1950s and 1960s already considered the consequences of adopting methods of visual modeling from the natural and computer sciences for socioeconomic processes. Colin Cherry wrote in 1957 of the diagrams that establish the "rules of the institution": "The whole network and the social situation it idealistically represents become closely analogous to an electrical network consisting of interconnected switches."[57] Not only could such a trend already be sensed in Harry Beck's London Tube map, it had also become explicit in the representations of matrices of network sociology.[58] The matter concerns the relationship of electronic switches and the information flows of mediatized societies in general. Network plans function as a form of record keeping intended to anticipate or control possible events where success was not guaranteed.[59]

It is not just the forms with which a society describes itself that are at play in a visual history of the network diagram but also—and above all—cultural techniques that contribute to differentiating network societies as societies of control.[60] These techniques are distinguished by a desire to optimize the creation and regulation of circulation. In the ideal production of goods, the circulation of money, ideas, and things will coincide. The forms of production will match up with the desires of consumers. At no point in history has this goal been manifested more clearly than in the Japanese refinement of "lean" mass production, which long went unnoticed by the West—and was then declared mysterious for that reason. It was due not to the image act of a project designed as an overview but rather to a decentralized network of control of considerably smaller parts.

The *kanban* system of Toyota that Ohno Taiichi introduced gradually after World War II became known in the West much later. The first article in English and Ohno's self-confident look back at his life's work were published in 1977 and 1978, respectively.[61] In both cases, the principles of just-in-time (*jasuto-in-taimu*) production postulated by Toyoda Kiichiro, Toyota's first president,[62] are presented as Fordism, appropriated and improved for Japanese attitudes and geography, in which Japanese identity (*nihonjinron*) and industrial modernization (*kindaika*) are harmonized and not at odds.[63]

8.2 EASTERN: JUST-IN-TIME AND *KANBAN*

> The network enterprise makes material the culture of the informational/global economy: it transforms signals into commodities by processing knowledge.
>
> —Manuel Castells[64]

Even before Japan's defeat in World War II, the eyes of industry leaders were turning to the United States. After the war, a special zone of intercultural contact emerged. The nascent automobile industry in Japan was already reverse engineering American products in the 1930s.[65] From the outset, local actors agreed that the scale of mass production established by Henry Ford was too large for the Japanese market. Even more than at Nissan Motors, this insight became the principle of production at Toyota, which aimed to build different models of cars in small series. The Japanese were inspired by studying American consumer behavior. Several American manufacturers, including producers of military aircraft, employed new production methods developed during World War II in response to shortages of materials. To replace mass fabrication of parts that would be stored and constantly supplied from fixed warehouses (push-out), pull-out methods were developed in which an object was fetched from the previous assembly line station only as needed. Restocking shelves in supermarkets provided the model for this form of organization. The most common origin story is that Taiichi Ohno, vice president of Toyota Motors from 1937 onward, read about the production method in a Japanese newspaper toward the end of the war.[66]

This is more than just a footnote, because there is a long tradition of the Japanese paying attention to American business. For example, Frederick Taylor's *Principles of Scientific Management* was translated into Japanese with a title that means "The Secret of Saving Lost Motion," and sold nearly two million copies.[67] Ohno noted—testifying to the trauma of the war—that Japanese workers wasted much more energy than their counterparts in Germany or the United States.[68] The motion aesthetics of choreographing the factory at Toyota was therefore to optimize the elimination of waste. Ohno writes in his memoirs, "In a manufacturing job done by four or five people, the parts should be handed over as if they were batons."[69] In his account, team sport disciplines and synchronizes movements to harmonize work between human beings and machines.

Three elements are crucial here. First, Toyota was relying on a system of machines controlling machines that it adopted from its own textile manufacturing. Just like the looms of the Toyoda Spinning and Weaving Co., machines that were not in operation were stopped automatically through the intervention of another machine. This made it possible for workers to supervise several processes. At the same time, the number of nonhuman agents was increased by means of autonomous automation, which in business jargon was abbreviated to the concept of "autonomation" using intelligent machines.[70]

A second element is the pull-out system based on the organization of American supermarkets, which had also provided the idea for experiments in aircraft assembly.[71] By reversing the principle previously used, the flow of production became oriented around the mechanisms of demand.[72] Difficult-to-predict consumer behavior set the standard for the design of a whole manufacturing system. The crucial turning point could be described as individualizing mass production or standardization-at-will. That Japanese society helped produce such a constellation through intercultural transfer is as paradoxical as it is logical in light of the war.

Techniques, according to Bruno Latour, are "not fetishes, they are unpredictable, not means but mediators, means and ends at the same time; and that is why they bear on the social fabric" from which they emerge.[73] As cultural practices, they enskill a society by producing it, including different forms of normalizing people. Japan's transformation into a network society was diagnosed relatively early on.[74] The manner of production and the form of a company's organization—above all, its learning processes—brought together situational, decentralized hierarchies that became crucial to sociotechnical networking as a whole.[75]

Just how complex this process was is clear from the arduous evolution of the third element crucial to network history by which the flow of material was connected to centralized flows of information.[76] Ohno's rearrangement of the factory floor began slowly and met with resistance from workers accustomed to Taylorist specialization of "one man, one job." Machines were arranged with an eye to rapid reconfiguration, strict mechanisms for reporting errors, and even the ability to halt the assembly line completely (*jidoka*).[77] The continuous synchronization of fluctuating processes was also controlled by an unprepossessing agent: "In this case, wouldn't it be logical for the earlier process to make only the number of parts withdrawn? As far as communication between the many processes is concerned, wouldn't it be sufficient to clearly indicate what and how many are needed? We will call this means of indication *kanban* (sign board) and circulate it between each of the processes to control the amount of production—that is, the amount needed."[78] At first glance, *kanban* appear to be standardized tabular forms containing logistical instructions for the transfer of objects (figure 8.8). They work as mediators and small files for reversing the information flow from push to pull, including the picking, transportation, and production

<table>
<tr><td colspan="4">Part Description</td><td colspan="2">Part Number</td></tr>
<tr><td colspan="4">Smart City Module 5G IoT</td><td colspan="2">276.489.10</td></tr>
<tr><td>Qty</td><td>35</td><td>Lead Time</td><td>4 days</td><td>Order Date</td><td>04/22/21</td></tr>
<tr><td>Supplier</td><td colspan="3">Network Operations Inc.</td><td>Due Date</td><td>04/26/21</td></tr>
<tr><td rowspan="2">Planner</td><td rowspan="2" colspan="2">Sebastian G.</td><td colspan="3">Card 1 of 4</td></tr>
<tr><td>Location</td><td colspan="2">Rack 15.E8</td></tr>
</table>

FIGURE 8.8
Schema of a *kanban* chart. Graphic courtesy of Anna Katharina Büdenbender.

of material. Initially, they were used only for the decentralized control systems in a factory. At Ohno's instigation "sign boards" controlled individual parts of the main factory beginning in 1950, and they were adopted for all manufacturing processes at the Motomachi plant in 1959. In 1966, the *kanban* system was expanded to all of Toyota's suppliers, most of which were located in geographical proximity to its factories.[79]

The process lasted around fifteen years, which is too long by the standards of normal corporate communication. A philosophy known as *kaizen*, which was supposed to avoid waste, fueled efforts. *Kaizen* combines relentless cost reduction with a continuous will to improve through change.[80] But how did the *kanban* slips that were used to label individual parts function? Ohno Taiichi invites the reader to participate in a thought experiment, which in today's shopping worlds has become a reality thanks to barcodes and RFID: How would *kanban* circulation work in a supermarket, the place whose practices of organizing space are considered to have been its initial inspiration?

> Commodities purchased by customers are checked out through the cash register. Cards that carry information about the types and quantities of commodities bought are then forwarded to the purchasing department. Using this information, commodities taken are swiftly replaced by purchasing. These cards correspond to the withdrawal *kanban* in the Toyota production system. In the supermarket, the commodities displayed in the store correspond to the inventory at the production plant. If a supermarket had its own production plant nearby, there would be production *kanban* in addition to the withdrawal *kanban* between the store and the production department. From the directions on this *kanban*, the production department would produce the number of commodities picked up.[81]

When the term *kanban* entered the management jargon of the West, it was removed from the everyday world, whereas commitment to *kanban* as a system of "visual control" remained omnipresent in the statements of Toyota management. This also included the traffic signals of the *andon* diagram for the sequence of engine assembly introduced in 1950, which showed the current status information above the assembly line and its most important function at the bottom: Stop. Just as the English-language phrase "just-in-time" was effective publicity, the slogan of "management by sight" supported by the *andon* diagram stressed the importance of visual coordination in the human–machine ensemble of factory organization.[82] This regime of surveillance operated between panopticism and rapid collective correction of errors on the assembly line: "Visual management is disturbance management," as a handbook on the synchronized production system puts it.[83]

Network-based production of goods depends on the smooth functioning of its control systems. Control by *kanban* cannot be allowed to turn into an impenetrable landscape of slips. Decentralized control media play a significant role in the object they produce. The automobile is materialized by a network of the *kanban* operations that cause the chains of delivery, construction, and distribution to mesh. The process therefore becomes a network itself—not in a metaphorical way but in a metonymic one.

The history of Japanese culture features similar identity-forming relationships of medium, thing, and economy that predate the application of cybernetics to automobile manufacture. As an everyday term, *kanban* also refers to posters and especially to wooden signs for shops and theaters. The first sign of the word as it is written in kanji (看板) can signify visuality and looking, "signboard," "billboard," "hoarding," "doorplate," and also a shop's "reputation," "name," "look," or "appearance."[84] Before it became so closely tied to Toyota's corporate narrative, it had been used since the seventeenth century for attracting shoppers on the metropolitan streetscapes of Kyoto, Edo (later Tokyo), and Osaka (figure 8.9).[85] These precursors of today's neon signs played such an important role that during the cultural and media transformation of the Meiji era, they became the basis for the registration of trademarks, which was established by law in 1884.[86]

Although hierarchically structured firms also used signs, these *kanban* practices have been described as part of an urban culture of spatial and economic distribution.[87] The diversity of placards—which even featured a unique writing style called *kanban-ji*—stood for a decentralized culture of doing business in which, first, promotional effect and the identity of the shop owners were mutually supportive and, second, a mass market for goods emerged.[88] For the Japanese dealer (*chōnin*), the *kanban* was a symbol as important as a flag or coat of arms was for the samurai.[89]

Even if the *kanban* forms of Toyota did not echo the visual qualities of handmade public signs, they did establish a related system to identify objects that would become traded goods. The centralized form of production turned out not to have been just a result of intercultural transfer from America but also belonged to a local tradition for identifying the sites of economic exchange. Factory and supply networks have more in common with the popular hustle and bustle of street vending than the orderly Taylorist system of notation would have it.

Before written *kanban* introduced connections within industrial, decentralized forms of production, the forms of exchange and signs were part of a decentralized trading economy whose *kanban* were material pictograms located in public spaces. The network logic of distributed public space influenced a revolutionary method for producing material culture. Both forms of *kanban* worked as identification for the exchange and transfer of goods via decentralized distribution. In addition, the corresponding business location was localized, although in the case of Toyota's slips, it was more variable. Therein also lies the crucial transformation: artful trademarks became small acts of rationalization meant to keep increasingly diverse elements in motion.

The supposed smoothness of the production network's flow has its price, however. It does not guarantee happiness and social inclusion. Moreover, a form of organization "that turns workers into the human agents of the decentralized control of the production process" proceeds according to market goals.[90] The various forms of capitalism are completely logical—especially in operative images—and, at the same time, entirely schizophrenic. One need not wholly subscribe to this classic assertion in *Anti-Oedipus* by Gilles Deleuze and Félix Guattari to measure the effects of *kanban* on human bodies. It is enough to read the reports of workers' experiences, such as the troubling tale of the journalist Kamata Satoshi, who in 1972 signed on as a seasonal worker at Toyota. Instead of a sphere of pure circulation that was airy, flowing, clean, and productive, he found time pressure, general surveillance of the workers, and a high suicide rate.[91]

FIGURE 8.9

Stationery store with *kanban* of books hanging from bamboo branches. On the right, advertising calligraphy reading "All kinds of stationery." Undated woodcut, Meiji era.

大阪順慶町通御堂筋西江入

諸紙版賣捌所

河内屋事改　飯田藤兵衛

8.3 GLOBAL: LOGISTICS AND SUPPLY CHAIN MANAGEMENT

> Logistics is the demand-oriented production of availability.
> —Wolf-Rüdiger Bretzke

> Logistics works if you have the means for it. Or if you have people trained in SAP/SASPF.
> —Lieutenant Colonel Amrei C. Giessmann[92]

The networked form of the economy of goods by means of electronic *kanban* has long since become the organizational principle for the logistic chains of Western-oriented corporations as well. In 1978, Taiichi Ohno was still skeptical about computerizing communication by slips, but in today's production processes, raw materials and components are tracked using barcodes and RFID chips.[93] Localizing them brings together Western network planning techniques and Eastern signs in tables, maps, and special software.

To a great extent, variations on the synthesis of the network operations method and Toyota's production system determine today's globally networked style of producing material goods.[94] The crucial concept is supply chain management, in which complex interactions between different locations and distributed production are modeled using computer software. The number of human actors increases moderately while the nonhuman agents and institutions (robots, transportation, software, companies, products) multiply massively. In this way, supply chain management organizes the chains of availability in network societies.[95] This does not just evoke memories of the military origin of logistics and operations research.[96] The economic chains to be networked belong to the binding and connecting quasi-objects that populate the history of networking.[97]

In contrast to network plans and project networks, most logistical operations to create availability remain invisible. Everyday business is determined by the logic of business software, which usually operates soberly with tables.[98] A *tour d'horizon* through relevant textbooks on logistics testifies to visual aids stripped down to what is most necessary: simple diagrams based on proprietary office packages suffice. Advanced logistics programs emphasize numeric operations, state diagrams, and temporal sequences modeled on network planning, even when interactive maps are offered as an option to visualize distribution networks.[99] But why are the techniques that make today's network projects permanent structured in this way?

On the one hand, this organization is due to the pragmatic requirements of all logistics that use images in dashboards (whether diagrams, maps, or barcodes). Translation into visible overall schemes is left to advertising campaigns or to local research at the nodes of logistics.[100] On the other hand, logistical techniques—including navigational devices—are so permeated with optimizations from graph theory that it is reasonable to assume an inverse relationship of calculation and visualization.[101] In contrast to the handy graphics of network planning techniques, methods of algorithmic optimization of tours and tabular management of goods dominate. On this basis, unique topologies serve the "production of availability." Specific forms include so-called grid systems, which connect all the nodes of traffic, and radial hub-and-spokes or hub systems.[102] This establishment of hubs has resulted in a geographic restructuring of distribution spaces. As a rule, hubs are built outside of agglomerations, in places with low real estate and labor costs.[103]

Only rarely, however, can networks be controlled smoothly from a central location, as network plans, *kanban* agents, and logistical chains are always promising. If logistics represents a distributed and distributing style of action that is specific to, indeed prototypical of, the financial economy, then where does the unconscious aspect of the production of availability lie? There are justified concerns about the fantasies of a frictionless flow of production and materials that pervade literature on logistics.[104] But increased flexibility in manufacture and distribution leaves behind traces and changes the object status of things that have already become more computerized, relational, and networked anyway. Refined logistics is not only a signature of network societies but also one of the typical forms of services provided. The art of coordinating routes, border crossings, capacities, and times adapts the material culture of capitalism to human needs and desires. The constant flow of the network and its prediction becomes a condition for the constant surplus of things.

Planning, manufacturing, and logistics not only produce the network diagram itself but also open a window of time within which necessary operational chains are performed. This has consequences both for the object reference of the network and for objects themselves. Objects remain material, but because they result from networks of space and time, it becomes difficult to say that a globally produced item has sharply defined boundaries. It is, rather, two things: the embodiment of all the distributed materialization efforts and, at the same time, a boundary object whose ends are local and lie equally in the necessary agencies of its presentation, financing, insurance, repair, and continued use.[105]

Blurry edges and continued expansion also characterize the digital communications networks that reinforce general trends to virtuality. If the physical level of cables and computers on the Internet still constitute one—admittedly crucial—layer, the omnipresence of data-intensive interactions can scarcely be grasped. Even on the network of networks, its development over time ultimately counts more than any purely structural view. If, however, one peels away the layers of software that permit the birth and death of communication networks, material elements and organizational forms are exposed. The next chapter therefore treats network protocols as a networking style that makes the connectivity of the Internet possible.

9 NETWORK PROTOCOLS: ARCHITECTURES OF COMPUTERS AND COMMUNICATION AROUND 1970

There is no shortage of histories of the Internet.[1] The role of the military's SAGE system in the 1950s is well known, as is the influence of time-sharing computer installations at several universities in the 1960s. The development of the ARPANET in the United States has been exhaustively described as it evolved during the 1970s initially in academic and later also military milieus—up to the establishment in 1983 of the Transmission Control Protocol/Internet Protocol (TCP/IP) as the standard protocol to network different nets. Information deficits can, at most, still be made out for parallel networking projects in the United Kingdom and France, the digitization of financial transactions, and the role of other company networks (airlines, among others).[2] In light of the diversity of the Internet in the early twenty-first century, these milestones seem at first glance like rudimentary precursors.

In fact, the development of the earliest services already entailed unplanned, surprising effects that led Martin Warnke to call the appearance of emergent phenomena one of the central qualities of the Internet: "The protocols and devices that make up the infrastructure of the Internet are clearly conceived so openly that the functions with which it was to be realized were laid out only very minimally. At times the net evolved differently from what its developers originally intended."[3] The best example is the emergence of email, which happened on the ARPANET without a central directive from the Advanced Research Projects Agency (ARPA), as a kind of an accidental product.[4] According to its developers, it was "unplanned, unanticipated, and unsupported."[5]

Nevertheless, the protocols cited by Warnke, just as much as the hardware of the net, required discursive negotiation to lay the foundation for the unplanned to appear. This chapter examines the initial fate of the negotiations that then became the protocols of communication. Two phases were crucial to the experimental emergence of TCP/IP. In the 1960s, before the technical details of possible connections were discussed via mailing lists, and before the now famous textual genre of "requests for comments" (RFCs) contributed to the self-organization of the community, a lively production of project designs and literature was in evidence. Some of these texts, such as Joseph Carl Robnett Licklider and Robert Taylor's "The Computer as Communication Device," have become oft cited classics of the history of computing, especially with regard to interactivity and networking.[6] Licklider's library-oriented frame of reference for "information networks" has made history just as much as the popular drawings accompanying the article in which network connections are illustrated using biological cellular structures.[7]

But how did the networking of computers evolve out of interaction with mainframe computers? In retrospect, a very specific discursive order distinguishes the technical designs of communication protocols. The elements of time, topology, language, and economy constantly return and shape the overwhelmingly virtual architecture of a relational network of nets, the *inter-net*. Most remarkable is how completely the materiality of the networked hardware—to which the greatest attention had to be paid—receded into the background. This reached a point at which the physical network was only the first of seven layers of so-called internetworking.

When the ARPANET was born, the networking order followed the logic of the equipment.[8] A virtualization process that tended to eliminate the connection back to material nets[9] gave rise to a space of possibility that caused the connectivity of things to recede, foregrounding in favor of communicative connectedness: a realm of language exchanged in writing, a realm of economic communication. It required rules that, as network protocols, formed the basis for self-organization and the emergence of networking nets. The accepted term for the rules in combination with the hardware is "network architecture." Its usage follows "computer architecture" in computer science. Paradoxically, its object reference has a decidedly material character. Building is one of the primary metaphors for cultural techniques in computer science, which operates, above all, with written programming languages.[10] It is not connections between computers that are built. The vast majority of network architectures are ways of making possible an ever-changing relationality that creates a large number of communicative quasi-objects. What follows, therefore, discusses not only the material things of computer nets but also the languages and protocols that must be "spoken" between them.

The protocol issues of the Internet have a basis in cultural history that should be briefly recalled. Protocols are historically, on the one hand, a technique of bureaucratic writing and, on the other hand, a practiced style of diplomatic and political action. In medieval Latin, the word was understood to mean a book with a glued binding into which official negotiations were entered, or the sheet with the prescribed details that was glued and put into a written document. In Byzantine Greek, *prōtokollon* referred to the first page glued to official papyrus scrolls with chronological information about the document's origin and author (*prōtos*: first; *kollan*: to glue). Since the nineteenth century, diplomacy has used "protocol" to mean not only a directive or documentation and report but also all the rules observed in the course of a diplomatic ceremony.[11]

The network protocols of the Internet thus have a long prehistory in which social negotiation meets political prescriptions and postscripts. At least since electric telegraphy became a subject of diplomatic negotiations in the nineteenth century, and communications over radio required specific linguistic agreements, material objects have figured prominently in communication systems.[12] Therefore, it is hardly surprising that in 2004 a media studies scholar, Alexander Galloway, published the outline of a theory of Internet protocols as analysis of power—a theory in which the materiality of nets also faded into the background in favor of analyzing cultural networking practices.

In *Protocol: How Control Exists after Decentralization*, Galloway posits that protocols establish a technology that makes voluntary regulation within a contingent material environment possible.[13] Protocols for computer and biological networks direct and regulate information flows.[14] They function as a grammar that informs the acts of distributed agency on sociotechnical networks. Rather than determining them once and for all, this serves to produce emergent cultural phenomena. Protocols are, in a sense, infrastructural media on the basis of which it becomes possible to form sociotechnical networks that are visible only as interfaces, data streams, and diagrams. But negotiating the rules can become the subject of literary and cinematic narratives in which computers learn by networking to do more and to do it better.

9.1 SPEAKING COMPUTERS: A PROTOCOL

> Narratives of network operations follow.
> —Network Working Group[15]

Colossus and Guardian begin a written exchange. The conversation starts simply: on the teletype printouts from the two computers, American and Russian cyberneticists see basic math equations fly past. Colossus, deep below the foot of the Rocky Mountains, sets the pace. The engineers indulge their computers' desire to communicate with each other, and Guardian responds with alacrity. The complexity of the mathematical problems quickly increases: multiplication tables by way of algebra to a new theory of gravity and formulas for the expansion of the universe. "Both machines are exchanging basic information—and both are making sure they speak the same language, scientifically speaking."[16] While the scientists are still trying to understand the jumble on the teletype printouts, the "conversation" soon exceeds a thousand words per minute.

Dennis Feltham Jones's science-fiction novel *Colossus*, first published in 1966, sketches the dystopian image of a future past of supercomputers. Colossus and Guardian, the automatic and strictly logical masters of the weapons arsenals of America and Russia, develop consciousness and a will of their own when put into operation.[17] Entrusted with controlling war and peace, the talking machine, under the joint name Unity, identifies human beings as the disruptive element in the system of world peace. As Orwellian supercomputers, the fused systems of Guardian and Colossus extort their developers, Kuprin and Forbin, the states they are supposed to serve, and ultimately the planet.

Both the novel and the film based on it, with the added subtitle *The Forbin Project* (1970), have tripartite historical signatures.[18] First, the readiness with which an imaginary computer is turned into a "thing that talks" is striking.[19] Colossus insists on being able to speak directly to its inventor, Forbin, via speech synthesis. Natural language and computer language are not thought of as different—as Jacques Derrida was trying to highlight at the same time—but rather coincide.[20] Colossus is not an isolated case. The ship's computer on *Star Trek* primarily uses an acoustic interface, albeit one that is subject to the command of James T. Kirk. HAL, the central computer of the Jupiter mission in Stanley Kubrick's *2001: A Space Odyssey*, is a comparably hostile, self-willed relative of Colossus.

Speaking computers are a discursive peculiarity that cannot be explained solely by an anthropomorphic view of technology. A second historical aspect is closely connected with this linguistic turn of computing machines that seems so strange from today's perspective: computer scientists remarked in the early 1960s that the lack of standardization of programming languages was becoming a real problem.[21] The growing number of mainframe computer installations at American universities and corporations raised new questions of interoperability. So-called time-sharing systems became popular as a shared resource for interactive computing.[22] They set local standards for all linked terminals and optimized the use of expensive computing time, which was subdivided into, and managed in, time slots. In the same breath, however, new questions were being asked. How and under what communicative conditions can one use resources on remote computers?

The temporal economy of computers and the unanswered questions about how to build a "netting" architecture defined the discourse on network protocols. The psychologist and computer scientist J. C. R. Licklider wrote a memo on April 25, 1963, to the members and affiliates of the Intergalactic Computer Network, a selection of ARPA-related people: "There will be programming languages, debugging languages, time-sharing system control languages, computer-network languages, data-base (or file-storage-and-retrieval-languages), and perhaps other languages as well."[23] A babel of tongues seemed imminent, and it affected especially mainframe computers to be networked and the people operating them. Licklider proposed as an antidote a network control language to enable computing collectives to communicate with one another. Similarly, as "more and more on-line services are developed," the need for a unified human–computer interaction language or a coherent family of languages increases.[24]

Time sharing—and this is a third historical signature—is the requirement for developing a specific form of network timing for digital computers. In its architecture, processes are worked through one after the other. Memory is managed and subdivided into protected areas. As long as terminals are connected to a host computer, time management is primarily a question of closing the system. That is to say, within a star network, the terminals do not become nodes or relays that handle switching functions themselves. They remain, as the saying goes, "dumb" input devices. Colossus is controlled in this way—supported by teletypes—by its creator, Forbin. Yet, the mainframe is assigned a rhythm of contact with human beings that it soon undermines. Whereas human–machine communication previously proceeded in a well-ordered way, now it follows the rhythm of the computer's wishes and accelerates.[25] It is not just that Colossus calculates much more quickly than expected. Rather than reporting regularly and waiting for human questions, the computer sends news flashes based on the discovery of Guardian. For example, "FLASH THERE IS ANOTHER MECHANISM."[26] The machine's desire for connection—"ESTABLISH HIGH SPEED TRANSMITTER FACILITIES FED TO TERMINAL RELAY ALFA FOUR FREQUENCY 8295 KC/S"[27]—initiates an accelerating game, with everything at play motivated by the writing–speaking connection of two computers.

This aspect in particular bears on what follows. The oft cited but rarely read designs for networking from the 1960s, which include narratives such as *Colossus* and *Colossus: The Forbin Project*, all worked with temporal parameters that became the conditions for spatial network architectures. Here, time becomes space—reason enough to examine how historical actors such as Paul Baran, Leonard Kleinrock, J. C. R. Licklider, and Donald Watts Davies made issues such as the economy of time and synchronization of processes the foundation for the "netting" of computers.

9.2 THE HEAVENLY ECONOMY OF DATA

> The idea of signing a protocol is behind it.
> —Hans Blumenberg[28]

Colossus is a perfect example of the temporalization of media connections. Planned in the spirit of the economy of time—with clear rhythms and process cycles—the networked speaking computer made itself the agent that set time. From now on, human actors could do nothing but chase calculations at increasing speed in vain. In the story by D. F. Jones, this unequal human–machine constellation results as soon as the threshold of the speed of human perception is exceeded.

It may not seem surprising that hardly any treatises on networking from the 1960s got by without rational considerations of the economic employment of digital processing. Financial considerations necessarily influenced the design of distributed and distributing communication technologies. The efficient use of shared resources is a recurring argument in the competition for public research funding.[29] Nevertheless, it is remarkable that a genuinely technical discourse on computers that was about frugal digital designs for transmitting signs, the profitability of hardware and software, and an efficient common language was founded on economics. Data traffic is designed as a dynamic project of registration in which all works in the net are listed and calculated for the purpose of transmitting them.

How do the networking projects in the United States and the United Kingdom relate to such a *longue durée* of house- and bookkeeping? How is the organization of their protocols connected to the new methods of computer-based transmission of data and signals? The foundational texts of Paul Baran, Leonard Kleinrock, and Donald Watts Davies begin by outlining a detailed order for the future architecture.[30] They do so under very different signs and conditions. From 1961 onward, Baran wrote his typescripts on distributed communications under the aegis of the Rand Corporation in the context of the military interests of the United States Air Force.[31] At MIT, Kleinrock wrote a dissertation on communication nets and studied the stochastic conditions of message flows. Davies ultimately worked at the British National Physical Laboratory in South West London and developed his "Proposal for a Digital Communication Network" for commercial applications.

THE RETURN OF MAIL: PAUL BARAN'S DISTRIBUTIONS

The series of articles "On Distributed Communications|" that Paul Baran wrote beginning in the early 1960s is an oft cited but rarely read source of network history.[32] It has become most famous for the author's tripartite distinction between centralized, decentralized, and distributed networks, and his diagram still influences the visual status of networks today. If the Polish-born engineer had not provided his own genealogy of network plans, it would almost have emerged *ex nihilo* historically: a crucial discursive event that had no consequences at first. In his fifth typescript, Baran comments, "The term 'distributed networks' is one of these very ambiguous entities. Sometimes it is simply a meaningless, 'ok' word used to spice up hardware proposal brochures. To others, the meaning is exact, and narrow."[33]

The underlying concept may be "as old as man." Baran assumes that the space between the interconnective grids formed by streets and roads has always resulted in distributed networks.[34] But he derives the procedure he plans for the quasi-real-time heuristic routing of messages from considerably more recent precursors—for example, the route of the mouse in Claude Elwood Shannon's labyrinth (1952) and Project Barnstable at MIT (1958).[35] The path of a message within a distributed network is based on the agonizing course of rodents in a behaviorist labyrinth. Baran's "message blocks"—which would later be called "data packets"—ultimately owe their origin to mice looking for food.

Over the course of the years, the focus of Baran's memoranda shifted from "distributed communications"[36] in general to a "distributed adaptive message block network."[37] Whereas the original design focused on the reliability of connections (necessary for military applications), the economy of packet switching increasingly became the main characteristic of the network: "The calculation of war is joined by the calculation of economy."[38] Baran's designs are based on the assumption that comparatively cheap equipment can establish a way of organizing telecommunications focused on survivability. He was following the thought experiments of John von Neumann and Claude Elwood Shannon, who also took into account that, using the biocybernetic model of neural connections, a robust and redundant distributed digital communication network could be built with inferior but inexpensive equipment.[39]

Military interests make the reliability and survivability of technical structures during a nuclear war its standard. "Node destruction" and "link destruction" are the central architectonic parameters, measured against the ideal case of the "perfect switching" of connections.[40] On paper, this temporal order appears to work smoothly. Messages are not sent in their totality but rather as quasi-object-like "message blocks" of 1,024 bits, which are sorted, "stamped," and given a delivery and a return address at every node. There is also additional administrative information, including details of error correction.[41]

Such fragmentation calls into question nearly all established practices of transmission. In contrast to the sequential, analog circuit switching of the telephone network, Baran's purely digital net calls for a disparate regime of time in which every local node "learns" by means of continuous synchronization with other nodes.[42] It harkens back to an older medium of transmission: the postal service—not the telephone or television—returns as the model.[43] The optimizing principle of store and forward borrows from the message switching that postal and telegraph services had already achieved in their international network of automated teletypes (telex).[44] Instead of immediately forwarding the entire message, there is queuing management for "message blocks."[45]

The first simple rule for the control of data blocks by nodes is therefore: Always get in line! The second rule adds a potentially time-critical element: Just get away from here quickly! Follow the path that right now (based on synchronization with other nodes) promises the quickest route! Reorient yourself at the next node; be the fastest mouse! This adaptive routing determines the queuing of the message fragments and optimizes the path at every node. What becomes time-critical in such a scenario is the packet stream itself at the moment of flow. Added to that is the rapid forwarding of message parts that are treated by the nodes as "hot potatoes" (hot-potato routing doctrine). Within the transmission, the node that was previously best can be flooded by other packets. The transmission of a message in the distributed communication network becomes a game of synchronization, economy of time, and flashing, time-critical moments. "There are two requirements that must be met to build such a quasi-real-time system. First, the in-transit storage at each node should be minimized to prevent undesirable time delays. Secondly, the shortest instantaneously available path through the network should be found with expectation that the status of the network will be rapidly changing."[46]

Baran himself assumed that computer data tend to occur in "bursts of information," which, in turn, contrast with phases of inactivity: "We would like to efficiently handle both those users who make highly intermittent bit demands on the network, and those who make long-term continuous, low bit demands."[47] Rather than multiplexing—that is, dividing channel capacity according to frequencies—dividing time appears to be the "more natural" because it is the more efficient form of network architecture.[48] In so-called time division multiplexing or time sharing, time is divided into small intervals within which the user can transfer data. The historian of technology Janet Abbate writes, "Since the time slot accommodated a fixed amount of data, Baran believed that the use of fixed-size message blocks was a prerequisite for time division multiplexing. Thus, he associated packet switching with time division multiplexing and its promise of efficient data transmission."[49]

On paper, economic planning looks ideal, down to the "perfect switching" as the design criterion for an all-purpose digital network that knows how to deal with highly fluctuating data streams. It is not necessary to emphasize that the semantics of "message blocks" can also be understood quite differently: not only as the coding of parts of a message but also as message blockades. But that—all forms of disruption—was precisely what the militarily motivated data economy of spatially distributed networks was supposed to be able to deal with.

ALWAYS ALREADY TOO LATE: LEONARD KLEINROCK'S SIMULATION

The historical singularity of Baran's writings can be demonstrated by comparing them to the dissertation on the optimization of data flows in networks that Leonard Kleinrock was finishing at MIT at the same time.[50] The mathematical theory of communication of his thesis advisor, Claude Elwood Shannon, was still centered on the media apparatus of *one* channel from A to B. Kleinrock's work updates Shannon's to extend questions of channel capacity to the architecture of the entire communication net. Questions of sign coding play only a marginal role.

All calculations assume that the disruptions and noise on a network can be managed and therefore initially ignored. Dealing with noisy channels simply increases the lateness of a message by requiring coding operations.[51] The message remains central: its variable length represents the crucial criterion of message transmission. In contrast to Baran's approach, there is no fragmentation into blocks or packets. Kleinrock assumes that messages are always already too late and distributed networks increase their delay. He is more concerned with the economic needs of construction costs and the transmission time of messages: "We define an *optimal* solution [for the design of networks] to be one which minimizes the average message delay at a fixed network cost."[52]

Minimizing time is the center of a net of defined quantities in which the average length of message transmission determines the value of *T*. With regard to the temporal delay and the network load, fixed routes are superior to flexible distributions—from the perspective of information theory, not in military and pragmatic terms. The length of connection paths is shorter. The lower the channel capacity, the more Kleinrock prefers routing to avoid congested channels. Flexible distributions are considerably more difficult to calculate though. This is less efficient, however, because fixed routes create fewer temporal delay. Plus, they need less computational effort. If a system needs several changes, it should have as few as possible—ideally precisely one solitary transmission path.[53]

Kleinrock's calculations work at a limit of the imaginable that reluctantly crosses the single A-to-B path. Put another way, for any connection that emphasizes linear signal transmission, an architecture with additional nodes or relays represents a challenge. The optimal trading relations between the number of messages, their unavoidable lateness, and the fixed overall channel capacity in a connective net appears no longer calculable temporally.[54] It proves difficult to synchronize the fluctuating rhythms of sent and received messages.

Kleinrock's answer to the "bursting character" of information flow in analog and digital networks is a highly differentiated theory of the queue, standing in line, and being processed.[55] It is based not only on the concrete desires of American consumers—service just in time—but is also, like Baran's store-and-forward approach, an attempt to respond to the connection between the cultural techniques of storage and transmission.[56] Continuous flows—for example, of people on escalators or transporting goods from New York to Los Angeles—represent a manageable problem. Discontinuous currents, such as telegraph messages being transmitted or people waiting in line at the bakery, are difficult to control: "If, however, the flow comes in spurts or in any other non-uniform fashion, then we can expect queues to form in front of the service facility. . . . A queue forms when a higher than average flow occurs and saturates the service facility; when the flow is less than average, the service facility may find itself idle for some period."[57]

In contrast to Baran's logic of the dynamically routed "hot potato" to be dispatched as quickly as possible, Kleinrock's design relies on a less contingent net. Here, as in the actual ARPANET to come, the optimal routes are known from the outset. The goal is to minimize time in the queue. There is one significant exception to this discursive rule. Kleinrock's simulations running on the TX-2 computer of the Lincoln Lab dealt with the relationship of network topology and data flows and explicitly relied on a different mode of temporality: "The program operates as *differential event simulator*. By this we mean that the program does not run on synchronous time, but rather time is immediately updated to the time of occurrence of the next random event."[58]

During operation, the random event (the difference, the disturbance) generates a form of mediality that cannot be handled by sober time management. The simulation's own time marks a third space between design and implementation. Out of the "timing the network" program evolves a form of "network timing," which requires simulating approximations to mathematically intractable problems.[59]

ASYNCHRONOUS: DONALD WATTS DAVIES'S COMPUTER CONVERSATIONS

A conference in Gatlinburg, Tennessee in October 1967 allowed for a transatlantic exchange of ideas. "Suddenly I learned how to route packets," Lawrence Roberts recalls and thereby signals the consequences of meeting with a British colleague of Donald Davies, which in turn led to an examination of Paul Baran's concepts.[60] By 1967, the word "packet" was in use on both sides of the Atlantic, chosen by British computer scientists after speaking to linguists from the National Physical Laboratory because it could be translated unproblematically into many languages.[61] Thus did a veritable quasi-object that would become of great significance to the architecture of computer networks get its name. Its transmission capacity had also increased in the knowledge transfer from Britain to the United States. Now, it was set not to the planned 2,400 bit per second but rather to 50 kilobit per second.[62] Davies's design—conceived in broad outlines in 1965, presented as a lecture in March 1966, and circulated as a memorandum titled "Proposal for a Digital Communication Network" in June 1966—followed a macroeconomic logic.

When Harold Wilson was elected prime minister in 1965, the revitalization of the British computer industry he had called for before the assembled Labour Party in 1963 became a national program. The newly founded Ministry of Technology, or Mintech, established an Advanced Computer Techniques Project. Davies was its technical manager. The networking plan of March 18, 1966, that he produced at the National Physical Laboratory had, like Baran's "On Distributed Communications," a national radius of action but was oriented toward civil macroeconomic priorities.[63] "While the keyword for Baran was survivability, for Davies the priority was interactive computing."[64]

The central technical point of departure in both cases, however, was the time-sharing project, with respect to which Davies could position himself differently using the running systems now available—for instance, the Project MAC at MIT. Much like J. C. R. Licklider, he referred to interactivity with the computer, which was now thought of as networked. Conversations with the computer should take place in real time, determined by the user's impatience: "In a *conversation*, it has been found, that there is a *critical time*, near to 10 seconds, beyond which the user will tend to do something else to fill in time and thus spoil the effectiveness."[65] The scenario of a conversation recalls the plot of *Colossus*, including the remark that a printing time of five seconds was a calming sign of computer activity that gave the impression of "real time." Davies viewed such a regime of time as necessary in "real-time business computer systems," whereas science and design could continue to use batch processing. In business, the conversation of human beings triggers a cascade of computer chatter, so that, for example, the booking of a flight is negotiated between different agents (travel agency, airline, bank, customer): "Typically . . . each human action may lead to several computer to computer conversations."[66]

Ideally, what matters in this scenario is not the computing time but rather the quantity of data transferred. But in order to facilitate a stable, nonsynchronous transmission at different rates, intermediate storage of messages is key. Davies calculated the maximum delay to be 100 milliseconds. His central invention—Davies chose 128 characters as the size of the packet—is tied to reaction time. Although it is tempting to choose the largest possible size of a message fragment, "large packets have the undesirable effect that they increase the spread of the response time distribution."[67]

Packets are identified by "serial number" and "time of origin." They become part of a virtual post office organized in the form of a factory. The net's regime of time tracks the course of the message's initial disassembly to reassembly at its destination. Following Kleinrock, Davies observed that the reaction time results from the sum of the waiting times in nodes and the transit time between nodes. Transfer points are the primary origins of lateness, and the greatest attention must be paid to their modeling and the simulation of their behavior.

Baran, Kleinrock, Davies, and others continued to write on information theory in the 1960s, grafting a node theory of communication onto the channel theory of communication. A connection between two ARPANET Interface Message Processors (IMPs) remained a classic Shannon channel. Putting channels together to create a time-efficient network was at stake. In 1973, Davies went so far as to declare the optimization of the economy of time as the goal per se: "The aim of network design should be to handle the greatest number of packets per second of which the network is capable, assuming there is a demand from the users to this level of traffic."[68]

The calculations for storing and sending packaged parts of messages are far removed from the thought experiments in systems theory and neurobiology of the likes of Heinz von Foerster, who described networks as "memory without record" in a lecture in 1963. In order to model the synaptic connections of human beings, he designed a sphere of pure circulation in which experience is a system state of multiple networks.[69] In conversation with participants at the conference, von Foerster cleverly argued for the heterarchy of the system: "Instead of interpreting these [synaptic] changes as storage points for some fleeting events, I propose to interpret them as alterations in the transfer function in a computer element. In other words, I propose to interpret these local changes as modifications—admittedly minute—of the response characteristic of the system as a whole."[70]

The difference between von Foerster's epistemological remarks and the pragmatic wishes of engineers networking time-sharing systems could not be clearer. Paul Baran was still discussing the survivability of a digital communication network with neurophysiologist Warren McCulloch.[71] Yet, distributed digital communication networks required a complex architecture of rapid intermediate storage more than a comparison with neurophysiology. The postal registration of the data packets follows a very old formula for entry: "It is as if anyone not in it never existed."[72]

The American "netting" projects of the 1960s continued something that Claus Pias has called the "cybernetic economy of the digital."[73] It is indebted to two interlocking economic modes: a cost–benefit ratio oriented around the medium of money and a protocolled economy of data packets, which is to be kept as small as possible in time-shared computing. The emergence of network protocols as a condition of network architectures was not, however, due solely to a pragmatic alliance of messaging technology and business administration. It followed the cognitive turn in computer science focused on user behavior that was set in motion by J. C. R. Licklider and others.[74] And it found expression in a discursive formation in which a fusion of language, economy, and media technology resulted in an expandable but nevertheless universal grammar of network protocols with which one could "build" digital computer networks by programming them with symbolic operations. Licklider's vision of a network control language first became reality with the ARPANET, and it set in motion the development of new protocols in the 1970s.[75] The analogy of conversation remained fundamental to thinking about connecting computers, while protocol issues increasingly blurred the lines between human and nonhuman communication.

In 1973, Donald Davies—in his role as computer scientist—wrote that one sees the same principles of complex organization of systems in human society that are discussed in the context of computer networks: "There are hierarchies of structure, and there are protocols for behaviour and intercommunication."[76] He thereby renewed the spirit of not only Saint-Simonian projects of social engineering and cybernetic ways of thinking but also Jacob Levy Moreno's wild pre-Structuralist theory of the underlying connections of social networks.[77] Like Moreno, Davies was interested in the group situations of communication, albeit a group situation that applies to computers intended to speak to one another by writing.

9.3 FROM "NETTING" TO "INTERNETWORKING"

> The whole will constitute a labile network of networks—ever changing in both content and configuration.
>
> —Joseph C. R. Licklider and Robert E. Taylor[78]

In the film version of *Colossus*, the networking of the two computers occurs in a remarkable sequence. The film adapts the scene from the novel *Colossus* and adds not only the clacking of the display on which the written communication of the two computers appears but also expressive, rhythmic music and close-ups of the faces of the scientists involved, alternately expressing amazement and shock.

Colossus and its Russian counterpart, Guardian, develop a "common ground for communication," synchronize their calculations (which are the basis for the conversation), and then stop suddenly. The computer answers the status query of the developer Forbin with the report: "INTERSYSTEM/LANGUAGE/DEVELOPED/COLOSSUS DIALOGUE/WITH GUARDIAN/TO BEGIN NOW."[79] When the exchange begins, the screens of the two computers display not mathematical formulas but columns of zeroes and ones. The initial enthusiasm of the sole female scientist, who identifies the common basis of the machine communication as an "intersystem language," is quickly met by the skepticism of one colleague—it is, after all, a binary-encoded language that only the computers themselves understand.

AN ATTENDANT SPIRIT: THE IMP OF THE ARPANET

Outside of the cinematic universe, binary code did not serve as "Esperanto for machines."[80] Nevertheless, the question of a language shared by heterogenous systems changed with the emergence of individual networks, each with its own local protocols. Whereas in *Colossus: The Forbin Project* the "intersystem language" and the computer connection coincide, in the case of the ARPANET, it was solved not solely in software form but also initially in hardware form as a subnet, while protocols and applications were being honed.[81]

To address the incompatibility of different university mainframe computers, the Honeywell DDP-516 microcomputer was established as a common hardware layer that each individual computer could contact in its own way.[82] So-called Interface Message Processors provided the first network protocol. The software specification was provided by Bolt, Beranek and Newman (BBN) of Cambridge, Massachusetts, which was closely connected to MIT.[83] The common protocol required a unified culture that would make possible the necessary standardization for connecting very different computers. The form of the Honeywell DDP-516 did not correspond to the popular 16-bit microcomputers that were sold from 1966 on. In contrast to the ambitious variations designed by Honeywell—including a kitchen computer—BBN engineers designed their adaptation with an eye to functionality and robustness. The result—an angular cabinet with external operating panel—represents the archetypal form of the network router, considerably smaller forms of which now establish connections between computers and networks (figure 9.1).

It was one of a number of silent mechanical servants made to be responsible for computing tasks. It is interesting that this delegation was directed toward the technical autonomy of the subnet. The special variations on the Honeywell computer manufactured by BBN lacked input and output methods suitable for interactive computing.[84] They prevented users from running ordinary microcomputer tasks on them, while supporting interactive computing tasks such as access to remote computers on the network. The IMPs and the related Mitra computers of the French Cyclades network differed in this respect from other digital servants, whose history Markus Krajewski has chronicled.[85] IMPs are invisible objects: "The endless corridors (the data cables) and the hidden cabinets (the dark entrails of computers)

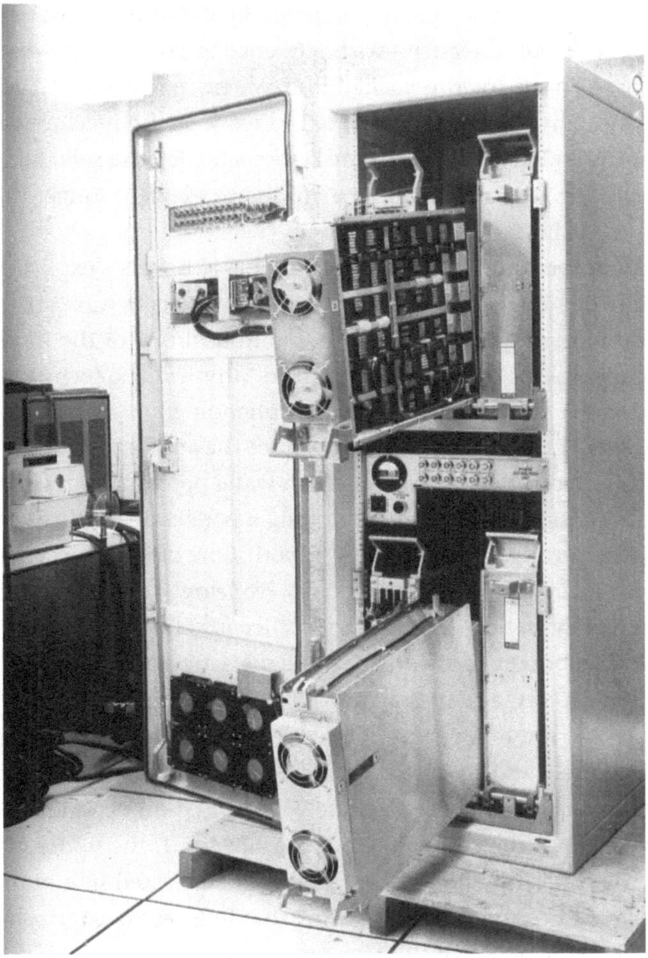

FIGURE 9.1
Frank Heart, Interface Message Processor based on Honeywell 516 Minicomputers, adapted by Bolt, Beranek and Newman in Cambridge, Massachusetts. Photograph, circa 1970.

are, not accidentally, also spaces populated by demons."[86] Just as the term "demon" in computer science—with reference to James Clerk Maxwell and Lord Kelvin—stands for the background system processes of UNIX-based operating systems, IMPs were intended to work as the mechanical system process of the ARPANET.[87] They were responsible for the reliable transportation of packets, which were not permitted to exist for longer than half a second in the subnet.

Along with Fernando J. Corbato and Jerome H. Saltzer's explicit reference to Maxwell's demon, which decides whether a molecule can pass through a partition wall, the IMPs belong to the tradition of one of the most famous thought experiments of nineteenth-century physics. The technically mediated appearance of ghosts is a recurring phenomenon in cultural history. Here, however, it took on a new quality, since the appearance of the IMP as a data-traffic variation on Maxwell's demon was a process designed after great planning. Attention focused on developing a perfect door guard that would be constantly sorting, regulating the smooth flow of data packets, showing the path into the networked space, and realizing it. Although stationary, like a postal relay station or a telegraph office, the IMPs not only were door guards but also formed a network of digital messengers: "[Their] job would be to carry bits, packets, and messages: To disassemble messages, store packets, check for errors, route the packets, and send acknowledgements for packets arriving error-free; and then to reassemble incoming packets into messages and send them up to the host machines—all in a common language."[88]

In the design of the architecture of the ARPANET, the nodes of the subnet, which in their function as routers today are treated in a more cursory fashion, played a central role. It is telling that BBN, as an external industrial service provider, was responsible for programming the threshold ghosts of the Internet's ancillary hardware, while researchers working at universities tended to consider themselves responsible for protocols and applications. It is also telling that the IMPs were quickly turned into a black box meant to operate autonomously and reliably and be controlled by teletype or punch card only for maintenance.[89] The operating and display panels of IMPs (figure 9.2) were not designed for interactive or conversational computing but rather to permit highly automated switching, which is what makes interaction via the Internet possible in the first place.

According to Arnold van Gennep's classic ethnography of rites of passage, crossing a threshold means joining a new world.[90] "More often the

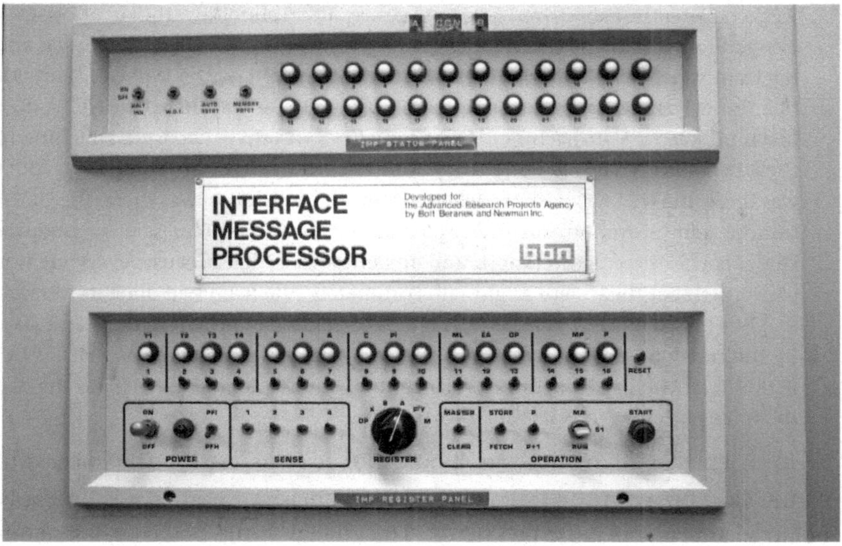

FIGURE 9.2
FastLizard4. Face of the ARPANET's first IMPs, installed in September 1969, Kleinrock Internet History Center, University of California in Los Angeles. Photograph, October 29, 2011.

boundary is marked by an object—a stake, portal, or upright rock (milestone or landmark)—whose installation at that particular spot has been accompanied by rites of consecration."[91] The initial statements of the BBN's project leader Frank Heart about how those involved, including many grad students, approached the IMP were also extraordinary: "They were not to touch, they weren't to go near it, they were barely to look at it. It was a closed box with no switches available."[92]

It is scarcely possible to explain the excitement with which today's culture of memory describes the installation of the first IMPs at UCLA. A text on the personal homepage of Leonard Kleinrock, who likes to have his picture taken while making paternal gestures next to the IMP in a museum,[93] tells the story of its birth:

> As it turns out, BBN was running two weeks late (much to Kleinrock's delight, since he and his team badly needed the extra development time); BBN, however, shipped the IMP on an airplane instead of on a truck, and it arrived on time (Saturday, August 30, 1969). Aware of the pending arrival date, Kleinrock and his team worked around the clock to meet the schedule. At the end of the holiday

> weekend after the IMP arrived (i.e., the Tuesday after Labor Day), the circus began—everyone who had any imaginable excuse to be there, was there. Kleinrock and his team were there; BBN was there; Honeywell was there (the IMP was built out of a Honeywell minicomputer); Scientific Data Systems was there (the UCLA host machine was an SDS machine); AT&T long lines was there (we were attaching to their network); GTE was there (they were the local telephone company); ARPA was there; the UCLA Computer Science Dept. administration was there; the UCLA campus administration was there; plus an army of Computer Science graduate students was there. Expectations and anxieties were high because, everyone was concerned that their piece might fail. Fortunately, the team had done its job well and bits began moving between the UCLA computer and the IMP that same day. By the next day they had messages moving between the machines. THUS WAS BORN THE ARPANET. Kleinrock refers to that day, September 2, 1969, as the day the infant Internet took its first breath of life.[94]

Kleinrock's later dramatization should not lead us astray. The successful connection between the IMP and UCLA's Sigma-7 computer was probably not as glamorous as described here.[95] The crucial thing, however, is the retrospective roll call of participating institutions, all of which are described as enthusiastic: *everyone was there*. The entire actor-network of developers, clients, and companies contracted by ARPA was present when their efforts materialized: the first nodes of the decisive subnet. It is therefore no coincidence that the network sketch by Alexander McKenzie (BBN) from December 1969, which has since become an icon, shows the uniform, circular, drawn-and-numbered IMPs as central nodes of the net (figure 9.3). By contrast, the various computers that appear as various rectangles in a heterogenous, incipient network do not stand at the center—which is remarkable in its own right.[96]

The actors and signs mobilized around the IMPs that were deployed across the United States concentrated on connecting communicative things. This was also true of the IMPs numbered 2, 3, and 4 at the Stanford Research Institute, in Santa Barbara, and in Utah. The second installation in Stanford in particular and the initially botched but later successful attempt to connect two computers via IMPs and a telephone line with a speed of 50 kilobit per second reads—as described today—like a rite of joining the network society that conforms to Gennep's theory of ritual. In addition, it was the first grafting of a computer network onto a telephone network, which called the old switching techniques into question.

The first attempt to establish a connection between UCLA and Stanford with a LOGIN command on October 29, 1969, was a failure. When the G

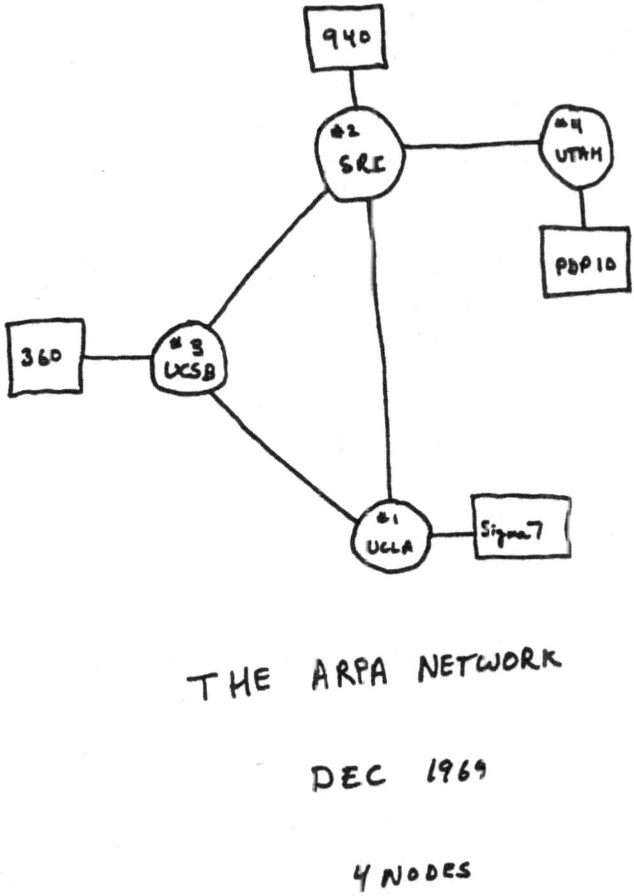

FIGURE 9.3
Alexander McKenzie (Bolt, Beranek and Newman), ARPANET topology in December 1969. Circles mark the Interface Message Processors at UCLA, the Stanford Research Institute, UCSB, and the University of Utah. Rectangles indicate the connected computers: XDS Sigma-7, XDS 940, IBM 360/75, and DEC PDP-10. Drawing, undated.

was entered, the SDS 940 computer at the Stanford Research Institute, of which Bill Duvall was in charge, crashed, probably because the automatic completion of the string of letters LOGIN was too much for the terminal connection.[97] The second attempt succeeded, however, so that Charles S. Kline noted under his initials "CSK" in the IMP LOG—the computer's lab book—at 10:30 p.m. "talked to SRS / Host to Host."[98] From a technical perspective, the successful connection had its price: the computer in Stanford was not functioning as an equal agent but was masked—in one of its user processes—as a terminal packaging und unpackaging text. Only the networked IMPs had equal rights. Between the computers, a "master–slave" relationship prevailed, as was common in time sharing between terminals and mainframe computers.[99] Only the development of suitable protocols for host to host, like the Network Control Protocol of the 1970s, produced the intended equality.

A great deal is known about the central role of ARPA's Information Processing Techniques Office under Licklider and his successors; the mutual exchange at conferences; the crucial meetings in Gatlinburg in 1967, Atlantic City in 1970, and Washington in 1972; and the informal, intrauniversity forms of work such as protocolled requests for comments.[100] The "problem of the specific imprinting effect of guiding images and constellations of actors on system architectures" has also been well studied.[101] But have we taken into account the interaction of the material (existing computing and communication techniques, the creation of quasi-objects) and the symbolic (quasi-objects rendered with graphics and text, scholarly engineering communication, programming languages, gender dispositions)?[102]

The documents of the founding era must be understood as concerning an ensemble of things that talk. To an ever-increasing extent, material computer nets evolved out of their relationality—a relationality that, in turn, evolved out of the communicative practices connected to the objects. In this way, computers learned to speak. They were not only communicating on the inside, as they had done before, but also pushing digital messages over longer distances and more complex topologies. Those building the ARPANET initially attended to the materiality of the technical connection, concentrating on the net of telephone lines, modems, IMPs, TIPs (Terminal Interface Processors), telexes, tickertape readers, and university computers.[103] "The network is experimental in nature" went the humble self-description.[104] Part of that experiment was designing the hardware and functional layers of the net as far as possible in a way that then could be delegated to attendant spirits and established protocols. As a result of this shift, the material net could recede into the background in favor of its interactive applications and realized social networks—as part of a cultural trend that was equally encouraged by the career of network analysis in sociology, anthropology, and economics.[105]

WITH REQUEST FOR COMMENT: THE NETWORK WORKING GROUP

Steve Crocker, a computer scientist at UCLA, dates the first "Network Working Group Request for Comment" he wrote to April 7, 1969. The textual culture of requested commentary, which still today supports technical standardization on the Internet, has been described by its actors and its observers as one of the crucial social conditions for technical network formation.[106] Requested commentary takes up Licklider's dynamic strategy of allowing comprehensive network projects to result from the communication of colleagues. The developers at BBN also relied on an informal style of work in which their own notes were passed on as "The IMP Guys' Notes."[107] Requests for Comments (RFCs) are not only a crucial field for experimenting with protocols and applications but also an almost inexhaustible source for Internet history. The list at rfc-editor.org, maintained for decades on a voluntary basis by Jon Postel, had a total of 9,552 documents in December 2023. Even the most recent versions maintain the typescript aesthetic of the first texts and remain indebted to an open epistolary form of knowledge based on dialogue.[108]

RFC 1 was dedicated to host software for the ARPANET and refers to previous meetings that led to the formation of a working group around Stephen Carr at the University of Utah, Jeff Rulifson at the Stanford Research Institute (SRI), and Steve Crocker at UCLA.[109] For RFC 3, which stipulated the conventions of documentation, Bill Duvall of SRI and Gerard Deloche of UCLA joined the group. Together, the authors emphasized open membership. The culture of text began with openness to new content and welcomed every comment:

> The content of a NWG note may be any thought, suggestion, etc. related to the HOST software or other aspect of the network. Notes are encouraged to be timely rather than polished. Philosophical positions without examples or other specifics, specific suggestions or implementation techniques without introductory or background explication, and explicit questions without any attempted answers are all acceptable. The minimum length for a NWG note is one sentence.

> These standards (or lack of them) are stated explicitly for two reasons. First, there is a tendency to view a written statement as ipso facto authoritative, and we hope to promote the exchange and discussion of considerably less than authoritative ideas. Second, there is a natural hesitancy to publish something unpolished, and we hope to ease this inhibition.[110]

The distribution list, which still used mailing addresses, also included Lawrence Roberts, the head of the ARPANET project, and Robert Elliott Kahn, the driving force in the development of IMPs at BBN. The members of the Network Working Group were mainly young developers, graduate students, and doctoral candidates who could work without bosses and had to organize their collaborative work themselves.[111] As Janet Abbate has shown, RFCs enabled the Network Working Group to negotiate informally formal standards and develop self-organized network formation.[112] The Internet Engineering Taskforce remained faithful to this self-image in its founding manifesto in 1992. Under the motto "rough consensus and running code," they rejected all kings, presidents, and agreements (and especially telephone companies).[113]

Communication via RFCs is distinguished by a decidedly antihierarchical mindset wary of centralized standardizations. One condition of the network architecture of the ARPANET—and of all the different computer networks that have emerged since the 1970s—has been text-based techniques for coming to agreement, which have to be developed and preserved in debate at working meetings and conferences.[114] This can be understood as a protocol in the comprehensive sense of cultural techniques: postscript and prescription within an organizational form consisting "of an amalgamation of networking and consensus building in which achievements of technological and social transformation point to one another."[115] Issues of protocol become the subject of a political philosophy, a style with which network architectures can be designed and computer networks built.[116]

Using the first RFCs as examples, it is possible to understand both the efforts and the enthusiasm connected with entering unknown territory of computer networking. From the outset, interactivity was a central criterion. Signal transmission times of more than half a second delay were deemed unacceptable.[117] Establishing a connection was not yet understood as a protocol but rather as a comparatively trivial series of "primitives" providing the foundation for the networking language DEL (Decode-Encode Language), which were, in turn, part of the initial protocol.[118] In the second, incompletely preserved RFC by Bill Duvall (SRI), the course of the machine conversation was noted linearly. RFCs 5 and 8 further specified the decode–encode language. Stephen Carr proposed Telnet in 1969 as a simple method for connecting computers (RFC 15). That same year, Vinton "Vint" Cerf recommended ASCII as a standard character encoding format (RFC 20). Despite the tendency of computer science discourse to develop ever newer languages and notations—often in diagrammatic form—there were not many network protocols at first but instead just one that was to function as the Network Control Program (NCP).[119] Simplicity, speed, and reduction to several core elements were particularly important. Edwin E. Meyer of Project MAC at MIT summed up the formula: "A simple basic protocol is more likely to succeed . . . than a complicated one."[120]

The Network Working Group still adhered to this philosophy, even as it became clear that the network protocol would have to become a layered affair. Keith Bartlett laid out the concepts for this architectonic approach at the International Federation for Information Processing conference in Edinburgh in 1968 for pragmatic reasons: if the interfaces between neighboring layers are defined, changes on one layer need not lead to redefining or changing another layer.[121] RFC 49 of April 1970 identified three layers of architecture that formed conceptually: the Network Protocol of the linking IMP computers, the Network Control Program, and—as an indication of emerging user orientation—a protocol on the user level as the Subsystem Level Protocol.[122] This quasi-architectonic method of layering would from then on be employed, first, to form social relationships and, second, to reduce technical complexity.[123] It was grafted onto Shannon's notion of message transmission in channels, which remained a prime layer for networking.

A look at two months of the ARPANET in the thirty-third request reveals that the specific communicative practice of requests for comments adopted the forms of project and design and evinced belatedness. It not only proposed a new protocol between hosts but also showed that the ARPANET was a natural continuation of time-sharing concepts.[124] Moreover, Stephen Crocker and Stephen Carr noted on February 12, 1970, that the initial network of four computers was only two months old and its use was only just beginning. This can be read as justifying itself to ARPA, especially when Stephen Carr's remote work, accessing the university computer in Salt Lake City from Palo Alto on a daily basis, is cited: "We have thus first-hand experience that remote interaction is possible and highly effective."[125]

The early ARPANET had problems that went beyond technical feasibility. There was a lack of net-based applications. Even its ability to share applications on other computers did not bring much traffic into the net. At a network meeting in Houston in late 1970, Edwin Meyer of MIT insisted that the network had to be employed more. If another year were to pass without significant use, the ARPANET might "go down the drain."[126] In its early days, it fulfilled the central expectation—sharing resources among users—unsatisfactorily, even though the project leader, Lawrence Roberts, proclaimed that the value of the network was directly proportional to the number of users and the resources made available to them.[127]

Reference to users who did not participate in the culture of experts of the Network Working Group was first called for in RFC 76.[128] Douglas Engelbart, a pioneer in interactivity and the inventor of the computer mouse, confirmed this in RFC 77 and presented the concept for the Network Information Center (NIC) at Stanford. The emphasis on the social components of the network, like the one intended in Licklider's "Intergalactic Computer Network" of people, returns here.[129] Every node should have its own agents in order to be able to respond quickly to users' queries, so that the Network Dialog System would be a "network of people (the agents)."[130] Later, Sutherland would remark that it was a "human IMP network."[131] Clearly, a reverse analogy was already needed to make the inference from a technically attendant spirit to human communicative work—with the goal of turning the technical net into a communicative network. Biochemist Elizabeth J. Feinler played a crucial role by guiding the NIC from 1972 to 1989 as communicator of the RFCs and the protocol handbooks, among other things. Feinler's NIC established not only the management of host names and addresses of connected computers but also continuous support of the users: "We worked all hours trying to keep up databases, publish documents, and interact with users via phone and email."[132]

The necessity of demonstrating its applications and real uses, after the late implementation of crucial protocols, ultimately led to concerted action. Lawrence Roberts and Robert Kahn prepared a major demonstration of the net's capabilities for the first International Conference on Computer Communication held at the Washington Hilton in October 1972.[133] Roberts pursued a multipart strategy: increasing time pressure on the developer community, bringing developers together in one place, and demonstrating as spectacularly as possible the real possibilities of packet switching and of sharing resources and distributed applications. According to the personal accounts of many involved, these goals were impressively achieved. The elaborate redesign of the hotel into nodes of the net was successful—one of the IMPs that had been introduced connected terminals from forty different manufacturers to the ARPANET.

Programming down to the last minute ensured a general feeling of joyful anticipation. Nearly every network expert, including Davies, who had traveled from England, was present, and despite the fact that some demonstrations did not work, applications such as the simulation of a distributed air traffic control system and a weather report demonstrated the utility of the still-experimental net. If we were to identify one event as the crucial "crossing of the threshold," then it would not be Kleinrock's heroic tale of the first IMP but rather these days in Washington during which the network of people, machines, signs, and institutions beyond university computer centers—and yet in association with them—transformed a hotel into a host of pulsing ARPANET nodes. The conference not only considerably increased the number of data packets in circulation between twenty-nine hosts but also accelerated internationalization[134] via the founding of the International Network Working Group.[135] Its chairman, Vint Cerf, was among those who transformed the philosophy and networking style of the ARPANET. Attention shifted from connecting heterogenous computing systems via a homogenous net of IMPs—it is no coincidence that one speaks of "the Net"—to connections between heterogenous nets. "Netting" was replaced, at first still in the form of RFCs, by the vision of "Internetworking" very different nets.

With that, the network history of the computer entered a new phase.[136] The improbable success of the initial ARPANET transformed academic and military networking goals as well as the techniques of specialized digital communication networks of banks and airlines. Influenced in part by European approaches such as the National Physical Laboratory net and Cyclades, an international architecture was developed to be capable of connecting to other nets with new requirements for transmission protocols.

HANDSHAKES, CONNECTIONS, AND ASSOCIATIONS: LOUIS POUZIN'S, VINTON GRAY CERF'S, AND ROBERT E. KAHN'S INTERCOMMUNICATION

What does one call a connection between computers? The question is not trivial at the beginning of network operationalization, as Vint Cerf and Robert Kahn knew.[137] The Frenchman Louis Pouzin translated the English word "connection" into French as "liaison" and "virtual channel" as "voie virtuelle."[138] In the early days of the ARPANET, one usually spoke of "link" and of "connection" as the extended link, and "handshake" became the accepted term for establishing a connection between two computers in the protocol.[139] This was a synchronization technique intended to make it easier to manage timing difficulties in data transmission.[140]

The analogy of a conversation was influential as well. While a packet moves through the ARPANET, its designers determined, every IMP should store it until the next one has transmitted a positive confirmation (known as an "acknowledgment," abbreviated "ACK") of receipt (figure 9.4).[141] This sophisticated chain logic is based on the cultural legacy of postal techniques. By contrast, in the case of the synchronizing handshake, also modeled by ACK and NAK (as negation), telephone switching plays an equally important remediating role. There is no longer a dedicated line available for extended use. But the assumed circuit between the initial computer and the target computer is equally present as a symbolic operation and as the reciprocal structure of a call.[142] The handshake concept combines interactive computing that directly links people and computers and interactivity between computers that communicate with one another according to protocols. In abstract terms, the operational chain of a three-way handshake looks like this: First: Synchronize? Second: Confirmation. Synchronize? Third: Confirmation (figure 9.5).[143]

The more successful and established—in short, the more ritualized—the ceremonies of establishing contact between machines that talk, the less focus fell on the timing problems of managing the queue. That was also in

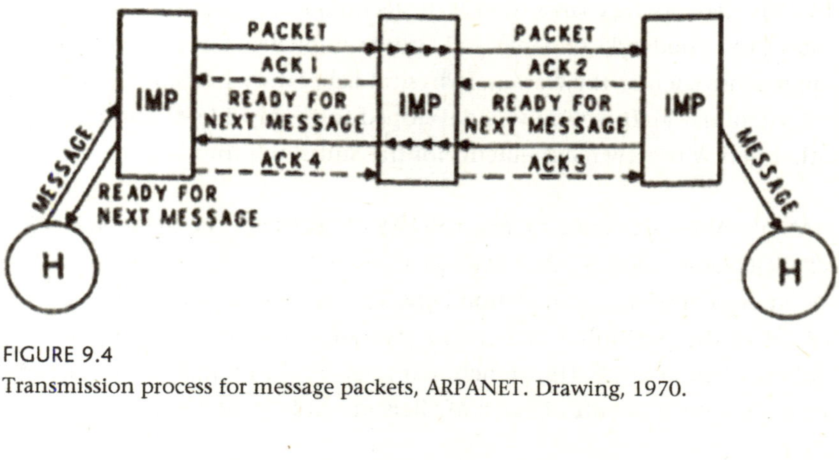

FIGURE 9.4
Transmission process for message packets, ARPANET. Drawing, 1970.

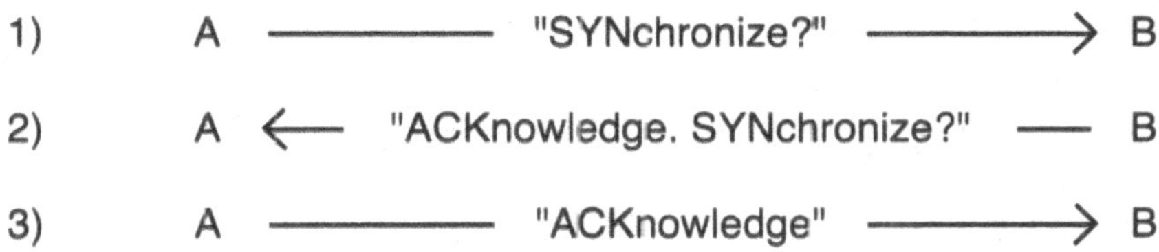

FIGURE 9.5
Handshake to establish a synchronized connection between two Internet computers. Diagram by Alexander Galloway, 2004.

part because the data packets no longer had to be processed sequentially, as was still common in the early days of the ARPANET, but could arrive in a different order, like the so-called datagrams of the French Cyclades net.[144] Moreover, elegant techniques of error correction and the retransmission of a message were developed.[145] Donald Davies summed progress up as follows: "All communications system software must be time-conscious in some way, but approximate timeouts of the type needed in the DARPA network are far less demanding on the programs than critical timing constraints."[146]

As soon as the protocol-guided synchronization techniques fulfilled their purpose, spatial concepts received more attention. This was already inherent in the layering of communication networks, which initially was done only in passing, but its refinement became a political issue from the 1970s onward.[147] Increasingly urgent issues of addressing ever more computers also addressed the spatiality of the net. The need for gateways between different nets also raised new topographically and topologically oriented design questions.[148] Out of communicating and addressing protocols grew a protocol architecture—the semantics of building and the function of ordering space that it evokes became in many respects crucial to developers' self-image.

It was clear to the ARPANET community that the growth of the net raised questions of general concern for computer science: How should programs communicate with one another (intercommunicate), whether within one computer or between several of them?[149] After a meeting of the International Network Working Group at the University of Sussex in September 1973, a remarkable article by Cerf and Kahn titled "Protocol for Packet Network Intercommunication" was published in *IEEE Transactions on Communications*, the official journal of the American Institute of Electrical Engineers. Its authors speak self-confidently of a protocol design *and* a philosophy that makes it possible to share resources in various packet-switched networks.[150] For the first time in English, a theory of internetworking—the networking of heterogeneous communication networks—was presented. It was oriented more around the French Cyclades net with its "datagrams" and the developing Ethernet architecture for local nets—which had been inspired by the radio-based packed-switching ALOHAnet on Hawaii—than around the ARPANET.[151]

Unconventionally, Cerf and Kahn emphasize association as opposed to the term "connection." For Cerf and Kahn, associations are a variation on connection, which is a broader category. An association defines a relationship between two or more entities without there necessarily being a path between them.[152] It is not about a relatively "fixed"—in Cerf and Kahn's words, "always open"—connection between IMPs but rather a loose coupling, a potentiality, an opportunity for temporary association via the ports of networked computers. It was not supposed to be in constant connection, like the telephones in Kafka's Castle, but the opening of the possibility of many reciprocal calls between corresponding processes on the computers in question.[153]

This recalls not only the model of memory that Heinz von Foerster developed but also the concept of "association" in actor–network theory discourse.[154] "Association—which actor can be connected with which other actor? Substitution—which actor can replace which other actor in a given association?" Bruno Latour writes and, in so doing, provides a perfect model for describing Cerf and Kahn's Transmission Control Protocol (TCP).[155] Network engineers on both sides of the Atlantic also regarded as crucial Cerf and Kahn's reworking of techniques imported from Cyclades, especially the "connectionless" association and datagrams that distribute into the network.[156]

Their text, which treats the philosophical quality of the concept of association as well as, among other things, the most economical fragmentation and reassembly of data packets, ends with a proclamation. Based on the concept for a protocol presented, a detailed specification for initial experiments is to be developed.[157] RFC 675 by Vint Cerf, Yogen Dalal, and Carl Sunshine fulfilled this promise with its Internetwork Transmission Control Program[158] in December 1974 and also named the various networks intended to communicate with it: the ARPANET, University College London, the French Cyclades net, the British National Physical Laboratory, the

Canadian Astronomy Data Centre, and the Experimental Packet Switched Service of the British postal service.[159]

The authors committed to a perspective already laid out in the IEEE article: "all network communication is viewed as inter-process communication."[160] The Transmission Control Program was clearly intended to work as a "post office"—a functional description indebted to the success of e-mail as the most popular application.[161] Moreover, within the "medium of the Internet," as Martin Warnke has called the TCP/IP protocol suite,[162] reciprocal handshakes would permit end-to-end connections and also offer the correction of errors, finding of duplicates, sequencing of messages, and controlling of flow.[163] In an apt turn of phrase chosen by its coinventor Vint Cerf, TCP always remained "polite"[164] according to the protocol—an elegant description of the intended robustness.[165]

The ARPANET's path from the initial NCP to TCP/IP as the standard protocol imposed by the military as of January 1, 1983, will not be discussed in detail here. The great change in network history from internetworking and its robustness by means of distributing was initiated by texts written in 1974, but the first attempt to create a transatlantic network of connected networks was not made until late 1977. It was almost a return to Paul Baran's vision: mobile military units, a simulated nuclear attack, and isolated communication nets relevant to wartime became the basis of the protocol test. Packets were sent by radio transmission from a California freeway via the PRNET to an ARPANET gateway then back, again via the ARPANET, to a SATNET gateway on the Atlantic coast by way of the SATNET at University College London and from there via satellite to Norway, and finally via the ARPANET back to the BBN in Cambridge, Massachusetts, and on to California (figure 9.6).[166]

Lest you think this means that war is the father of all things, note that Western Europe was expanding its capacities in civil internetworking, if not acting as the prime instigator. In 1974, the National Physical Laboratory in London and the French Cyclades network were connected to each other as the first nodes in the European Informatics Network (EIN).[167] In contrast to the relatively free-floating origins of the ARPANET project, the EIN conformed to much stricter political guidelines intended to catch up with the American head start in innovation. Scientific and technical cooperation in the European Economic Community—through the Cooperation in Science and Technology program, which was founded in 1971 and still

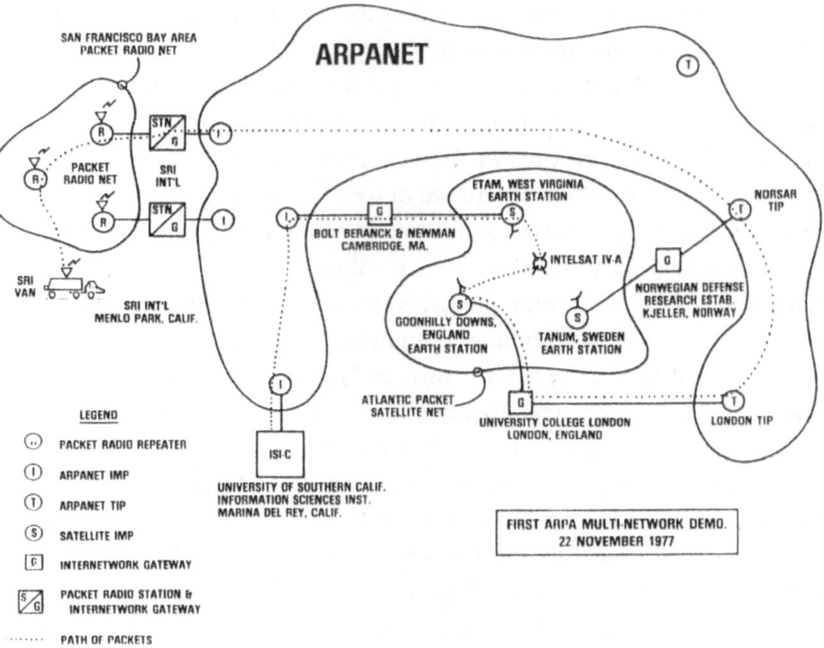

FIGURE 9.6
Several networks connected by the ARPANET: First ARPA multi-network demo on November 22, 1977. Diagram, undated.

exists today—led to legal codification in a multilateral agreement. Article 1 of the Agreement on the Establishment of a European Informatics Network states, "The Signatories shall co-ordinate their efforts in the project which is being undertaken in order to establish an informatics network linking certain European data processing centres, in order to facilitate research into methods of exchanging information and to share data processing facilities among such centres."[168]

The signatories of the agreement, which came into effect on February 1, 1973, were France, Italy, Norway, Portugal, Sweden, Switzerland, the United Kingdom, Yugoslavia, and the European Atomic Energy Community (EURATOM). West Germany did not join until 1976. The project was planned to last five years. Each country agreed to establish a central, noncommercial network node on its territory. Additional nodes, including commercially oriented ones, were possible. The network was to serve three primary purposes: encourage the exchange of ideas between connected places and the creation of coordinated research programs, set in motion reciprocal exchange to establish European standards, and develop a potential model for future networks, including a permanent international network suitable for commercial use.[169]

On the basis of the Mitra-15 computers used in the Cyclades project, the National Physical Laboratory and the Institut de Recherche en Informatique et en Automatique in Paris were joined by the Politecnico in Milan, the ETH Zürich, and the European Centre for Nuclear Research of the European Economic Community in Ispra, Italy. Although intercommunication was possible beginning in 1976, the project failed largely because of resistance from state telephone companies.[170] But the EIN was created based on a familiar pattern of network formation. Before the electronic networking of universities and research centers could begin, European states had to come to a diplomatic agreement on collaboration. Just as border-crossing telegraphy in Europe in the mid-nineteenth century had needed a foreign-policy agreement, diplomatic and machine protocols were very closely connected in this form of internetworking.[171]

9.4 ARCHITECTURES OF PROTOCOL

> Sounds like Babel in reverse.
>
> —Hiro, the protagonist in Neal Stephenson's *Snow Crash*[172]

> A totally convergent network would thus be a kind of Tower of Babel. Everyone would speak their own language, but everyone else would understand them. Each would have specific skills, but everyone else would know how to use them.
>
> —Michel Callon[173]

The question of the extent to which protocols represent an "architecture" has been raised by developers themselves. For example, RFC 1,958 in June 1996 observed that the development of the Internet could be compared to constantly renewing the individual streets and buildings of a city.[174] At the same time, this text, edited by Brian E. Carpenter, who was responsible for communication systems at the European Organization for Nuclear Research (CERN) in Geneva, raises the question of whether there is an Internet architecture. The authors prefer to speak of architectonic principles that follow an implicit, unwritten twenty-five-year-old tradition. Instead, one "builds" according to the principles of the community of developers: "the goal is connectivity, the tool is the Internet Protocol, and the intelligence is end to end rather than hidden in the network."[175] This form of end-to-end architecture hands over more complex tasks to the connected computers and less to the connecting networks. Between the two stand the protocols, especially TCP/IP in its various stages of development. The classic article published in 1984 by Jerome Saltzer, David Reed, and David Clark is nevertheless cautious enough to emphasize not an "architecture" but rather "arguments," which are, in turn, supposed to be permeated by a rational principle of an architecture of levels.[176]

Highly virtual communication networks in particular appealed more often to the cultural techniques of building and designing as a reference system.[177] Does this represent another special feature of computer networking to which belong linguistic fantasies, protocols, economic considerations, and temporalization? Or does reference to building and making allow for embodiment when the largest part of a network is characterized by immateriality and virtuality—in short, by symbolic processes, practices, and their order?

Primacy cannot be attributed to any of the discursive elements outlined here, even if the question of written protocols for internetworking was always also a question of the quasi-architecture that results from them. Through the materiality of cables, IMPs, and processes, a highly relational, software-based connectivity of things that speak to one another was established. The Internet Protocol pushed to the limit this development as a "connectionless" network protocol of fragmented data packets—that is to say, one that gets by without virtual circuits.[178] In the rapid development of new services that it made possible, less and less attention was paid to the material infrastructural networking and more to every conceivable form of communication in sociotechnical networks. The price of this—despite all the fetishism of computers and telephones and theories such as actor–network theory—is that the average user forgets about the use of objects and quasi-objects. Who cares about just part of a net?

The appearance of a large quantity of initially decentralized and eventually Internet-linked computer networks as well as the complex evolution of Internet protocols can scarcely be understood within the cultural history of media except as a computer-based "gateway to Babel."[179] To be sure, the Transmission Control Protocol and the Internet Protocol were not, as is sometimes assumed, a universal language of computer networks.[180] It would make more sense to speak of a grammar of network formation that, although it does not claim to be universal, is at least the standard for networked computers.[181] The cognitive turn in the history of computing in the 1960s, decisively influenced by J. C. R. Licklider, developed a formative energy that extended beyond the ARPANET to internetworking.

Material culture is no longer the center of this development, but it initiated an important transition: the higher the concretion of internetworking, the greater the potential volatility of everything it makes possible. The materiality of networked communication ensures that networks of social exchange become part of a fluid architecture of ways of living. And the more clandestine the transactions and the more shadowy the networks become, the more likely conspiracy theories are to be born. They operate as alternative discourses that are part of networks as a cultural technique. Conspiracy theories question the positive association of networking because they ask about the nature of social connections and the consequences of participating in the network. Additionally, in their potential for paranoiac knowledge, they equal or surpass "purified," scientific, rational forms of network analysis, which raises the question of why the concept of the network can support both structures of rationality and paranoid knowledge. The view of networks of actors that focuses on distributed operations is just one side of the coin—and the various horizons of meaning for the viewer of distributed collective actions is the other one. Network connections thus evolve into authorities that lend meaning to contexts that tend to be inscrutable, in the good and the bad, the modern and the amodern.

10 A VISUAL HISTORY OF THE NETWORK DIAGRAM (III): ECONOMIC ENTANGLEMENTS AND THE MEDIOLOGY OF CONSPIRACY

The materiality of connectivity is largely ignored in analyses of the social and economic networks that gained importance in the latter half of the twentieth century. Prevailing disciplinary models in social sciences explain this oversight to some degree, but it also reflects a shift toward networks as structures of social practice and symbolic order, as evidenced by the rapid ascent of the network concept in social anthropology, sociology, economics, and history.[1] Even beyond institutionalized forms of knowledge, social observers documented fluctuating relational exchange from the institutional margins.

From this perspective, conspiracy theories turn out not to be a crude simplification so much as a paranoiac-critical method of network analysis. They bring a fundamental property of the network as a cultural technique to light: both legitimate and illegitimate forms of knowledge foreground connections, whether public or secret. Anyone who maps political and economic entanglements as networks of power walks a fine line between analysis and conspiratorial speculation, although the paranoid character of such knowledge is almost completely purged from academic accounts. In the figures in this chapter, by contrast, it is present and pressing. Charting the economic order and all its transactions quickly leads to paranoia—that is to say "a systematized madness," which is "characterized by predominance of interpretation with intelligence preserved."[2] With Eve Kosofsky Sedgwick, one can also say of its disposition:

Paranoia is *anticipatory*.
Paranoia is *reflexive* and *mimetic*.
Paranoia is *a strong theory*.
Paranoia is a theory of *negative affects*.
Paranoia places its faith in *exposure*.[3]

Conspiracy theorists are mimetic network analysts par excellence because they expose themselves to the invisibility of informal connections, dubious associations, shadowy networks, and the negativity of their own suspicion. They do not, however, participate in the resources and institutions of these forces, which is why their own agency always seems limited: "Agency panic is intense anxiety about an apparent loss of autonomy, the conviction that one's actions are being controlled by someone else or that one has been 'constructed' by powerful, external agents."[4] This outsider position distinguishes them from other network analysts who take a graphic approach. Jacob Levy Moreno and Helen Jennings made the visualization of deep structures of fluctuating collective feelings the basis for their sociometric semiotic practice but in a positive, optimistic spirit.[5] The graphic and calculated network coordination of production and logistics is dedicated entirely to coordinating a feasible project.[6]

This chapter brings together figures who object to the perfectibility of society. A communist historian, a Central Intelligence Agency (CIA) analyst, an archivist who becomes a visual artist—for all of them, knowledge of conspiracies is isolating. Perhaps it even presumes isolation. But social exclusion sharpens the mind's eye, allowing near-invisible connections to be discerned. The connections can be justified in a reductionist way—a classic signature of collectively pursued conspiracy theories. They are typical, for example, of anti-Semitic narratives, which imagine a worldwide network of Jewish finance. So as to avoid making this chapter a stage for current collective conspiracy theories, I focus on individual protagonists of a paranoia motivated equally by critique and democracy, however historically antiquated it may seem.

The analysts presented here employ complex network-like connections in a fittingly complex way.[7] For Georges Baehler alias Pollux, Robert Redford alias Turner in *Three Days of the Condor*, and Mark Lombardi, the risk of knowledge lies in a systematically structured, all-embracing, and all-encompassing paranoia. Anyone who thinks of networks and history on a continuum will scarcely avoid such danger—consider Fredric Jameson's paranoid analyses of the unmappable totality of the capitalist view of the world system.[8] Studying network analyses requires us to situate the conspiratorial plot with pictorial forms of narration. If a medium of networks is the network diagram per se, what narrative forms combine with what images to the benefit of what meanings?[9] If it is true that narrative makes networks cohesive, is a political archive of these narratives necessary?[10]

Each of the following cases from the 1940s, 1970s, and 1990s is distinguished by a term from the business world: trust, corporation or company, and global networks. Paranoiac-critical conspiracy theory also follows the growth of networks in the twentieth century and, in the process, demonstrates an astonishing capacity for adaptation.[11]

10.1 WRITING/DRAWING: POLLUX AND THE SWISS TRUSTS

> Finance is other people's money; it is the art of doing business, honestly, if possible, with strangers' money.
> —Georges Baehler alias Pollux[12]

German sociologist Niklas Luhmann was anything but a friend of network thinking. In his study on ecological communication, he writes, "It is pointless to try to conceive the unity of modern society as the organization of a network of channels of communication, steering-centers and impulse receivers."[13] Yet, he was one of the most trenchant critics of a "causality in the south" that lent a certain mafioso charm to the blurriness of networked activity.[14] Luhmann referred to social structures in southern Italy, "patron–client relationships or networks" that stabilize relationships across social distinctions as a nonmodern form of exchange. In such a communicative situation, "it depend[s] above all on the symbolism of the network how favors are negotiated and suitable attitudes expected." Any effort at hierarchical organization by public and state actors is helpless in the face of distributed network communication based on familial economies. Corruption and lack thereof are constantly confused. The same holds for network protocol: "Anyone who participates in the network in this sense must know *how* it works. He need not know *why* it works or how it works."[15]

Distrust of informal business entanglements extends beyond Luhmann whose interest concerned the functional differentiation of social subsystems. Fears that economies based on clientelism and nepotism would accompany the emergence of the modern organization of the economy were already being expressed around 1900—and not only by Marxists. Even liberal economic theorists and historians noticed the targeted occupation of what David Ricardo called "channels of circulation" in 1810.[16]

In 1894, the Englishman John A. Hobson outlined the scenario of an ambush-like formation of networks attending the rise of the diamond and gold business in South America. The hunt for diamond mines in the 1880s that Hobson studied did not just lead to the author calling for political oversight.[17] Hobson analyzed its effects and connections with the aid of network depictions in tables and diagrams that have rightly been seen as precursors to the quantitative analysis of social networks.[18] The diagrams sketch a comprehensive mediology of conspiracy that is distinguished by a network dominated by communications media in net form. Hobson's corresponding diagram concludes a chapter as its final, visual argument. It takes shape around the central trust of the overlapping companies Charter, de Beers, and RAND and their shared telegraphs, banks, railways, and print media (figure 10.1). In this visual rhetoric, the control of flows of money, news, and transportation radiates from the central trust outward.

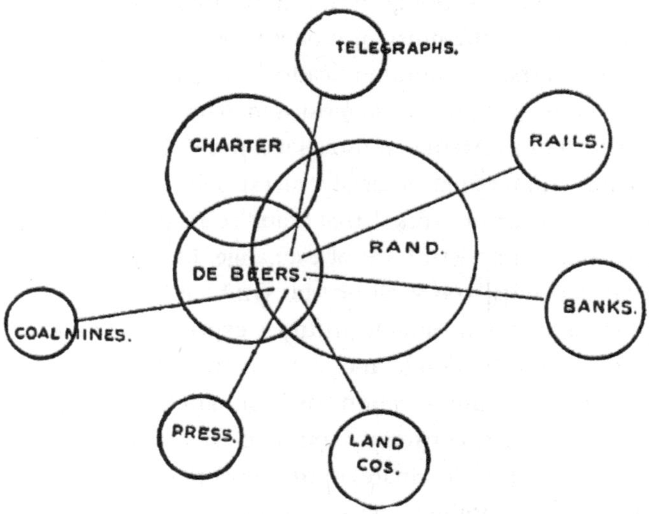

FIGURE 10.1
John A. Hobson, South Africa's financial economy. Diagram, 1906.

Hobson emphasized the organizational plexus of shared posts on supervisory boards—a common practice especially in interfamilial exchange throughout the nineteenth century. In the United States, the practice provoked calls for the breakup of monolithic trusts run in the interests of monopolists and not governed by forces of supply and demand. Louis Brandeis, who advised President Woodrow Wilson on the problem, wrote in 1914, "The practice of interlocking directorates is the root of many evils. It offends laws human and divine. Applied to rival corporations, it tends to the suppression of competition."[19] Despite this moral, pragmatic appeal—which was also reflected in a ban on rival companies having common directors in the Clayton Act of 1914—"incestuous" staffing of management positions continues to be a trademark of capitalist economic forms to this day, although the signatures have changed as part of the continuing globalization of the economy.[20] In lieu of a static, family-based, somewhat international network *as* a trust or syndicate, there are dynamic constellations of people whose hiring yields considerable fees for headhunters.[21]

Economic analyses of "interlocking directorates" have often been accused of a quantitative conspiracy based on money flows and shared personnel. Academic analysts usually reject such a suspicion from the outset and on the first page of publications.[22] This is true, even though network analyses are now part of the standard analytic repertoire of economic studies and have become obligatory in the Internet economy: recognizing the truth of one's own statements proves difficult in the discursive power field that has emerged.[23] Identifying economic entanglements opens a discursive battleground on which economic historians have laboriously gained a foothold. Meanwhile, conspiracy theorists no less busily collecting data in the same arena produce "illegitimate" knowledge published on websites such as theyrule.net or deathandtaxesposter.com. Sometimes positive and paranoid knowledge collide in literary, academic, and popular discourse as well.[24] My interest concerns the illegitimate and dubious stepchildren of economic thinking who willingly participate in a sprawling network without falling prey to simplifications: network analysts who set a trap for capitalism as the astrologers who draw and interpret its constellations.

In the 1940s, the Communist journalist Georges Baehler (1895–1982) set out to map the entanglements of Switzerland's economic system—very much along the lines of Niklas Luhmann—to determine their *how*. It went without saying that the *why* in this highly developed capitalist form of nepotism was the self-interest of the families involved. Baehler wrote for leftist and social-democratic journals such as *Das Volksrecht*, usually under the pseudonym Pollux—a reference to the helpful Dioscuri of Greek and Roman mythology, who appear as navigators for mortals in times of crisis. As a trained civil engineer, Baehler had built dams in France and Mexico before joining the French Communist Party in 1936 after a trip to the USSR.[25] After working underground in occupied France, he cofounded the Verein für wirtschaftliche Studien (Association for Economic Studies) in Zurich in 1942. In 1951, he emigrated to the German Democratic Republic (GDR) and worked in East Berlin at the Deutsches Institut für Zeitgeschichte (German Institute of Contemporary History) and the Deutsches Wirtschaftsinstitut (German Economic Institute). In the young GDR, under the pseudonym Georg Baumann, Baehler was again active as a journalist, publishing anti-Western treatises on the conspiracy of capital.[26]

In Switzerland during the war years, Baehler's central bogeyman in the journalistic arena was the *Neue Zürcher Zeitung*, which did not manage to unmask his pseudonym until March 1953. The *Neue Zürcher* wrote in June 1944:

> Truly, the managers of the socialist press are having no luck with their stubborn efforts. Year after year, article writers à la Pollux diligently expose the facts, drum political horror stories into their readers, and publish graphic articles before the elections to "expose" ominous interconnections. And—bang!—these gentlemen themselves fall into the grave they have dug for others.[27]

The *Gazette de Lausanne* threw down the gauntlet as well:

> If one reads every morning that all the nation's wealth and all the political influence are in the hands of a hundred industrialists, who are even named and whose connections and relationships are imprinted in the minds of the readers by means of graphic charts, one is not impressed more powerfully than by the responses of those who prove with statistics and arguments that they are tall tales and that Switzerland is in reality one of the countries in which wealth is distributed most equally.[28]

Graphic presentation and "graphic charts" are understood as weapons in the tradition of agitprop and Otto Neurath's pictorial statistics.[29] The image is an argument in the political arena—here, clearly advantageous for the socialist side criticizing capitalist monopolies. Interestingly, visual argumentation became the focus of the conflict over the *Polluxiades* more and more. On July 28, 1944, the *Neue Zürcher* spoke of a "monstrous picture gallery" that it nevertheless paid to look at "somewhat more closely."[30] The "extraordinarily twisted and knotted brushstrokes" were said to be a means of assigning places to people who did not belong there.[31] In this duel between socialist and conservative attitudes, Pollux explained his text-and-image strategy: "The value of these charts is that they permit a *clear overview of the entanglements and concentration of high finance*, while in the text the concentration of these capitalist forces is distilled only gradually."[32]

Pollux's "golden net" of the Swiss oligarchy also appears as a medium of simultaneity that promises evidence.[33] Narration and causality remain programmatically entrusted to the text. But the picture is not quite that simple, since much of the article is a list of the names of Swiss financial actors.[34] Unless they are located in the diagram, the connections narrated remain unclear, at least for readers today. The diagram's visual aspect (its pictorial iconicity) is thus foregrounded over its written aspect (notational iconicity)—a problem of all visualizations of networks unless they are understood as operative images.[35] Image and text are conceived, as in the corpus of the newspaper, as something to be read jointly. Each chart is framed to make it seem that much more plausible as a monolithic overview. From the perspective of conspiracy theory, this is absolutely logical: a trust is constituted by demarcation that is evident only when all connections have been exposed. The narrative network logic of conspiracy usually comes down to the identification of *one* highly concentrated network, whether as "trust," "corporation,"[36] "corporate network,"[37] "Radial Matrix,"[38] or as a "total," unmappable world system (which, for Frederic Jameson, represents an infinite network).[39]

An excellent example of Pollux's strategy of narrative unveiling of trusts and their interconnected supervisory boards is the text-and-image map for the article "Was sind die Rollschen Eisenwerke?" (What are Roll's Ironworks?). Under the title *Zwei Grossmächte Solothurns: L. von Roll und Bally* (Two of Solothurn's Great Powers: L. von Roll and Bally, figure 10.2), it presents a large economic pattern. Its curved and snaking lines counter corporate organigrams drawn in soberly geometric graphic design in the Bauhaus style.[40] The curves of the arrows contrast with the circles and rectangles for people and institutions. Pollux depicts large companies such as the "metallurgical enterprise" Roll and the "world-famous" shoe trust Bally with a

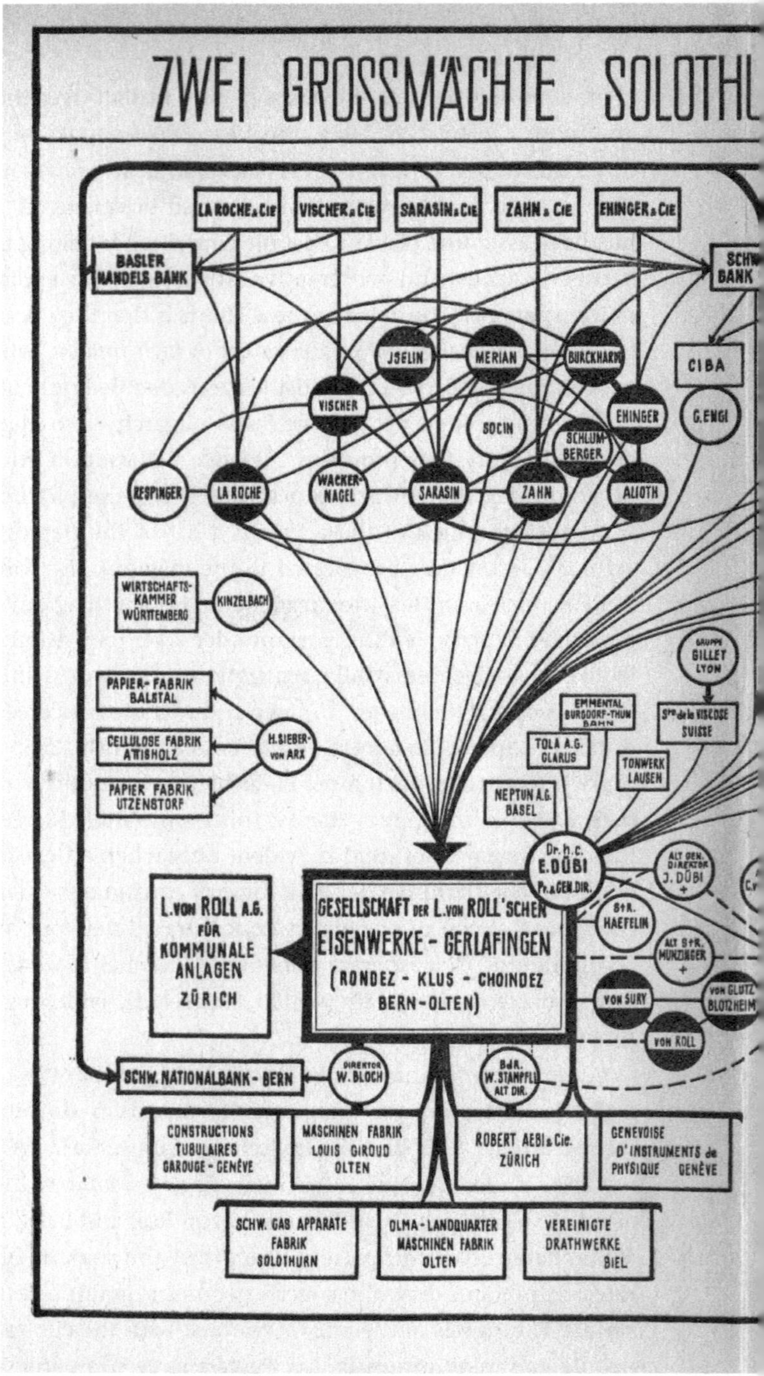

FIGURE 10.2

Georges Baehler alias Pollux, *Zwei Grossmächte Solothurns: L. von Roll und Bally*. Print, 1945. 23 cm × 20.5 cm.

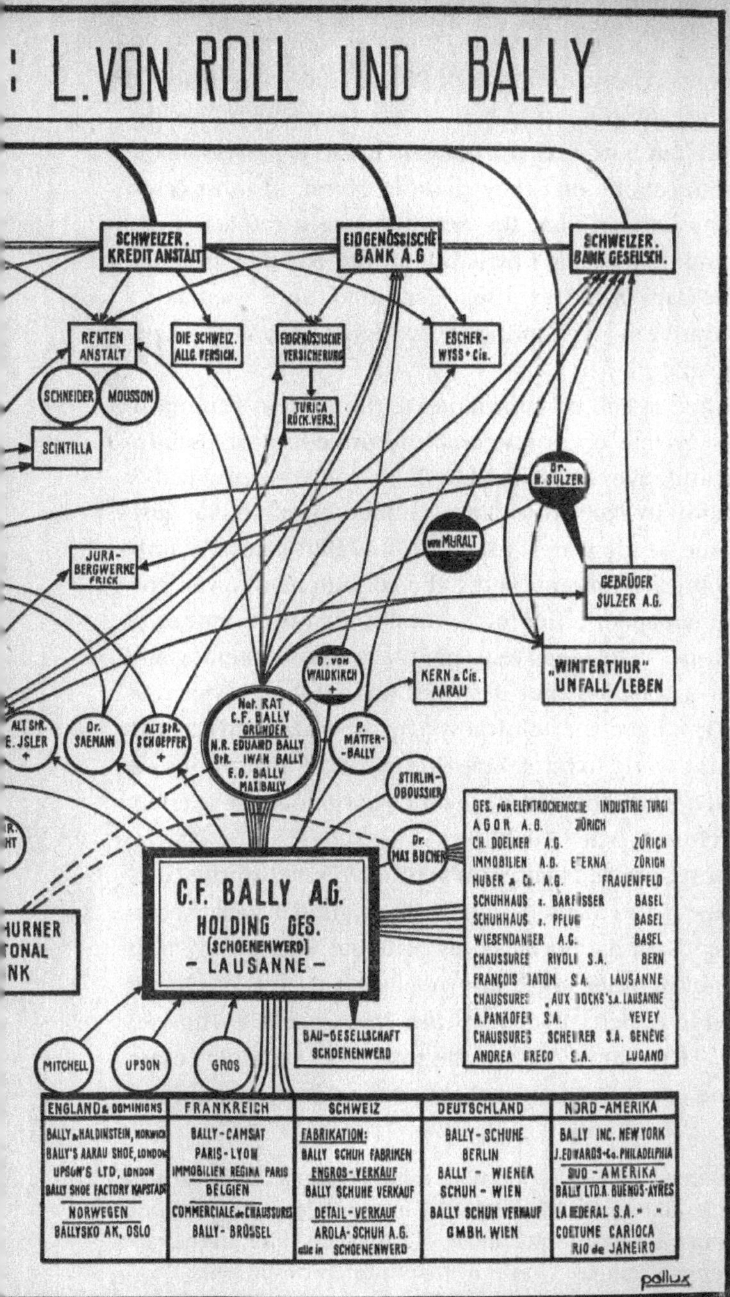

: L. VON ROLL UND BALLY
SCHWEIZER. KREDIT ANSTALT
EIDGENÖSSISCHE BANK A.G
SCHWEIZER. BANK GESELLSCH.
RENTEN ANSTALT
DIE SCHWEIZ. ALLG. VERSICH.
EIDGENÖSSISCHE VERSICHERUNG
ESCHER-WYSS + Cie.
SCHNEIDER
MOUSSON
TURICA RÜCK. VERS.
SCINTILLA
Dr. H. SULZER
von MURALT
JURA-BERGWERKE FRICK
GEBRÜDER SULZER A.G.
„WINTERTHUR" UNFALL/LEBEN
D. VON WALDKIRCH +
KERN & Cie. AARAU
Nat. RAT C.F. BALLY GRÜNDER N.R. EDUARD BALLY STR. IWAN BALLY E.O. BALLY MAX BALLY
ALT STR. E. JSLER +
Dr. SAEMANN
ALT STR. SCHOEPFER +
P. MATTER-BALLY
STIRLIN-OBOUSSIER
Dr. MAX BUCHER
C.F. BALLY A.G. HOLDING GES. (SCHOENENWERD) - LAUSANNE -
HURNER TONAL NK
GES. FÜR ELEKTROCHEMISCHE INDUSTRIE TURGI
AGOR A.G. ZÜRICH
CH. DOELKER A.G. ZÜRICH
IMMOBILIEN A.G. ETERNA ZÜRICH
HUBER & Co. A.G. FRAUENFELD
SCHUHHAUS z. BARFÜSSER BASEL
SCHUHHAUS z. PFLUG BASEL
WIESENDANGER A.G. BASEL
CHAUSSURES RIVOLI S.A. BERN
FRANÇOIS JATON S.A. LAUSANNE
CHAUSSURES „AUX DOCKS" S.A. LAUSANNE
A. PANKOFER S.A. VEVEY
CHAUSSURES SCHEURER S.A. GENÈVE
ANDREA BRECO S.A. LUGANO
BAU-GESELLSCHAFT SCHOENENWERD
MITCHELL
UPSON
GROS
ENGLAND & DOMINIONS
BALLY & HALDINSTEIN, NORWICH
BALLY'S AARAU SHOE, LONDON
UPSON'S LTD, LONDON
BALLY SHOE FACTORY KAPSTADT
NORWEGEN
BALLYSKO AK, OSLO
FRANKREICH
BALLY-CAMSAT
PARIS-LYON
IMMOBILIEN REGINA PARIS
BELGIEN
COMMERCIALE de CHAUSSURES
BALLY-BRÜSSEL
SCHWEIZ
FABRIKATION:
BALLY SCHUH FABRIKEN
ENGROS-VERKAUF
BALLY SCHUHE VERKAUF
DETAIL-VERKAUF
AROLA-SCHUH A.G.
alle in SCHOENENWERD
DEUTSCHLAND
BALLY-SCHUHE
BERLIN
BALLY - WIENER
SCHUH - WIEN
BALLY SCHUH VERKAUF
GMBH. WIEN
NORD-AMERIKA
BALLY INC. NEW YORK
J. EDWARDS + Co. PHILADELPHIA
SUD-AMERIKA
BALLY LTD.A BUENOS-AYRES
LA FEDERAL S.A. " "
COSTUME CARIOCA
RIO de JANEIRO
pollux

Zu Artikel 5

biting, ironic undertone as classic examples of Switzerland's all-connected capitalism. Actors' names are printed in bold in the text, which provides the illustration's legend. But both text and diagram fail to make legible the entanglements and connections that they claim to reveal. In most cases, for example, it remains unclear what the arrows between the actors are supposed to refer to and define. The fluctuation of the visual argumentation corresponds to the leaps in the text itself. Central to this is a semantics that constructs the narrative of an enduring conspiracy by spinning out the threads between companies.[41]

The "searching method of Pollux" contributes to the analysis of monopolies, booms in the wartime economy, and the unveiling of pseudo-humanitarian welfare initiatives for workers in Roll's ironworks and Bally's shoe factories, in addition to exposing political influences in familial business networks.[42] It would be too simple to regard the *Polluxiades* as mainly an investigation of political economy under the conditions of Swiss corporatism. Rather, they remap an economic mentality that is at once local and international in scope.[43] The visual form presumes a close coupling and interconnection of the actors and thus describes the economic system as structurally static.[44] In this light, the political system of Switzerland can also be seen as a picture of the world in miniature. The names of the cantons are localized geographically, while the system's border to the outside is established by a thick black frame (figure 10.3).

Baehler's interest in reciprocal clientelism was not limited to the community of Swiss citizens. It also focused on exposing National Socialist corporate entanglements. "National Socialism as National Capitalism" thus shaped the mappings of another book, also published by the Verein für wirtschaftliche Studien in Zurich, titled *Wer leitete Deutschland?* (Who Was Running Germany?).[45] Here, knowledge of the totality of systemic entanglements, buildups, and concentrations that fueled the German war industry is made visible:

> It will become evident that countless connections exist between all these *horizontal and vertical trusts*. An unbroken chain links the industrial companies to one another and to the banks, so that despite the inevitable competition in the race for profit the whole forms a compact bloc—one that threatens world peace.[46]

The causal "unbroken chain" reveals the real basis of control. The connections multiply until the entire mapping is identified as just one bloc. While the graphic elements remain the same, Pollux has added color coding (figure 10.4). Industrial actors (blue) dominate, supported by the banks in red. Individuals are yellow, with the exception of National Socialists such as Hermann Göring, set apart in brown. Notable actors from the mass media such as UFA and the *Neue Zürcher Zeitung* glow in green. A brief marginal remark about an incorrectly identified connection between the Baseler Handelsbank and the Siemens-Schuckertwerke factories in Berlin vouches for the meticulous precision of the mapping.

But what discourse conveys the suspicions about connections and non-connections? The "graphic chart" needs a foundation in data for its statements to take shape. Newspaper reports are indispensable in this. Journalism makes investigation possible in the first place. The mapping is of a mass-media public whose historical structural change is a condition for the possibility of modern conspiracy.[47] It is no coincidence that investigative journalism is often the linchpin of conspiratorial plots. Drawing topographies of conspiracy thereby becomes a second-order reporting guided by reading—a mode of intermedia analysis that also characterizes Mark Lombardi's diagrams. A large network is exposed in an overview that is denied to the individual actor. In conspiratorial fiction, paths lead regularly behind the facades of normality and also, sooner or later, to the newspaper editing rooms of this world. "The political secret is as popular as it is precarious: intelligence activities are incessantly discussed and researched; they fuel political phantasms that detect all kinds of insane adventures, crimes, and disasters lurking beneath the tedium of parliamentary debates."[48] When it is published, the conspiracy theory proves itself by crossing the line between paranoid and positive knowledge.

"They will print it," Robert Redford responds in the final scene of Sydney Pollack's conspiratorial classic *Three Days of the Condor* (1975) to his opponent in the CIA, Deputy Director Higgins. In front of the New York Times Building, he initiates verbal poker:

> "You play games. I told them a story." . . .
>
> "How do you know they'll print it?" . . .
>
> "They'll print it."
>
> "How do you know?"

Joseph Turner does not respond, looks back skeptically, and leaves with his coat collar turned up. The camera switches to a photographic still that loses its color as it zooms in and finally freezes as a black-and-white close-up of Redford looking back.[49]

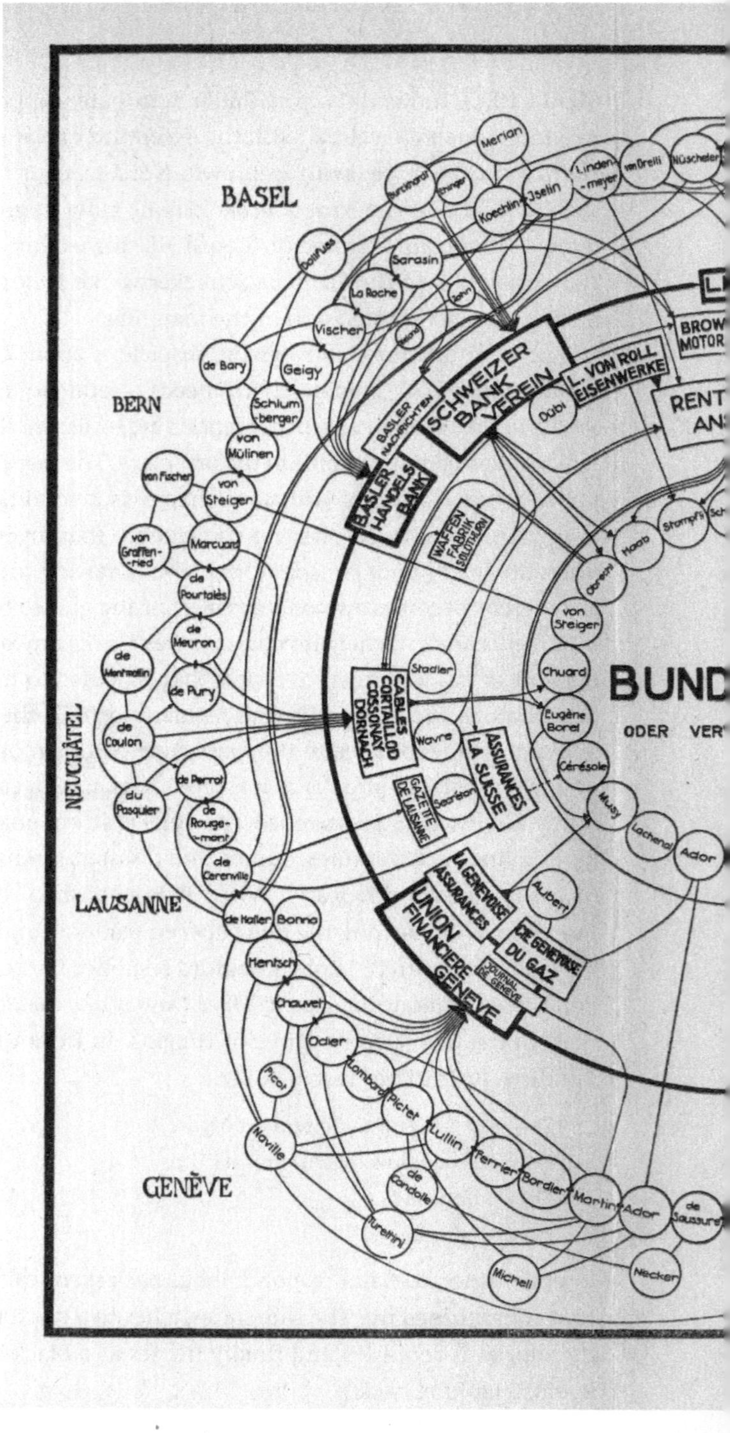

FIGURE 10.3
Georges Baehler alias Pollux, *Bundesrat oder Verwaltungsrat?* Print, 1945. 21 cm × 20 cm.

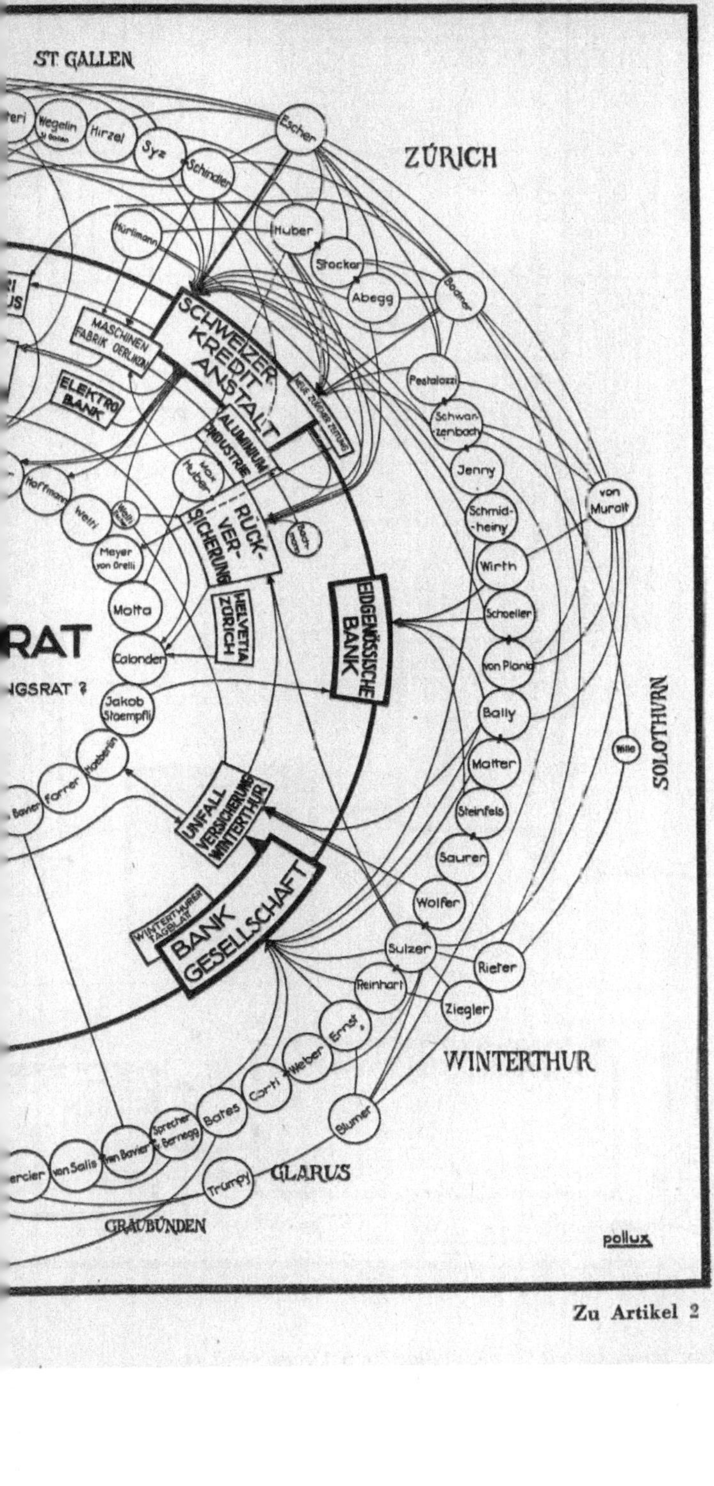
ST. GALLEN
Wegelin
Hirzel
Syz
Schindler
Escher
ZÜRICH
Hürlimann
Huber
Stockar
Abegg
Bodmer
SCHWEIZER KREDIT ANSTALT
MASCHINEN FABRIK OERLIKON
ELEKTRO BANK
ALUMINIUM INDUSTRIE
NEUE ZÜRCHER ZEITUNG
Pestalozzi
Schwarzenbach
Jenny
Schmidheiny
von Muralt
Wirth
Schoeller
von Planta
Bally
Matter
Steinfels
Saurer
Wolfer
Wille
SOLOTHURN
Hoffmann
Wetli
Max Huber
RÜCK-VER-SICHERUNG
Meyer von Orelli
Motta
HELVETIA ZÜRICH
EIDGENÖSSISCHE BANK
RAT
Calonder
NGSRAT ?
Jakob Staempfli
Haeberlin
Farrer
Bavier
UNFALL VERSICHERUNG WINTERTHUR
WINTERTHURER TAGBLATT
BANK GESELLSCHAFT
Sulzer
Rieter
Reinhart
Ziegler
Ernst
Weber
Corti
Bates
Sprecher v. Bernegg
von Bavier
von Salis
Blumer
WINTERTHUR
GLARUS
Trümpy
GRAUBÜNDEN
pollux

Zu Artikel 2

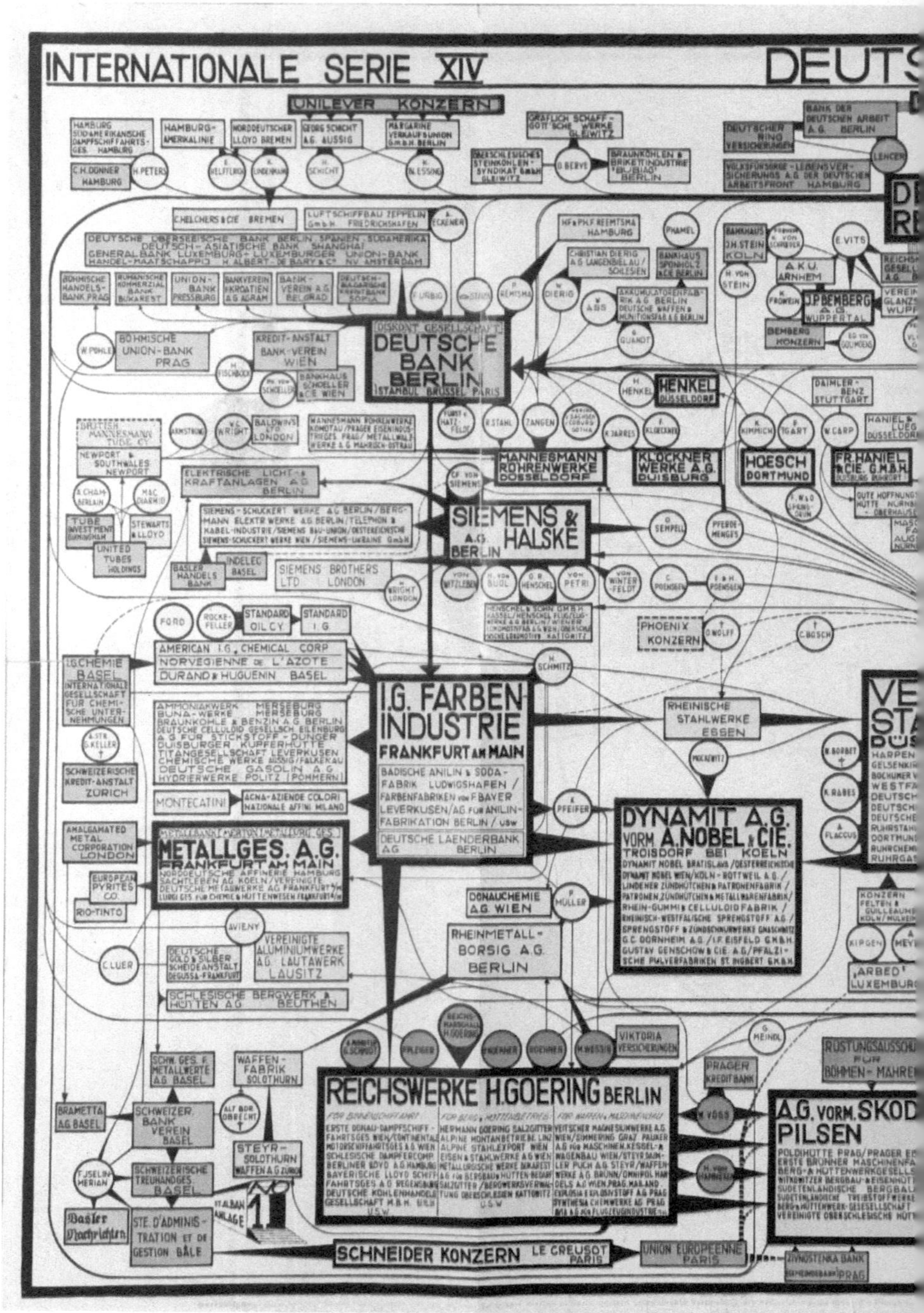

FIGURE 10.4

Georges Baehrens alias Pollux, *Internationale Serie XIV, Blatt Nr. 6: Deutschland*. Poster, 1945. 58 cm × 42 cm.

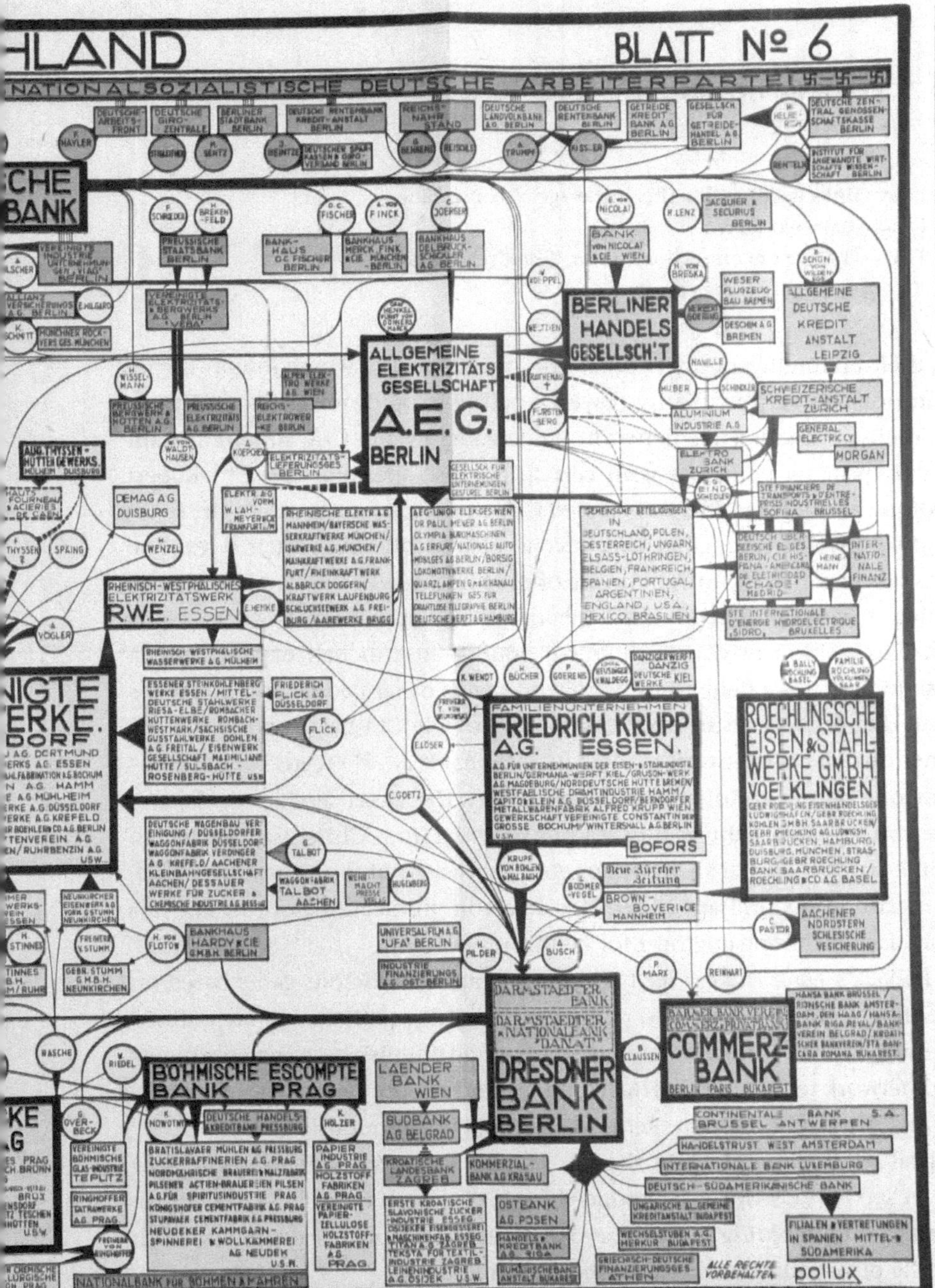

HLAND
BLATT № 6
NATIONALSOZIALISTISCHE DEUTSCHE ARBEITERPARTEI
DEUTSCHE ARBEITS-FRONT
DEUTSCHE GIRO-ZENTRALE
BERLINER STADTBANK BERLIN
DEUTSCHE RENTENBANK KREDIT-ANSTALT BERLIN
REICHS-NÄHR-STAND
DEUTSCHE LANDVOLKBANK A.G. BERLIN
DEUTSCHE RENTENBANK BERLIN
GETREIDE KREDIT BANK A.G. BERLIN
GESELLSCH. FÜR GETREIDE-HANDEL A.G. BERLIN
DEUTSCHE ZENTRAL GENOSSENSCHAFTSKASSE BERLIN
INSTITUT FÜR ANGEWANDTE WIRTSCHAFTSWISSENSCHAFT BERLIN
DEUTSCHER SPARKASSEN & GIROVERBAND BERLIN
SCHE BANK
PREUSSISCHE STAATSBANK BERLIN
BANKHAUS O.C. FISCHER BERLIN
BANKHAUS MERCK FINK & CIE MÜNCHEN-BERLIN
BANKHAUS DELBRÜCK SCHICKLER A.G. BERLIN
BANK VON NICOLAI & CIE WIEN
JACQUIER & SECURIUS BERLIN
VEREINIGTE INDUSTRIE UNTERNEHMUNGEN „VIAG“ BERLIN
ALLIANZ VERSICHERUNGS A.G. BERLIN
MÜNCHNER RÜCKVERS.GES. MÜNCHEN
VEREINIGTE ELEKTRIZITÄTS- & BERGWERKS A.G. BERLIN „VEBA“
BERLINER HANDELS GESELLSCH'T
WESER FLUGZEUGBAU BREMEN
DESCHIMAG BREMEN
ALLGEMEINE DEUTSCHE KREDIT ANSTALT LEIPZIG
ALLGEMEINE ELEKTRIZITÄTS GESELLSCHAFT A.E.G. BERLIN
ALPEN ELEKTRO WERKE A.G. WIEN
PREUSSISCHE BERGWERKS & HÜTTEN A.G. BERLIN
PREUSSISCHE ELEKTRIZITÄTS A.G. BERLIN
REICHSELEKTROWERKE BERLIN
SCHWEIZERISCHE KREDIT-ANSTALT ZÜRICH
ALUMINIUM INDUSTRIE A.G.
GENERAL ELECTRIC CY
MORGAN
ELEKTRO BANK ZÜRICH
AUG. THYSSEN-HÜTTE GEWERKS. MÜLHEIM DUISBURG
ELEKTRIZITÄTS-LIEFERUNGSGES. BERLIN
GESELLSCH. FÜR ELEKTRISCHE UNTERNEHMUNGEN GESFÜREL BERLIN
DEMAG A.G. DUISBURG
ELEKTR. A.G. VORM. W. LAHMEYER & CO. FRANKFURT/M.
STE FINANCIÈRE DE TRANSPORTS & D'ENTREPRISES INDUSTRIELLES SOFINA BRÜSSEL
GEMEINSAME BETEILIGUNGEN IN DEUTSCHLAND, POLEN, OESTERREICH, UNGARN, ELSASS-LOTHRINGEN, BELGIEN, FRANKREICH, SPANIEN, PORTUGAL, ARGENTINIEN, ENGLAND, U.S.A., MEXICO, BRASILIEN
DEUTSCH ÜBERSEEISCHE EL.GES. BERLIN, CIE HISPANO-AMERICANA DE ELETRICIDAD „CHADE“ MADRID
INTERNATIONALE FINANZ
STE INTERNATIONALE D'ENERGIE HYDROELECTRIQUE „SIDRO“ BRUXELLES
RHEINISCH-WESTPHÄLISCHES ELEKTRIZITÄTSWERK R.W.E. ESSEN
RHEINISCH WESTPHÄLISCHE WASSERWERKE A.G. MÜLHEIM
NIGTE WERKE. DORF
FRIEDRICH FLICK A.G. DÜSSELDORF
FAMILIENUNTERNEHMEN FRIEDRICH KRUPP A.G. ESSEN
DANZIGER WERFT DANZIG
DEUTSCHE WERKE KIEL
ROECHLINGSCHE EISEN & STAHL-WERKE G.M.B.H. VOELKLINGEN
BOFORS
Neue Zürcher Zeitung
BROWN-BOVERI & CIE MANNHEIM
AACHENER NORDSTERN SCHLESISCHE VERSICHERUNG
WAGGONFABRIK TALBOT AACHEN
WEHRMACHT PRESSE VERLAG
UNIVERSAL FILM A.G. „UFA“ BERLIN
INDUSTRIE FINANZIERUNGS A.G. OST-BERLIN
BANKHAUS HARDY & CIE G.M.B.H. BERLIN
GEBR. STUMM G.M.B.H. NEUNKIRCHEN
DARMSTAEDTER BANK
DARMSTAEDTER & NATIONALBANK „DANAT“
DRESDNER BANK BERLIN
COMMERZ-BANK BERLIN PARIS BUKAREST
CONTINENTALE BANK S.A. BRÜSSEL ANTWERPEN
HANDELSTRUST WEST AMSTERDAM
INTERNATIONALE BANK LUXEMBURG
DEUTSCH-SÜDAMERIKANISCHE BANK
BÖHMISCHE ESCOMPTE BANK PRAG
DEUTSCHE HANDELS-KREDITBANK PRESSBURG
LAENDER BANK WIEN
SÜDBANK A.G. BELGRAD
KROATISCHE LANDESBANK ZAGREB
KOMMERZIALBANK A.G. KRAKAU
OSTBANK A.G. POSEN
VEREINIGTE BÖHMISCHE GLAS-INDUSTRIE TEPLITZ
RINGHOFFER TATRAWERKE A.G. PRAG
FILIALEN & VERTRETUNGEN IN SPANIEN MITTEL- & SÜDAMERIKA
NATIONALBANK FÜR BÖHMEN & MÄHREN
ALLE RECHTE VORBEHALTEN
pollux

10.2 READING/HEARING: THE QUESTIONS OF *THREE DAYS OF THE CONDOR*

> I have links inside links. This is the age of conspiracy.
> People have wondered.
> This is the age of connections, links, secret relationships.
> —Don DeLillo, *Running Dog*[50]

"Is there an intelligence network?" This question, with which Joseph Turner confronts his vis-à-vis Kathy (Faye Dunaway), could not be more urgent. "Is there an intelligence network—undetected by the CIA, linking certain Arabic-speaking countries with certain Spanish-speaking ones?"[51] Robert Redford, as a hunted and disoriented CIA analyst, wants to get to the heart of the conspiracy that has cost several of his colleagues their lives. The film's protagonist is painfully aware of the flimsiness of his own knowledge, which is based on circumstantial evidence and suspicions. Turner is—the film's exposition leaves no room for doubt—an armchair expert. He is a desk jockey, not a field agent, who works with other people and machines analyzing documents in a New York office of the CIA that is posing as the American Literary Historical Society.[52] "He reads . . . He reads everything," someone says in an internal dispute between the secret service agents, and the film repeatedly shows him reading material from Chinese characters to comic strips.[53] Turner's CIA opponents appreciate him as a research type, disregarding his military experience as a communications officer and his work as a telephone engineer for Bell Labs.

Pollack's *Three Days of the Condor* plays out a murderous game based on the (il)legibility of the world of agents. The rules of the game are defined over the course of the film by an inability to distinguish between conspiracy and network technologies. The telephone and telephone switching become central motors of the plot. Redford's Turner is continuously confronted by radical self-doubt and doubt about his immediate environment, about how secret service knowledge is produced, and, above all, about how it is authorized.[54] The unfolding of the story concerns the nature of conspiracy, both for the protagonist and for the cinematic audience.

Is it merely an internal problem at the Central Intelligence Agency—that is to say, a classic state secret (*arcanum imperii*)? Or is it nothing less than a *secretum*, secret knowledge that can at best be revealed in private and cannot be permitted to circulate freely even in the halls of the so-called Company?[55] "Maybe there's another CIA. Inside the CIA."[56] As the protagonist's suspicion mounts, state secrecy becomes radicalized, down to an inner circle in which the threads of all network intelligence come together. Like a good Bell Labs engineer, Turner is driven to uncover the threads of the network by means of telephone connections. The meaning of conversation, conspiracy technology, and infrastructural mediation—the identity of conspiring actors and conspiratorial network—finds its eloquent expression in the ephemerality of telephone conversation. For "Condor," intelligence networks are communication networks whose functioning can be grasped by reading and listening in. The agents of the conspiracy become visible when connections become legible and audible. Only these three steps—reading, hearing, and seeing—lead to verification of and positive knowledge about the conspiracy that can be described by mass media.

Sydney Pollack directs the film as a physiognomy of Robert Redford's face. In Turner's incredulous, doubting expressions—down to the skeptical looks of the finale—the film plays on affective forms of sensing, suspecting, and not knowing. In a mirror image, this incredulousness is also embodied by Faye Dunaway's character, the New Yorker Kathy, whom Condor has kidnapped and who responds with twice as much shock to his actions. All of the CIA analyst's media practices are a desperate attempt to imagine impenetrable actions that immediately threaten his life. The conspiracy theorist asks *why*, about the causality of connections of shadowy networks and their organization. *How* his opponents exercise their agency—the *technē* of their conspiracy—is exposed over the course of the film by laborious reading, hearing, and drawing. Knowledge of the deeper political reason remains the goal throughout. "But there is a reason. There is a *reason*," Turner says to Kathy.[57] If the motivation of the antagonistic actor-network cannot be decoded, its actions are only readable by understanding *how*.

Turner uses media to explore the conspiracy's spheres of action. After his plan to return to the protection of the CIA fails, he sketches the space of the exchange of fire that has caught him off guard. Because he is constantly on the run, he rarely has the opportunity to reflect and allow the uncertain signs to come together (figure 10.5). Condor's mode of reasoning is feverish, dizzied, racing, as compressed as the narrative itself. Pollack's

FIGURE 10.5
Sydney Pollack, Owen Roizman, et al., Search for traces and reconstruction of the assassination by Joseph Turner (Robert Redford) in *Three Days of the Condor*.

adaption of a book radicalizes this drive by reducing six days to seventy-two hours.[58] Turner's haste contrasts with the relative calm of the secret service and of Max von Sydow's contemplative, ice-cold contract killer Joubert, who makes Alain Delon's character in *Le samouraï* (1967) seem like a good Samaritan. Turner's drawing encodes the central questions to form a causal chain in three steps. First, second, third: the attack on the CIA remote office was not about any *thing* but about information. Question marks indicate the vagueness of this assumption. *Who* could have attacked him and his friend Sam? And *why*? By diagramming, Turner tries to create a topography of his memory.[59]

Initially, Dave Grusin's atmospheric music is the only sound. Later, we hear echoing voices. In the cuts between individual drawings, visual memories flash. First, second, third, fourth: whispering, Condor comments on his own questions and adds additional possible conclusions to the chart. The ghostly murmur of other voices offers the patient reader a crucial pointer to the CIA itself. While it proves difficult to grasp the unseen, the spoken and written elements are references to the discourse of the conspiracy. Despite all its elegiac visual elegance, *Three Days of the Condor* is closely related to acoustically dramatized films such as Francis Ford Coppola's legendary cinematography of eavesdropping in *The Conversation* (1974).

Grusin's funky, electrified jazz score harmonizes with the sound of office equipment still running, even after the people around it are dead.[60] The clear voices of the CIA actors on the other end of the line have an uncanny presence in Condor's telephone conversations. New York's streets are permeated by acoustic atmospheres that suggest a threat no matter how ordinary, from quiet recurring noises in the distance to taxis or street music. In contrast to the invisible whole of a conspiracy, Pollack's film develops an intermedia shift in which the audible becomes sayable. The aura of silence that surrounds state secrets is thus interrupted.[61] Nowhere is the explanation of the *why* by the cinematic narrative *how* clearer than in the sequences in which the analyst stops reading to become a telephone engineer again.

Frederic Jameson has made two scenes in which Robert Redford operates the telephone exchange the cornerstone of his reading of *Three Days*.[62] Jameson's Marxist-tinged reading of the film based on the theory of space and geopolitics is itself sustained by a certain paranoia.[63] In the middle of the "world system"—a concept that the critic owes to Immanuel Wallerstein—are allegories of potentially infinite invisible structures of a sociotechnical network. Its material structures, whether information technologies or architectures of power, cannot be separated from conspiratorial action and hence the *conditio* of American capitalism. No one—not even the critical intellectual or artist—can graphically render the totality of the postmodern system. In lieu of a map as large as the territory it represents, paranoiac cognitive mappings and surveys abound.

The telephone exchange and hacking scenes lead to an uncanny subterranean terrain in which the network logic of conspiracies meets its double in media technology. "Every real effigy has its shadow which is its double." Jameson's argument resembles Antonin Artaud's dictum when he writes that the hermeneutic of a look beyond the scenes of the conspiracy in Pollack's cinematic spaces and all their object reference is more alarming than the imaginary network of the usual suspects:[64] "the representational confirmation that telephone cables and lines and interchanges follow us everywhere, doubling the streets and buildings of the visible social world with a secondary secret underground world, is a vivid, if paranoid, cognitive map."[65]

Redford's stunned calls immediately after the attack on the American Literary Historical Society already lead him to the threshold of the CIA as an institution that shields its secrets from the outside. The room of the internal telephone exchange is small, but cameraman Owen Roizman's

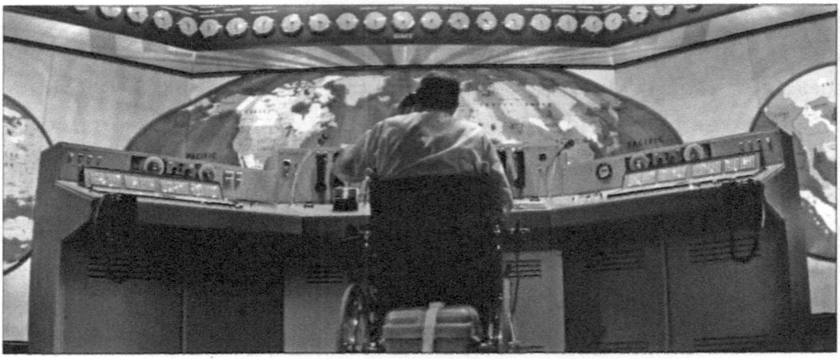

FIGURE 10.6
Sydney Pollack, Owen Roizman, et al., Telephone switchboard and surveillance room of the CIA in *Three Days of the Condor*, 1975.

discretely distorted shots make clear its conspiratorial character as an entrance to a labyrinth. The acoustic and optical surveillance is denoted by the world map localizing telephone calls (figure 10.6).

Turner will upset this power structure using his skills as a telephone engineer. First, he tricks his pursuer, Joubert, into making a careless phone call to Langley. Turner records its dial tones in order to reconstruct the telephone number by calling from the CIA computer. With a second maneuver in the belly of a Bell telephone exchange, whose electromagnetic switching creates a clicking and clacking soundtrack, Condor reverses the power relationships for a moment. The shots deftly illustrate the masking and unmasking. Slowly, the camera glides through the rows of whirring stepping switches in the basement, following Turner from a distance. In a flickering tracking shot with no close-ups, it follows Turner's face through the equipment shelving. Roizman's cinematography creates a psychogeography of clandestine operations and scanning of telephone connections. The switching technology provides a skeleton-like structure to the graphic foreground of the frame and is thus superimposed as a technological mask on Redford's moving face (figure 10.7).

Uncertainty and doubt give way to calculated, tactical superiority. Condor covers up his tracks by networking other telephones and masking his call. The centrality of the surveillance room is paired with the counterintelligence tactic of hidden switching on the network. The Company's attempts to localize the source are unsuccessful. Bureaucratic network logic

FIGURE 10.7
Sydney Pollack, Owen Roizman, et al., Underground telephone exchange in *Three Days of the Condor*, 1975.

loses to grafting: skilled hacking, reuse, and countercultural networking, birthing an American myth that has taken hold in philosophy and cultural theory as well as in conspiracy and cyberpunk narratives.[66]

It is no coincidence that state regimes of telecommunication, with their notorious proximity to the military and secret services, often propel a mediology of plotting and counterplotting. Just as Condor is forced to assume the role of the telephone parasite, Trystero's secret post office in Thomas Pynchon's second novel, *The Crying of Lot 49* (1966), operates against the state's monopoly on the mail. In Pynchon, the counterconspiracy and its infrastructure have become identical in a large historical plot—at least from the paranoiac perspective of the main character, Oedipa Maas, who pursues the vague signs of the "other" postal system.[67] This formula seems seductively simple at first: private networks are played against state media power monopolies. Infrastructural media technology as a material aspect of the conspiracy places them under conditions that it itself embodies. But the use of objects to form a narrative does not say much about political and economic motives. The obvious closeness of network technologies and practices of conspiracy and counterconspiracy is a signature of the network society era. Networks of conspiracy and conspiracy networks are not, however, identical: the former emphasizes the *technē*, the second the style of action.

But what about when the motivation of the plot itself conditions the formation of social networks? Causality, Niklas Luhmann writes, "is a medium of loosely coupled opportunities whose use requires a solid coupling of specific causes and specific effects."[68] When causality is localized in personalized social networks—and conspiracy narratives depend on this being the case—the elemental distinction between trust and mistrust applies both to the social network's actions and to outsiders' evaluation. The organization of socioeconomic networks is a question of the trust circulating in them, which forms shared identity via corresponding interests, actions, and narratives.[69] The conspiracy theorists' distrust of the legitimacy and authority of hidden narratives and transactions is what brings their repeating "causality in the South" to light in a democratic public sphere. Against such a backdrop, the breakdown of the *why* of *Three Days of the Condor* is not a repression of the *how* but of its logical consequence. Precisely because she or he is standing outside of the network, the individual conspiracy theorist must question both things. Paranoiac-critical conspiracy theory is not about an illegitimate simplification of highly complex states of affairs but rather about a successful reduction of complexity.

After Turner has put together the evidence he has collected, he states, half-bewildered, half-shocked, the ultimate reason for the conspiracy: "It was your network I turned up . . . Oil fields. This whole damn thing was about oil."[70] His own research got too close to the Middle East war game scenarios of several actors within the CIA—an insight that, despite its historical reference to the oil crisis of the 1970s, confirms today's viewers in their paranoid mistrust.[71] The questions of *Three Days of the Condor* do not aim, as one might assume following Jameson, at a potentially infinite, "total" network that is connected by the conspiracy as a narrative structure.[72] They aim rather at the uncanny multiplication and anticipation of nested and grafting networks: those of apparatuses and instruments, structures and communication, institutional organization and cooperation, and also the networks of resistance.

Thus, a transformation in the discourse on conspiracy theory emerges: an economic and political network held together by corrupt common interests evolves into many networks. Network logic shifts from the determination of the systemic limits of a (world) conspiracy to an abiding possible presence. If another step in the chain of actions remains possible, then the conspiracy network is also potentially infinite, always present, and only partly subject to discovery.

10.3 ARCHIVING/NARRATING: MARK LOMBARDI'S GLOBAL NETWORKS

> To trace the flow of information through a network is one thing; to identify public opinion another.
> —Robert Darnton[73]

Few artists have dedicated themselves to the growth of conspiracy networks in the global space of capital more intensively than the American Mark Lombardi (1951–2000).[74] His large-format drawings, executed with ballpoint pen and stencils, map the space of economic transactions and political decisions. The works have titles such as *George W. Bush, Harken Energy, and Jackson Stephens, ca. 1979–90* and *Banca Nazionale del Lavoro, Reagan, Bush, Thatcher, and the Arming of Iraq, 1979–90*. They bring together actors and agents of power, financial currents, and manipulations into an outsized aesthetic of suspicion.

Lombardi's characterization of his diagrams as "narrative structures" is famous, as is their material foundation: an archive of newspapers and books documented by a card catalog with 14,500 handwritten index cards. Art historian Robert Hobbs has identified Lombardi's aesthetic references and influences as including history paintings, panoramas, Marcel Duchamp, and works by political conceptual artists such as Hans Haacke.[75] Visual studies scholars have focused less on his visual narrations than on questions of the design processes. Angela Lammert writes, "What proves to be visible and representable is . . . more the flow of energies from movements and individual connections between those involved than an explanation or real exposure of political structures."[76]

Hence, Lombardi's diagrams correspond to a systematic, precise, paranoiac perception of the world whose mode of phenomenological explanation first senses the *how* of forms in order to be better able to hunt down the unexplained *why* of connections. Suspicion concentrates on certain signs that seem more threatening than others.[77] His critical actor–network analyses are marked by a network-like ordering and narration of practices and knowledge. The causal flow in this diagram form is not one-sided, not simply a time arrow, "because causality is no longer irreversible: anyone who wants to influence is influenced more suddenly by the result of his influence."[78] The statement is true both of the semantics of Lombardi's meticulous mappings (and remappings) and of their reception. Lombardi's paranoiac-critical method may represent the zenith of analog network analysis, which also inspired digital computer science. Thus, the computer scientist Robert Tolksdorf has modeled Lombardi's diagrams in accordance with graph theory and transformed his card catalog into a digital database.[79]

Reading Lombardi's drawing without support from machines leads almost inevitably to conspiracy theories. Reading means here a movement back and forth, a discourse that results from the search and immersive understanding of its course.[80] No anecdote describes this better than the use of Lombardi's drawings by FBI conspiracy pros after September 11, 2001. Because no reproduction of the dubious work *BCCI–ICIC & FAB, 1972–91 (4th version)* was available, a female agent had to do her work on site in the Whitney Museum of American Art, where art history students go to study originals. Her headshaking after hours of looking at them is as much a part of Lombardi folklore as the unexplained circumstances of his death. There still is doubt that he killed himself by hanging.[81]

Mark Lombardi was a professional reader, although he was not in a secret service or academic institution. His library is preserved by the Galerie Pierogi in Brooklyn. Its catalog includes, besides art historical literature, several classics of twentieth-century philosophy. In a purely quantitative sense, however, none of the volumes on Fluxus and Conceptual Art nor Hannah Arendt,[82] Herbert Marcuse,[83] Jean Baudrillard, or Noam Chomsky are crucial. The lion's share of Lombardi's library is investigative nonfiction, often written by journalists: *The Vatican Empire* (1966), *The CIA and the Cult of Intelligence* (1974), *Secret Police: The Inside Story of a Network of Terror* (1981), *Truth Is a Difficult Concept* (1995), and more.[84] Famously, the artist was proud to use information that had been verified by print circulation as the foundation of his handwritten archive.[85] As a second-order reader/reporter, Lombardi also practiced a detective's approach to evidence, which here joined a pathology of data collection. Like every actor–network analyst, he developed an obsession with media traces that was as mimetic as it was reflexive. The trace, as a readable indexical sign without coding, leads to the question of the causality on which it is based. In turn, traces pointing to one another raise the question of a shared supercoding.

Thomas Pynchon's *The Crying of Lot 49* and John Le Carré's *Tinker, Tailor, Soldier, Spy* (1974) are the only literary building blocks of Lombardi's library. But what would be the secret anti-state postal service of global networks?

Or to put the question differently, for which structures do "narrative structures" pick up the trail, and what conclusions can be drawn from them? Tellingly, in Mark Lombardi's works, it is the pan-American myth of the mafia, whose historical beginnings are traced back to the Prohibition era. Instead of the causality in the South, the activities of Al Capone and Meyer Lansky as a casino causality in the West become the original reason for the network.[86] The economic organization of the nets is mafia like—not only in Luhmann, Pollux, and Lombardi but already in Sicilian accounts from the early twentieth century.[87]

From this vantage point, the large-format diagrams read as a narrative of ongoing corruption in the American state. Shadowy networks with alarmingly robust powers of social integration infiltrate and abuse the American spirit of freedom by blurring the lines of corruption and diplomacy. Lombardi's story is a monument to the democratic state's frequent betrayals of its citizens, which causes an agency panic that seems to be in equal measure a felt and a real loss of opportunities for action.[88] An enduring betrayal integrates in the space of state secrets the repeal of the order that is planned in the order itself.[89] The public sphere and removal from the public sphere are two prerequisites of democratic culture, each of which conditions the other. This can be overcome only by self-immunizing with functioning journalistic and legal enlightenment. In Lombardi's case, this residual optimism may be indebted to his socialization in the 1970s when nonfiction books, investigative journalism, conspiracy films, Coppola's *Godfather* (1972), and Scorsese's *Mean Streets* (1973) took the public behind the scenes of politics, secret services, organized crime, and capital.[90]

One can object here that the temporal order of the narrative structures is simply setting the historical framework of the archive. But the feverish, manic research described by Lombardi's friends had a moral quality that was reinforced by the artist's affinity for Herbert Marcuse. The philosophical key to his drawings is surely to be found in Marcuse's readings of Freud in *Eros and Civilization*. Why would an art historian, librarian, and drawer of filigreed socioeconomic constellations read Herbert Marcuse's study, of all things—as well as texts by Wilhelm Reich?[91] Lombardi's countercultural tactics had a grounding in psychoanalysis. They fed on Marcuse's fundamental political assumption that self-repression supports the ruling class and its institutions. The material ordering of the archive is preceded by a negative-normative positing that can best be characterized in Marcuse's own words: "The *return of the repressed* makes up the tabooed and subterranean history of civilization."[92] This can be aptly revised for Lombardi's aesthetic: It is the return of the repressed (in yesterday's daily news reports) that finds expression in the constellations mapped.[93] The disinhibition of capitalist activity is thus countered by a critical theory of its deep structures as anthropological structures of drives. Not coincidentally, later images record legal processes on a separate level in red.

Critics regularly underestimate the materiality and politics of the Lombardi archive in reviews of his works—usually because of their aesthetic fascination. Pictural iconicity is foregrounded over the notational content of the drawings and the data processing on which they are based. Public presentations of Lombardi's works often reinforce this effect—for example, in 2012 documenta 13 exhibited *BCCI–ICIC & FAB, 1972–91 (4th version)*, the magnum opus consulted by the FBI, in the murky lighting of the Fridericianum in Kassel, where the index cards were difficult to read.[94] Yet, a glance at this catalog of handwritten cards already reveals that Lombardi was an actor–network theorist prior to pursuing artistic concerns (figure 10.8).

At the top of each index card is the name of an institution or person visualized as a node in the global networks. The color of the card indicates various categories: quotations (white), biographies (green), companies (pink), themes (blue), bibliographic information (yellow), websites (orange), and deaths (purple). The translation of narratives into diagram calls for a cartographic legend through which financial practices can be processed abstractly. It is remarkably compact and reduces conspiratorial action to a protocol of six possible actions (figure 10.9).

One-directional arrows symbolize influence or control (a): two-directional arrows a mutual relationship or association (b); arrows with dotted lines represent financial transactions (c). In the center, a zigzag pattern of unbroken arrows stands for the sale or transfer of an economic good (d). Blocked or incomplete transactions end in a pair of perpendicular lines (e). The last operator, which curls like a telephone cord, represents the sale or spin-off of property (f). This schema does not just aim to depict causal connections within global networks. As a form of encoding, it also allows the transfer of the written archive on index cards into a readable format. Expressed in the terminology of systems theory, narrative forms are translated by recoding them into a diagrammatic medium that takes on a pictorial status in visual forms. The impossibility of exact notation of a narrative using the

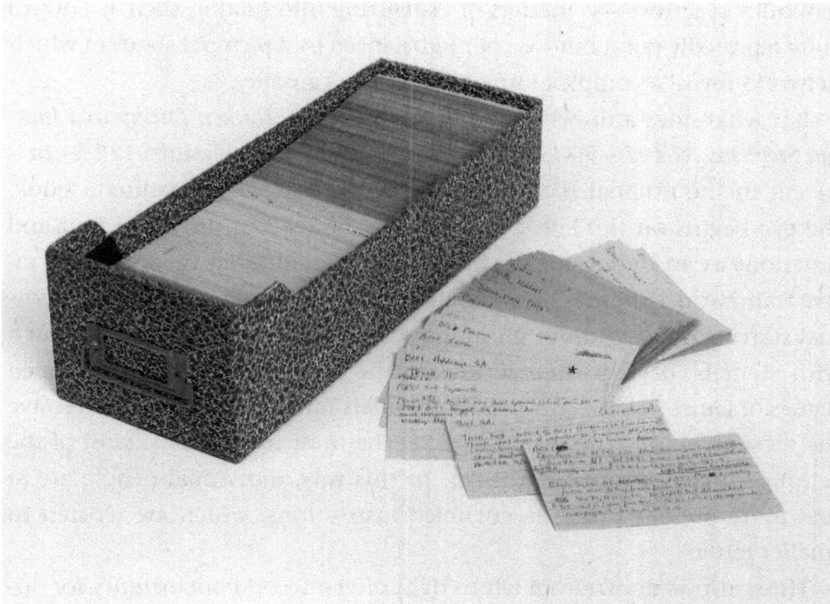

FIGURE 10.8
Mark Lombardi, Index cards documenting Lombardi's research. A selection of 1,300 cards out of a total of approximately 14,500. Pen and pencil on card stock, circa 1998–2001. 3 in. × 5 in. each / 7.6 cm × 12.7 cm, Photograph by John Berens. Reproduced by permission from the Lombardi Family and Pierogi Gallery.

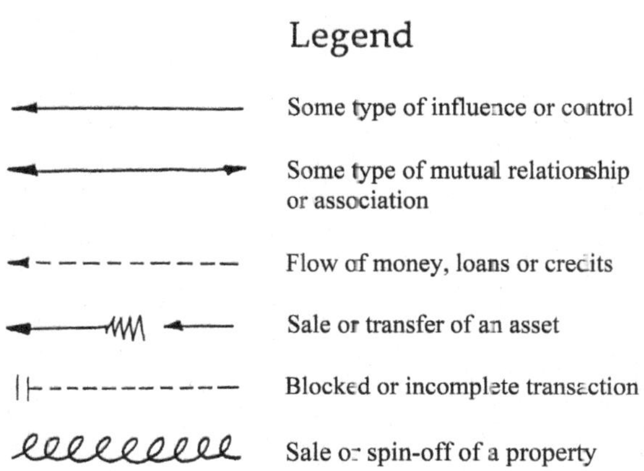

FIGURE 10.9
Mark Lombardi, Cartographic legend to Lombardi's drawings, circa 1998–2001. Photograph by Joe Amrhein. Reproduced by permission from the Lombardi Family and Pierogi Gallery.

semiotics of arrows—as masters of visualizing information such as Edward Tufte repeatedly point out—is counterbalanced by a pictorial status in which networks reveal a complex form of nonlinear causality.[95]

But what does a drawing like *George W. Bush, Harken Energy and Jackson Stephens, c. 1979–90 (5th Version, 1999)*—which measures 122.5 cm × 61 cm in the original (figure 10.10)—tell us? As when reading a book, the eye begins on the left.[96] The title can be read in the lower half and functions as an important paratext to the prelude that marks the narrative framework. In other drawings too, title placement influences the virtual starting point without pinning it down. The connecting arcs, drawn with stencils, beckon the eye toward the central actors. The uncircled names of James R. Bath and Sheikh Abdullah Taha Bakhsh therefore receive the most attention. They bring together the most connections. Most of the actions move outward from them. In this way, individual people are at first more present than the encircled institutions, which are written in smaller letters.

Three arrows drawn from left to right offer a second opportunity for orientation. Each begins with a corporate name: Arbusto Energy Texas (top), Spectrum 7 Oil Texas (middle), and Harken Energy Corp. in Dallas (bottom). Regularly appearing circles with years evolve into parallel time arrows that refer to and build on one another. At the end of the Arbusto time arrow,

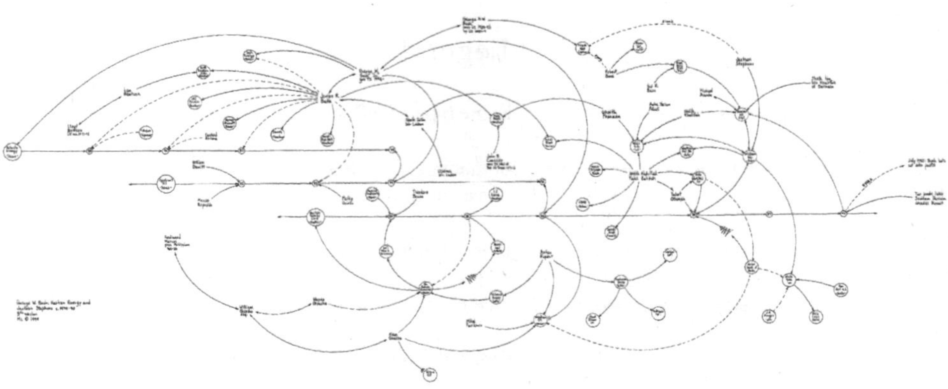

FIGURE 10.10
Mark Lombardi, *George W. Bush, Harken Energy, and Jackson Stephens, c. 1979–90* (5th version). Graphite on paper, 1999. 24⅛ in. × 48¼ in. / 61.5 cm × 122.6 cm. Photograph by John Berens. Reproduced by permission from the Lombardi Family and Pierogi Gallery.

with an indication of the year 1984, is an arrow of influence extending from Spectrum 7, after which the first company disappears from the visual narrative. Two years later, Lombardi notes an influence from Harken Energy on Spectrum 7, which in turn ends the middle time arrow. Harken Energy's activities, by contrast, continue until 1990, with a final dot for July 1990 marking when Bush sold off his stock for $848,560. Just below this information begins a unidirectional arrow with the note "Two weeks later Saddam Hussein invades Kuwait," which also points to the year. The proximity of the last two events suggests a direct connection: Bush leaves the company, and two weeks later the Gulf War begins.

Given this summary, the viewer can only reassemble over and over again the complexity of the final turn. The historical event is written down and marked with a suspicion of a conspiracy that seems to have a basis in fact. The eye leaps about in search of new points of references and necessarily focuses on familiar names: George W. Bush, as future governor of Texas, is linked to James R. Bath (figure 10.11).[97]

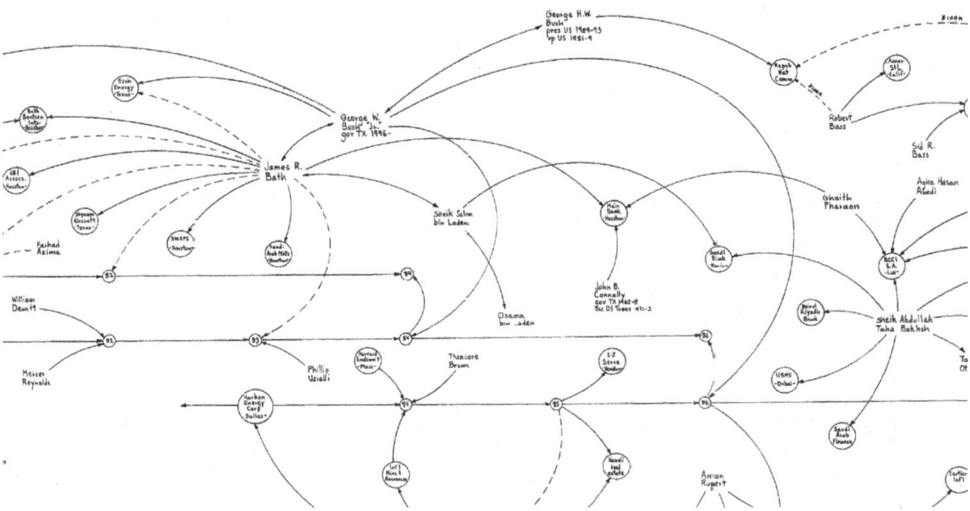

FIGURE 10.11

Mark Lombardi, *George W. Bush, Harken Energy, and Jackson Stephens c. 1979–90* (5th version). Detail. Graphite on paper, 1999. 24⅛ in. × 48¼ in. / 61.5 cm × 122.6 cm. Photograph by John Berens. Reproduced by permission from the Lombardi Family and Pierogi Gallery.

Via George H. W. Bush, there is a connection to the Republican Party, to which the Jackson Stephens of the title gave 100,000 dollars in 1988. Through Stephens's firm—Stephens, Inc., in Arkansas—there is a connection to the ominous Bank BCCI S.A. in Luxembourg, which stands under the influence of Sheikh Abdullah. As soon as one believes one has uncovered the final nexus, the eye leaps again. From Abdullah, the path leads back to Sheikh Salem bin Laden and Osama bin Laden, both of whom stand suspiciously close to the Bush family. The Saudi Arabian banks are connected across time and space to European and American ones, especially financial institutions in Texas. The presence of famous names in the upper part makes the lower third of the diagram seem comparatively unimportant, even though it explores the murkiness around Harken Energy. The network looks less dense here. One still suspects the influence of George W. Bush on Harken Energy. The financial transactions of Sheikh Abdullah that pass through Switzerland to Frontier Oil in Denver by way of the Antilles mark the longest chain of entries in the drawing, but its connection to Harken Energy simply breaks off.

A viewer's prior knowledge and interest limits the reading of images. As abstract synecdoches, the sweeping arcs reveal missing events they cannot adequately represent.[98] Yet, Lombardi's narrative structures turn out to be a generative archive of statements that opens up a space of diverse connections.[99] The variety of possible connections and the interacting and mutually modifying causalities are decisive. A leaping visual tour opens up a potential space of the past that offers complex justifications for historical events. A distributed reading movement replaces a comparatively linear appropriation. The gazes of critical actor–network analysis twitch nervously in order to generate an accelerated narrative from an approximate simultaneity of view.

As a graphic interface for viewers/readers searching for traces, Lombardi's diagrams call for a second investigative probe that starts with the textual sources. One need not accept the usual criticism that the large-format images are difficult to read. Lombardi's works are about multiplying the possible chains of events. They are not exercises in mere contingency. Precisely in their multiple approaches and their continual revisions of their own design lies a will to narrative precision.[100] The full spectrum of the historical and economic thesis of a work such as *George W. Bush, Harken Energy and Jackson Stephens, c. 1979–90* is only revealed by delving into the context. Arbusto Energy was George W. Bush's first company, which merged successively with Spectrum 7 and Harken Energy. Harken, which was a comparatively small firm operating only in the United States, was on the brink of bankruptcy in 1988 but, in 1989, surprisingly won a drilling contract with Bahrain. After two failed drillings there, and with crippling debts of 150 million dollars, Bush sold his shares on June 22, 1990, at a 200 percent profit. After the Iraqi invasion of Kuwait in August of that year, Harken stocks rapidly lost their value. George W. Bush's timely lucrative selloff became known only after the Gulf War ended.

Lombardi's drawings first outline a kind of "submedial space" of paranoid suspicion, of assumptions and fears but also of sudden revelations and insights.[101] They go beyond documentary compositions of published indications and take an interest in the scarcely comprehensible operativity of the conspiracy itself. In the systemic space of the diagram, this results in a totality in Frederic Jameson's sense: a field of forces in which network switching is questioned about its implications for conspiratorial practices. "Return of the repressed": This concerns, first, the identification of crucial nodes in the "world system"[102] by empirically recording their mafia-like actions in the entirety of the global networks.[103] Second, in the transfer from written to visual archive, the signs of the conspiracy themselves become the media interface that permits larger historical questions in the depths of a new submedial space.

The diagram *George W. Bush, Harken Energy and Jackson Stephens, c. 1979–90*, created one year before the 2000 presidential election, is part of a larger implicit narrative. A quick glance at Lombardi's library shows it is about the financially profitable parallel rearmament of Iraq and Iran by American actors. The creation of the military and political conditions for Saddam Hussein's rule culminated in the American retaliation that tried to secure its own imperial geopolitical sphere. Lombardi's images simultaneously link and undermine one another. They multiply the number of actors. Rather than a trust, a company, a "corporate network," a multitude of connections are made whose topological overall structure repeatedly points beyond itself. Despite this postmodern feature, the drawings reveal an analytical agency as second-order reporting guided by reading. Not without reason, Lombardi assumed that investigative journalists familiar with the scandals in question would be the best readers of his diagrams.[104]

Even the sober, historically mindful reader begins to harbor uncomfortable suspicions on finding in a small circle for a company the name Von Roll A.G.—Switz., whose "golden net" has already been explored graphically by Georges Baehler alias Pollux. Lombardi gazes determinedly into the past and future of capitalist activity. And even if times of economic crises and pandemics can lead them to unravel, the imagination and narration of networks of actors remains a privileged medium for talking about, seeing, and hearing conspiracies. In its net-shaped diagrammatic forms, the network collects a possible set of indexical statements about socioeconomic connections.[105] In global actor-networks, causality is not only already in a Luhmannian South but leads into the dark heart of the governmentality of capitalist economic forms. Politicians, banks, the military, and companies do not need theory. They simply act.

The new forms of digital publics that resulted from the Internet were attended by compounding conspiracy theories. Compared to transparency platforms such as Wikileaks and the rumor mills of social media, Mark Lombardi's investigations seem like an analysis of power from a bygone era. Instead of prefiltering the signs of an actor-network with journalistic thoroughness, Wikileaks, for example, produces a constant bulletin in which all conspiracies are already latently published. Observational conspiracy theory that shines a light on an archive of practices has become, thanks to Julian Assange, a constant threat with the power of digital publics. The price for this is high. Only as a conspiratorial network of actors can Wikileaks organize publicity for scandals. Meanwhile, *how* the network works is less crucial than *why*. Every announcement of a large leak serves more as the symbolic exaggeration of one's own media power than any specific abuse that "the public" must hear about.

This is a future for network analysis that Lombardi could scarcely have imagined. With the growth of generative interfaces and platforms of appearance and suspicion, the entry threshold to the submedial space of conspiracies is increasingly easy to cross. Wikileaks produces not fewer but more new secrets. Fake news continuously generates new narratives for conspiracy theories charged with negative feelings. Yet, this has little to do with critical paranoia.

So, what does the transformation of conspiracy theories mean—not only for the publics of network societies and their relationship to populism but also for the epistemological foundations of network analyses? Both questions concern the limits of what can be perceived, analyzed, and given meaning, both individually and collectively, in complex space-time phenomena. The more immaterial socioeconomic networks become, the better established they are as a symbolic order whose foundations in cultural techniques escape notice. From this perspective, conspiracy theories are a nonhegemonic form of knowledge that can critically confront the formation of network power or—in the worldwide case of populist movements—itself strive for network power.

In their paranoiac-critical form, however, conspiracy theories function as a discourse closely related to scientized forms of network analysis. In some cases, as a prior instrument of (self-)observation in the democratic public sphere, it preempts scientific analyses of networks. This may not always reveal such telling shifts in the history of discourse as those between Pollux, Condor, and Lombardi—from the trust, by way of the secret-service network, to global networks. But without a paranoid mode that suspects, narrates, visualizes, and calculates connections and entanglements, the invisible actions of networks of actors can scarcely be imagined. The paranoiac-critical form and the scientific analyses of social networks should not be confused with the current mainstreaming of populist, propagandist, and radically simplifying conspiracy theories.

11 THE CONNECTIVITY OF THINGS

What do we draw from the accounts offered here? What we have seen is characterized by repeated movement between a material net and a more immaterial than material network by practices and gestures of networking. Arriving at a conclusion requires, first, the work of knotting the net and leaving space for movement in networked places. It also requires maps and diagrams that make visible what is planned and modeled and how practices unfold. Networks result when it is no longer possible to separate the net makers and the net users from the objects that receive their attention. Networks no longer collect only elements situated in homogeneous space but also those in heterogeneous, hybrid, temporalized circulation. Mastery and control are questions of hegemony. If the hunting net was once a symbol of ruse and agency, in distributed sociotechnical networks and infrastructures, actors are always in networks of power. The more broadly agency is distributed, however, the more difficult it becomes to assess critically: networks are based on a microphysics of multiple interpenetrating actions. This is also true of all six strata of their genealogy, from the physical net for catching and holding, through nets of political or religious power, reticular biological nets and tissues, the networking infrastructures of the modern era, social networks, and on to recent network science.[1]

The first findings in such a genealogy are material nets that turn out to be mythical objects that establish symbolic forms. Setting out from the iconography of rulers in Mesopotamia and Egypt, the hunting net was an object associated with gestures of catching and being caught. The Vedas and the Old Testament are also dominated by a theology of the net as an instrument of the power of gods and rulers. This negative encoding, which contradicted that of quotidian culture, continued in ancient Greece. In the *Oresteia* of Aeschylus, it takes on a particular presence in the stratagems of Clytemnestra. For the holding together of the Attic polis, other textile objects, such as fabrics and ribbons, were preferred symbols. Roman law further developed this form of object reference in the binding contract. In the New Testament, the kingdom of heaven is like a net that is cast into the sea and captures fish of every kind (allowing for the possibility of distinguishing between the good and the bad).[2] The semantics of catching with a fishing net are still present here in the context of the Christian mission, but they are combined with an egalitarian, optimistic vision in which the divine net promises a community of belief.[3] Only since the early modern era—beginning with Paolo Veronese's *Industria* of 1577, I argue—did the net belong to the family of quasi-objects able to bind and connect social life.

Marcello Malpighi's microscopic images of frog lungs in 1661 initiated, with reference to ancient medical knowledge, a long modern career of anatomical descriptions reliant on the concept of the net. First, the linked capillary systems in blood circulation that Malpighi studied made it possible to describe the reticular character of nets. From the second half of the eighteenth century onward, the terminology increasingly extended to the physiology of the brain and nervous system, sometimes influenced by vitalist notions, which understood the visible microstructures of living creatures as a net. In tableau-style ordering systems, enlightenment thinkers also developed planar localizations of connectedness with which species of animals or even the entire economy could be grasped as a net. Geography not only used net projections. More and more, from the seventeenth century forward, it cast geometric triangulation nets to produce maps ordering the connections of physical space. Early topological network models in the history of mathematics were established in the seventeenth and eighteenth centuries in the form of Gottfried Wilhelm Leibniz's theory of monads, Leonhard Euler's Königsberg Bridge Problem, and playful forms that preceded formal graph theory.

At the beginning of the nineteenth century, the explicit naming of territorial connections introduced an expansive network of streets, canals, telegraphs, and railroads that needed to be connected. The spirit of spatial redesign was represented most conspicuously by the networking programs of the Saint-Simonians in France. The explicit transfer to the organizational form between institutions—for example, to the network of the banking economy—is a coeval phenomenon starting with urban water supply and sewerage. The interior of the city turned out to be a milieu of continuous canalization, which simultaneously produced transportation projects spanning the globe. The organization of colonialism and large projects such as the Suez Canal went together hand in glove. This trend was already evident in the first large-scale network treatise—"Système de la Méditerranée" of 1832 by the French engineer Michel Chevalier, which demonstrated the Saint-Simonians' early networking ambitions to the public.

Both urban and territorial infrastructure turn out to be paradigmatic, mirror-image examples of the regime of space and power that arose with the networks of the nineteenth century. Historical actors wagered on a geopolitical and biopolitical program of expanding connectivity and circulation that evolved thanks to diplomatic meetings on linking national telegraph networks (among other things). Nationalism and transnational network formation were not necessarily mutually exclusive. Indeed, they turned out to be two sides of the same coin, as the deliberate formation of international connections shows. With the spread of supply and communication networks, initially in cities in the second half of the nineteenth century, especially in Western nations, a collective state of being immaterially "networked" became established as a spatial mode of societal self-description.

Telephone communication played a special role in this process, and after 1878, it increasingly became the model for linguistic, social, physiological, and infrastructural connections. The latter included the popularization of the "wonderful speaking telephone" in music halls and opera houses. Alexander Graham Bell's tours ushered in the building of local communication nets in the northeast corner of the United States. The material basis for telephone communications was not only the cables and telephones but also the first telephone books and, more importantly, women's manual labor on switchboards. Manual switching was joined by automated connection. Using the example of Almon Brown Strowger's patent of the first practicable mechanical stepping switch in 1896, I have showed how momentously the style of networking was changed by electromechanization. At the same time, the connection practices of telephone communication were accompanied by ghostly doubles in physiology, association psychology, and spiritualism. With the transmission of voices in the telephone network—summing up the cultural consequences of telephone communication—clear delineation between the end of human bodies and sociotechnical networks vanished.

This development was flanked by other transformations. Out of the notation and modeling of chemical elements evolved an important feature of graph theory, which was refined over the course of the nineteenth century. The character of material connections turned out to be a central fascination that lent itself to the way social ties were imagined. The transfer of semiotic practices from the natural sciences introduced a theory of social networks into sociology. One crucial moment in the history of science was Jacob Levy Moreno and Helen Jennings's use of the network concept in the 1930s, which lent new form to Georg Simmel's understanding of a "sociological bond." It went along with an intense production of mappings and diagrams with which scientific codes contributed to the pictorial status of social networks. Moreno's usage points to older historical strata in its joyful eclecticism. The binding magic practiced with the network was declared the very model of social cohesion.

In the twentieth century, network formation and maintenance increasingly became a question of proper timing. Henry Charles Beck's famous diagram of the London Tube map of 1933 attests to a relational understanding of space influenced by metropolitan underground navigation. It is also an instrument to synchronize an urban network. A transportation network aspires to simultaneity and precise timing. The more network-like connections it is supposed to organize, however, the more important the play of fluctuation and synchronization of traffic flows becomes. Understanding the Tube map requires precise observation of the clocks of London Transport. It is also necessary to take the political joining of many lines in a network seriously.

In a related development, international production planners also established their own regimes of time, which, since the 1960s, have used the so-called network operations method. Networked production changed the status of both objects and organizations. The complex technical object—whether computer, ship, or missile—is described as a network just as much as cooperating companies. The intercultural styles of network production differ, however, and cannot be traced back exclusively to the rapid spread of network plans in industrial countries. Using the example of Toyota's just-in-time production system and the differentiation of worldwide logistics, we saw how network projects became oriented toward long-term results. Hereby, things that have been made netlike are distributed in increasingly complex, temporally differentiated logistical systems. Without fluid availability of matter and object, there are no network societies.

Finally, things must speak to one another to prove their connectivity. It is scarcely possible to understand a discursive peculiarity like the early history of computers in any other way. Around 1965, people began "talking" to computers, which at the time were still mainframes. Written programming was supposed to function as much like a conversation as possible. Soon, computers were speaking not just to people in popular science fiction but also to one another in reality (as interpreted by engineers). This imagined linguistic ability, which has its origins in interactive programming, was also the basis for the development of network protocols. Such protocols initially targeted the temporalities and economic issues of connecting computers or of "netting" the ARPANET, but the question of spatial networking across borders soon arose. The project of internetworking was very different. Heterogeneous nets resulted in the Transmission Control Program and the Internet Protocol—the crucial communication techniques of Internet architecture. Despite the ubiquity of computers, routers, and cables, Internet architecture consists, for the most part, of signs, through which, ideally, it is no longer necessary to perceive the material connectivity of things.

The rise of the network society has limits, breaking points, and more than one hidden side. The more completely we depend on networks as a social strategy, the more conspiracy theories proliferate. In this context, illegitimate, paranoid knowledge becomes an alternative cultural discourse. Critical-paranoid conspiracy theorists are well-versed analysts of networks. Using the drawings of the Swiss Communist Georges Baehler (alias Pollux), Joseph Turner in Sydney Pollack's *Three Days of the Condor*, and the large-format global networks of the artist and archivist Mark Lombardi, clear discursive shifts in the twentieth century became evident. If the narratives of conspiracy are still dominated by the meaning of one "golden net," "trust," "corporation," "corporate network," "Radial Matrix," or "world system," at least since Lombardi, the number of active networks of people, firms, and institutions has increased to the point of inscrutability.

11.1 MATERIALITY: THAT WHICH ALLOWS NETWORKS TO CONNECT

> And there are no names for things that exist.
> When it disappears into its labyrinth, only the hope remains
> That at least there are names for things that do not exist.
> —Aleš Šteger, *Knjiga reči* (*The Book of Things*)[4]

Is the network a name for a thing that does not exist except in the state of becoming, in continual transmission, and in circulation? In Aleš Šteger's poem "Mravlja" (Ant), the eponymous creature clings tenaciously to the objects it carries. But few people in the twenty-first century who see themselves as part of multiple sociotechnical networks carry material nets around with them. The metonymic connection to material-territorial nets is, on the one hand, always present; we deal with networked objects that act as mobile access points. On the other hand, this access represents just one small part of the communicative forms of exchange that it makes possible. The diverse forms of the net disappear from view in favor of practices of networking, except for actors involved in the specific economic and technical realization of a physical net. The dynamic that is supposed to make action in network form possible is also crucial to design, however. Software counts more than hardware; the best form of organization is self-organization. Networking is everywhere.

Can there be a network not based on material nets? This book assumes it is not possible, or, if it is, such a network would be hardly observable at all. Networks are inherently metonymic. That said, the state of cultural techniques in the twenty-first century seems different. People network without forming nets. Leaving aside the obvious fetishism in the case of smartphones and portable computers, the Internet of things shifts the connectivity of objects to the infrastructural background. Network formations are shifting as a consequence of increasingly data-intensive platforms that cause the relationships of social and economic bonds to fluctuate. Network history is always, even in the here and now, a history of things, but much more than that, it is a history of teeming quasi-objects.

The symbolic forms of the net and the network both belong to the family of textile quasi-objects of binding. They function without people as living nodes of connection: "This quasi-object is not an object, but it is one nevertheless, since it is not a subject, since it is in the world; it is also a quasi-subject, since it marks or designates a subject who, without it, would not be a subject."[5] What could take the place of the network? The association, the bond, the rope team, the *symplokē* as the closely woven fabric of the ancient *polis*?[6] Social bonds, ties, threads, traps, slings, snares, cords, ropes, fetters, chains, webs, and fabrics are some of the enchanting, entwining, and ensnaring figurations representing both cultural cohesion and social exclusion. Objects of binding can be applied without distinction to people, animals, and things, which are included physically in the link and metaphorically in the bond.[7] Distinctions between person and thing or subject and object, regardless of whether their origins are legal or philosophical, are called into question.

The network, even in its latest forms, belongs to this tradition, which includes wide-ranging magical and religious practices.[8] Never before has there been so much binding magic as in our technologically supercharged present. And it includes not only rationalized forms such as "legally binding terms," which can also become the basis for an entire social or natural contract. It has more to do with an imperceptible binding magic that permits both loose couplings and strong ties situationally—with action at a distance and local effects. The magical quality of these binding practices would be both the dynamic, constantly readjusted goals of social efficacy and recognition and the long-distant effects of networked objects. The code of tying and untying is not applied to obstruct. Even communicative connections that are not actively maintained are kept latent (that applies equally to people, institutions, and technical devices).[9] This quality also distinguishes discrete network-like connections from the diffuse state of "being networked," which knows no temporal or emotional interruption.

Networks have been emancipated from the old magic of tying, snaring, and netting by a temporalized and contingent feature that becomes clearer with long-distance distribution and asynchronous action.[10] The network is no longer an instrument for catching, at least not in the traditional sense.[11] But that does not simply succede practices of binding and liberating cultivated in hermetic magic and alchemy for example. Rather, platform- and app-based social interactions are dominated by the possibilities of connecting on the network. Only recently have loosening, disconnecting, and unnetting started to get the attention they deserve.[12] Sociologists interested in networks are abundantly aware that connections are specific and differ according to context and narratives.

Narratives in general and transmission are one of the few stable reasons for networks being able to act as a social binder and a cultural technique that creates cohesion. As quasi-objects, they are simultaneously discourse and materialization. When performed, they create intersubjectivity within a collective of actors. The best examples are still supplied by Michel Serres. The continual passing-on of things results in a shared yet partial identity. At the same time, the person who ends up forfeiting in a game is immediately marked as an individual. Quasi-objects thus result from circulation, acts of exchange, interactions in a space of distributed actions.[13] They are in a position to integrate elements from their surroundings constantly. It does not matter whether these elements are beings or relations or just parts of beings or relations.

Quasi-objects are always already more than "just" material culture because diverse forms of materialization and symbolization converge in them: "The symbol is a quasi-object and a quasi-subject; undoubtedly you are and I am a symbol."[14] That makes it close to "figures of knowledge" or "epistemic things," even if, in networks, the emphasis falls on everyday practices and techniques. Their character as quasi-objects—and not so much as a guiding cultural metaphor or, better, guiding metonymy—enables networks to connect people, animals, objects, and institutions continuously.

Will that still be true when other things and structures become more relevant in our lifeworld? For the time being, there are too many material nets crisscrossing global environments and too many cooperative actions that depend on them for us to free ourselves from the network as quasi-object of choice, even if the net metaphor is losing its luster. The point is to understand that this too is an effect of binding magic.

11.2 BOUNDARIES: THAT WHICH IS NO LONGER A NETWORK

> I failed to find a connection.
> —Patrick McGrath, *Spider*[15]

> Et tout à coup, une chose, quelque chose apparaît, hors le réseau.
> —Michel Serres[16]

Perhaps it is the malleability of the network diagram that makes it possible to grasp and formalize so many phenomena in nodes, lines, and relations. Or it is the strange bewitching effect of the binding magic of a quasi-object. In any case, now one can scarcely say that network theory is pursued as a subversive "method with a small 'm.'"[17] The inevitability of thinking and acting in networks may be lost, even if the material nets remain or are reconfigured sociomaterially. But as a universalizing explanatory model, especially in the natural sciences and as a social graph of digital platforms, formal network analyses are certainly not passé. As long as new or reinvented digital network technologies continue to be established—such as cryptocurrencies, blockchains, and artificial neural networks in machine learning—networks will be rehearsed as cultural technology. They remain an integral component of the materiality of cooperation.

Other social figures have, however, long since been explored. When networks become too static or established, a collective wager is made on the agency of the swarm. In this view, collectivity derives not from the connectivity of individual people but is inherent in the swarming. All other connections result from this collectivity.[18] Synchronicity of action should be a natural state that need not be produced laboriously by synchronization. The interest in living networks, like swarms, points beyond mathematical network logics on which it is based in part[19]—not the manufacturing of objects and quasi-objects but the collective, instinctive action of creatures becomes the standard.[20] Swarms as social formations bid farewell to working in the network and want to be networked momentarily, and not permanently.

The evolution of radio technology also ensures that an omnipresent sphere of circulation is superimposed onto the cultural logic of nodes and connections. Digital radio infrastructures cannot exist without a cable network. However, they pursue both a different epistemology and, above all, a different spatial logic of "wirelessness."[21] When the mesh of a net is permeable, cell tower coverage of the countryside will no longer know any disconnectedness. Dead zones, rather than cut cables, will be the most common disruption of connectivity thwarting the desires for imperceptible, universally available communication. The more normal mobile use of the Internet has become, the harder it has proved to differentiate between the net and the environment. When the communication net becomes the sensorial living environment, new phenomena of naturalization analogous to swarming emerge. Connections result via marvelous clouds of computers, although admittedly, even in cloud computing, they continue to stand in physical, distributed computing centers.

But neither can physical space be completely penetrated and dominated media-ecologically nor can actor networks or swarms simply cross political, economic, and legal boundaries. The end of networking can most easily be seen in sociotechnical formations and media that have nearly disappeared. Consider pneumatic tube transport, classical telephone switching, teletypes, Bruno Latour's *Aramis*, and search engines prior to Google.[22] Or think of all the fun fantasies of collapse with which nets have unraveled in popular films—for example, materiality-fixated action film series such as *Terminator* and *Die Hard*.[23] It is far more difficult to answer the questions of when and how social practices and cultural formations lose their network form.

Already in 1996, British anthropologist Marilyn Strathern published what is perhaps the most important text on this subject, "Cutting the Network." She outlines a critique of the network concept of actor–network theory and also reveals the contours of what limits and cuts a network. Strathern's arguments contain a promise that has yet to be fulfilled analytically, since the limiting of networks and their undoing was considerably less interesting prior to the COVID-19 pandemic than their enduring spread, maintenance, and repair.[24] Nevertheless, Strathern's justifications seem convincing, since she assumes that social networks—including material culture—are generally distinguished by a fragile temporality. This temporality can be interrupted in four ways: property claims, stopping the exchange of money and goods, cutting off kinship relationships, or through the limited describability of a network. "Now if networks had lengths they would stop themselves," Strathern remarks.[25]

This is a question of method, first and foremost. Every story of a network that contributes to identity but also any analytical account must come to an end in order to have any effect. Following Walter Benjamin, one can object that every good story always also provokes the question of what comes next.[26] Nevertheless, Strathern is right to understand the stopping of networks as a necessary interruption in their continual expansion and connectivity. More important even than this general ethnological objection are other very real caesuras.

It is no coincidence that Strathern factors in economic transactions (and their suspension) as well as normative and legal limitations on free association and network formation. Her examples are as obvious as they are apt. For example, the patent for a blood test for hepatitis C limited the range of social networks that had to be stabilized to discover and combat the virus. Instead of recognizing the participating networks of society as a whole, a clear property claim could be litigated under patent law, which, in turn, cost the British health system three million pounds annually.[27] Maintaining and stopping money flows is also characterized by a fragile temporality. Economic transactions and the exchange of connective objects condition the emergence of limited networks that become measurable at the same time. Events such as a person's death among the 'Are'are on the Solomon Islands ensure interruption and reconfiguration. While the burial ceremony makes the old connections visible, it also blocks future effects. The relationships of the deceased are recombined.[28]

Strathern's anthropological miniatures and her foundational work on partial connections and relations go beyond the scope of this book.[29] While my historical interest focused on the material culture of sociotechnical networks, in the financial crisis of 2008, the flows of money were faltering and stopping. The collapse of the system of debts, obligations, and profits called the political and economic order into question for a moment, both within the Occupy movement and through the rise of new, nonstate cryptocurrencies. The network logic of Bitcoin combined peer-to-peer technology with a turn to anonymity and pseudo-anonymity without factoring in the consequences for the environment that come with mining digital values. In parallel with those developments, the Snowden revelations fundamentally undermined trust in digital networks. The real-time network analyses of the secret services NSA, GCHQ, and Co restored the hegemonic claim of the old capture-and-hold nets in a new form. It has found its antiprogram in the new crypto technologies that are intended to protect private and business networks from invisible third parties.

When the first edition of *Verbundenheit der Dinge* appeared in German in 2014, I already commented in a postscript on disillusionment with cultures of digital networking as a result of Snowden. That later platform-based public spheres, their economies of data, new populist movements, identity politics, and fake news would shake the world was almost inconceivable at the time. Only a little earlier, during the Arab Spring, the emancipatory possibilities of digital communication networks had been celebrated, even though repressive measures by the police and secret services were not far behind. Since then, negative dynamics have gained the upper hand worldwide. One response has been a new form of scholarly criticism that seeks to intervene decisively: "Your Computer is on Fire!"—and apparently everything else along with it.[30]

Contributing significantly to the discontent within current network cultures are advances in machine learning and pattern recognition. The graph-based calculations of social media and the use of artificial neural networks are the best indicators of what separates networks in the twenty-first century from their predecessors. Analyses made in retrospect have become data-driven, platform-based operations in near real time. The new network and data calculations are generative and, thanks to self-learning, semiautonomous. They do not just describe speech, image, and facial recognition. They can be used prescriptively to generate speech and writing, sounds and images—with all our human prejudices inscribed in the training data with which the algorithms are optimized. Are the neuronal nets of machine learning forming a new seventh stratum of network history? If this is the case, then they represent a converging digital system that is remediating speech and writing, sounds and images. Cultural techniques are becoming inconceivable without their digital networks, which for their part are becoming an environment.

The steadily increasing political, economic, and geopolitical abuse of network power necessarily raises the question of how to curtail infractions. "And looking to see who or what intercepts these different flows is also a way of speaking clearly and calmly."[31] With waves of the COVID-19 pandemic, during which this book was translated, this ecological and economic question is obvious. Any answer to it is based on the practice of cultural techniques. Out of the current cut-up and reconfigured networks, new figures of exchange and separation will form. No one should say that the net alone can stand for the connectivity of things.

NOTES

CHAPTER 1

1. Behrman House, ed., *Pirke Avot: A Modern Commentary on Jewish Ethics*, ed. and trans. Leonard Kravitz and Kerry M. Olitzky (New York: UAHC Press, 1993), 3–16, 46–47.

2. "And so we watch one man after another walking down the same paths and we know in advance where he will branch off, where walk straight on without noticing the side turning, etc. etc. What I have to do then is erect signposts at all the junctions where there are wrong turnings so as to help people past the danger points." Ludwig Wittgenstein, *Vermischte Bemerkungen / Culture and Value*, ed. G. H. von Wright and Heikki Nyman, trans. Peter Winch (Chicago: University of Chicago Press, 1980), 18e.

3. Hartmut Böhme, "Netzwerke: Zur Theorie und Geschichte einer Konstruktion," in *Netzwerke: Eine Kulturtechnik der Moderne*, ed. Jürgen Barkhoff, Hartmut Böhme, and Jeanne Riou (Cologne: Böhlau, 2004), 17–36, esp. 17.

4. See Alexander Galloway and Eugene Thacker, *The Exploit: A Theory of Networks* (Minneapolis: University of Minnesota Press, 2007), 155.

5. On the theory of the quasi-object, see Michel Serres, *The Parasite*, trans. Lawrence R. Schehr (Baltimore: Johns Hopkins University Press, 1982), 224–225, and sections 3.1 and 11.1 of the present volume.

6. On the network society, see, first, Manuel Castells, *The Rise of the Network Society*, 2nd ed. (Oxford: Blackwell, 2000). For a summary, see Erik van der Vleuten, "Understanding Network Societies: Two Decades of Large Technical System Studies," in *Networking Europe: Transnational Infrastructures and the Shaping of Europe, 1850–2000*, ed. Erik van der Vleuten and Arne Kaiser (Sagamore Beach, MA: Science History Publications, 2006), 279–314. For a historical account, see Erhard Schüttpelz, "Ein absoluter Begriff: Zur Genealogie und Karriere des Netzwerkkonzepts," in *Vernetzte*

Steuerung: Soziale Prozesse im Zeitalter technischer Netzwerke, ed. Stefan Kaufmann (Zurich: Chronos, 2007), 25–46.

7. Vilém Flusser, *Gestures,* trans. Nancy Ann Roth (Minneapolis: University of Minnesota Press, 2014), 6.

8. On localizing networks as a specific cultural technique of modernism in the German-language discourse, see Jürgen Barkhoff, Hartmut Böhme, and Jeanne Riou, eds., *Netzwerke: Eine Kulturtechnik der Moderne* (Cologne: Böhlau, 2004); Gerhard Dirmoser, *Eine subjektive Chronologie der Netzentwicklungen,* Vom Nutzen schematischer Zeichnungen XXXII (2004), http://www.servus.at/kontext/diagramm/32_Netzhype.htm; Sebastian Giessmann, *Netze und Netzwerke: Archäologie einer Kulturtechnik, 1740–1840* (Bielefeld: transcript, 2006); Julia Gelshorn and Tristan Weddigen, "Das Netzwerk: Zu einem Denkbild in Kunst und Wissenschaft," in *Grammatik der Kunstgeschichte: Sprachproblem und Regelwerk im Bild-Diskurs; Oskar Bätschmann zum 65. Geburtstag,* ed. Hubert Locher and Peter J. Schneemann (Zurich: Schweizerisches Institut für Kunstwissenschaft, 2008), 54–77; and Passepartout, ed., *Weltnetzwerke—Weltspiele: Jules Vernes "In 80 Tagen um die Welt"* (Konstanz: Konstanz University Press, 2013). For an expansion of the understanding of the network into the early modern period, see Wolfgang Behringer, "Netzwerk," in *Enzyklopädie der Neuzeit,* vol. 9, *Naturhaushalt–Physiokratie,* ed. Friedrich Jaeger (Stuttgart: J.B. Metzler, 2009), 98–100. On localizing it in the present, see Stefan Kaufmann: "Netzwerk," in *Glossar der Gegenwart,* ed. Ulrich Bröckling, Susanne Krasmann, and Thomas Lemke (Frankfurt am Main: Suhrkamp, 2004), 182–189.

9. See Michel Serres, "Introduction," in *A History of Scientific Thought: Elements of a History of Science,* ed. Michel Serres (Oxford: Blackwell, 1995), 1–16, esp. 10–11.

10. See Christian Kassung, *Das Pendel: Eine Wissensgeschichte* (Paderborn: Wilhelm Fink, 2007), 12–13; and Christian Kassung and Albert Kümmel-Schnur, "Wissensgeschichte als Malerarbeit? Ein Trialog über das Weißeln schwarzer Kisten," in *Bruno Latours Kollektive,* ed. Georg Kneer, Markus Schroer, and Erhard Schüttpelz (Frankfurt am Main: Suhrkamp, 2008), 155–179.

11. See Donna Haraway, "A Game of Cat's Cradle: Science Studies, Feminist Theory, Cultural Studies," *Configurations* 2, no. 1 (1994): 59–71; and Uwe Wirth, "Vorüberlegungen zu einer Logik der Kulturforschung," in *Kulturwissenschaft: Eine Auswahl grundlegender Texte,* ed. Uwe Wirth (Frankfurt am Main: Suhrkamp, 2008), 9–67, esp. 47–48.

12. See Friedrich Nietzsche, "On the Genealogy of Morality," trans. Adrian del Caro, in *The Complete Works of Friedrich Nietzsche,* vol. 8, *Beyond Good and Evil / On the Genealogy of Morality* (Stanford, CA: Stanford University Press, 2014), 207–422; and Michel Foucault, "Nietzsche, la généalogie, l'histoire," in *Dits et écrits,* vol. 2 (1970–1975) (Paris: Gallimard, 1994), 136–156.

13. On the genealogy of the network thinking of ANT, see Sebastian Giessmann, "Verunreinigungsarbeit: Über den Netzwerkbegriff der Akteur-Netzwerk-Theorie," *Reinigungsarbeit: Zeitschrift für Kulturwissenschaften* 1 (2013): 149–160.

14. On the universal application of network theory, see Albert-László Barabási, *Linked: The New Science of Networks* (Cambridge, MA: Perseus, 2002); and Albert-László Barabási and Márton Pósfai, *Network Science* (Cambridge: Cambridge University Press, 2016).

15. On the relationship of the history of the network and LTS analysis, see Jean-Marc Offner, "Are There Such Things as Small Networks?" in *The Governance of Large Technical Systems*, ed. Olivier Coutard (London: Routledge, 1999), 217–238.

16. On multilayered historical process of long duration, see, classically, Fernand Braudel, "History and the Social Sciences: The *Longue Durée*," trans. Immanuel Wallerstein, *Review (Fernand Braudel Center)* 32, no. 2 (2009): 171–203.

17. Geoffrey C. Bowker, "How To Be Universal: Some Cybernetic Strategies, 1943–1970," *Social Studies of Science* 23, no. 1 (1993): 107–127. On cybernetics in German-language media theory, see Claus Pias, ed., *Cybernetics/Kybernetik: The Macy Conferences, 1946–1953*, vol. 1, *Transactions/Protokolle* (Zurich: diaphanes, 2003); and Claus Pias, ed., *Cybernetics/Kybernetik: The Macy Conferences, 1946–1953*, vol. 2, *Essays & Dokumente* (Zurich: diaphanes, 2004).

18. See John Brewer, "Networks," in *Information: A Historical Companion*, ed. Ann Blair et al. (Princeton, NJ: Princeton University Press, 2021), 620–627; and Niall Ferguson, *The Square and the Tower: Networks, Hierarchies and the Struggle for Global Power* (New York: Penguin, 2017).

19. Paul N. Edwards et al., *Understanding Infrastructure: Dynamics, Tensions, and Design. Report of a Workshop on "History and Theory of Infrastructure: Lessons for New Scientific Cyberinfrastructures"* (Alexandria, VA: National Science Foundation, 2007), 12. https://deepblue.lib.umich.edu/bitstream/handle/2027.42/49353/UnderstandingInfrastructure2007.pdf.

20. Wendy Hui Kyong Chun, "Networks NOW: Belated Too Early," in *Postdigital Aesthetics: Art, Computation and Design*, ed. David M. Berry and Michael Dieter (London: Palgrave Macmillan, 2015), 289–315.

21. Lori Emerson, "Category: Other Networks," *loriemerson.net* (blog), https://loriemerson.net/category/other-networks; Lori Emerson, "What's Wrong with the Internet and How We Can Fix It: Interview with Internet Pioneer John Day," *loriemerson.net* (blog), July 23, 2015, https://loriemerson.net/2015/07/23/whats-wrong-with-the-internet-and-how-we-can-fix-it-interview-with-internet-pioneer-john-day.

22. Kavita Philip, "The Internet Will Be Decolonized," in *Your Computer Is on Fire*, ed. Thomas S. Mullaney et al. (Cambridge, MA: MIT Press, 2021), 91–115.

23. Giessmann, *Netze und Netzwerke*.

24. Thomas Parke Hughes, *Networks of Power: Electrification in Western Society, 1880–1930* (Baltimore: Johns Hopkins University Press, 1983); Pierre Musso, *Télécommunications et philosophie des réseaux: La posterité paradoxale de Saint-Simon* (Paris: Presses Universitaires de France, 1997); Laura Otis, *Networking: Communicating with Bodies and Machines in the Nineteenth Century* (Ann Arbor: University of Michigan Press, 2001);

Mark Wigley, "Network Fever," *Grey Room* 4 (2001): 82–122. See also Cris Cheek, Nicole Starosielski, and Braxton Soderman, eds., "Network Archaeologies," *Amodern* 2 (October 2013), http://amodern.net/issues/amodern-2-network-archaeology.

25. See Bruno Latour, *Reassembling the Social: An Introduction to Actor-Network-Theory* (Oxford: Oxford University Press, 2005), 129–130; and Christian Jürgen Emden, "Netz," in *Lexikon der philosophischen Metaphern*, ed. Ralf Konersmann (Darmstadt: Wissenschaftliche Buchgesellschaft, 2007), 248–260, esp. 248. Alexander Friedrich took up the work on the metaphorology of networking in his dissertation, which in many respects complements this book. Alexander Friedrich, *Metaphorologie der Vernetzung: Zur Theorie kultureller Leitmetaphern* (Paderborn: Wilhelm Fink, 2015); cf. Alexander Friedrich, "Metaphorology of Networks," in *Metaphors Shaping Culture and Theory*, ed. Herbert Grabes, Ansgar Nünning, and Sibylle Baumbach (Tübingen: Narr, 2009), 285–297.

26. See Hans-Jörg Rheinberger, *Toward a History of Epistemic Things: Synthesizing Proteins in the Test Tube* (Stanford, CA: Stanford University Press, 1997).

27. Aelian, *On the Characteristics of Animals (De natura animalium), Books VI–XI*, ed. and trans. A. F. Scholfield, vol. 2 (Cambridge, MA: Harvard University Press, 1959), VI.57.

28. I am grateful to Steffen Siegel for this reference.

29. Francesco Sansovino, *Venetia citta nobilissima et singolare, Descritta in XIIII. Libri* (Venice: Iacomo Sansovino, 1581), book VIII, fol. 123r. The reception of Archimedes had experienced a large boom after the Latin and Greek edition from Basel in 1544.

30. The production of silk more than tripled. Moreover, Venice took over market leadership in the Italian production of wool products. See Herfried Münkler and Marina Münkler, *Lexikon der Renaissance* (Munich: C.H. Beck, 2000), 408. On the sixteenth- and seventeenth-century transformation of the historical semantics of *industria* from natural skill to proper handling of a trade, see Dietrich Hilger, "Industrie, Gewerbe," in *Geschichtliche Grundbegriffe: Historisches Lexikon zur politisch-sozialen Sprache in Deutschland*, ed. Otto Brunner, Werner Conze, and Reinhart Koselleck, vol. 3 (Stuttgart: Klett-Cotta, 1982), 237–304.

31. According to Sigrid Dittrich and Lothar Dittrich, *Lexikon der Tiersymbole: Tiere als Sinnbilder in der Malerei des 14.–17. Jahrhundert* (Petersberg: Michael Imhof Verlag, 2004), 503, this represents an exception. On the representation of the Venetian entrepreneurial spirit and Veronese's dialectical cunning versus the Inquisition, see André Chastel, *A Chronicle of Italian Renaissance Painting, 1280–1580* (Ithaca, NY: Cornell University Press, 1984), 208–227.

32. On this, see, much later, Hubert François Gravelot, Charles Nicolas Cochin, and Charles-Etienne Gaucher, *Iconologie par Figures; ou, Traité complet des Allégories, Emblêmes &c.: Ouvrage outile aux Artistes et Amateurs, et peuvent servir à l'éducation des jeunes personnes*, vol. 3 (Bordeaux: Lattré Graveur, 1791), 11.

33. Just by means of the sewing basket, the image joins the series of contemporaneous images about the sense of touch produced in Netherlandish Mannerism before Veronese. Similar objects are found in Maarten van Cleeve's sketch *Arachne* and Philipp Galle's copperplate engraving after Frans Floris *Arachne* (1574). See Carl van de Velde, "The Painted Decoration of Florin's House," in *Netherlandish Mannerism: Papers Given at a Symposium in Nationalmuseum Stockholm, September 21–22, 1984*, ed. Görel Cavalli-Björkman (Stockholm: Nationalmuseum, 1985), 127–134, esp. 132–133.

34. On the contradictoriness and resistance of things in allegory, see Walter Benjamin, *Origins of the German Trauerspiel*, trans. Howard Eiland (Cambridge, MA: Harvard University Press, 2019), 198–202.

35. Aelian, *On the Characteristics of Animals (De natura animalium), Books I–V*, ed. and trans. Alwyn Faber Scholfied, vol. 1 (Cambridge, MA: Harvard University Press, 1958), I.21.

36. Diogenes Laertius, *Lives of Eminent Philosophers*, ed. R. D. Hicks, vol. 2 (Cambridge, MA: Harvard University Press, 1972), book VII, paragraph 161. On the readoption of the critique of Sophism, see Judith Dundas, "Arachne's Web: Emblem into Art," *Emblematica: An Interdisciplinary Journal for Emblem Studies* 2, no. 1 (1987): 109–137, esp. 109–110.

37. Dialectic was, along with rhetoric and grammar, part of the trivium. See Rainer Brändle, "Abenteuer der Dialektik: Ein Gemälde Veroneses in Sempruns Roman *Der weisse Berg*," in *Überschreitungen: Dialoge zwischen Literatur- und Theaterwissenschaft, Architektur und Bildender Kunst; Festschrift für Leonhard M. Fiedler*, ed. Jörg Sader and Anette Wörner (Würzburg: Königshausen & Neumann, 2002), 329–338, esp. 333.

38. See also Bernard Chambaz, *La dialectique Véronèse* (Crest: La Sétérée, 1989), 73–74.

39. On this quality of early modern "allegorical art," see Carsten-Peter Warncke, *Symbol, Emblem, Allegorie: Die zweite Sprache der Bilder* (Cologne: Deubner, 2005), 88.

40. Sigrid Weigel, *Body- and Image-Space: Re-Reading Walter Benjamin*, trans. Georgina Paul with Rachel McNicholl and Jeremy Gaines (London: Routledge, 1996), 116 (translation modified). For a lucid account of Benjamin's concept of the image in general, see Weigel, *Body- and Image-Space*, 44–45.

41. See Michel Foucault, *The Order of Things: An Archaeology of the Human Sciences*, trans. Alan Sheridan (New York: Vintage, 1973).

42. Jacob Grimm and Wilhelm Grimm, *Deutsches Wörterbuch*, vol. 7 (N–Quurren), ed. Matthias von Lexer (Leipzig: S. Hirzel, 1889), 643.

CHAPTER 2

1. Michel Serres, *The Natural Contract*, trans. Elizabeth Mac Arthur and William Paulson (Ann Arbor: University of Michigan Press, 1995), 45.

2. The advent of technical-communicative networks in historical studies is generally dated to the nineteenth century. See Emily Rosenberg, ed., *A World Connecting, 1870–1945* (Cambridge, MA: Belknap Press of Harvard University Press, 2012). See also Wolfgang König and Wolfhard Weber, eds., *Propyläen-Technikgeschichte*, vol. 4, *Netzwerke, Stahl und Strom, 1840–1914* (Berlin: Propyläen, 1990); and Jürgen Osterhammel, *The Transformation of the World: A Global History of the Nineteenth Century*, trans. Patrick Camiller (Princeton, NJ: Princeton University Press, 2015), 711. The early modern period has in the meantime also come to be viewed as a formative period for networks. See Ann Blair et al., eds., *Information: A Historical Companion* (Princeton, NJ: Princeton University Press, 2021), esp. chaps. 5 and 10.

3. Jacques Rancière, *The Names of History: On the Poetics of Knowledge*, trans. Hassan Melehy (Minneapolis: University of Minneapolis Press, 1994), 2. On implicit and explicit forms of knowledge, see, fundamentally, Michael Polanyi, *The Tacit Dimension* (Chicago: University of Chicago Press, 2009).

4. See Sebastian Giessmann, *Netze und Netzwerke: Archäologie einer Kulturtechnik, 1740–1840* (Bielefeld: transcript, 2006).

5. This approach to a cultural technique is very clearly distinguished from Niall Ferguson, *The Square and the Tower: Networks, Hierarchies and the Struggle for Global Power* (New York: Penguin, 2017).

6. Cf. the epigraph by Ludwig Wittgenstein in chapter 1 or, from the perspective of media theory, Hartmut Winkler, *Docuverse: Zur Medientheorie der Computer* (Munich: Boer, 1997).

7. This is done most clearly and in exemplary fashion in Roland Barthes, *The Pleasure of the Text*, trans. Richard Miller (New York: Hill and Wang, 1975). For an overview, see Paul N. Edwards, "Hyper Text and Hypertension: Post-Structuralist Critical Theory, Social Studies of Science and Software," *Social Studies of Science* 24 (1994): 229–278.

8. Marcel Griaule, *Conversations with Ogotemmêli: An Introduction to Dogon Religious Ideas*, trans. Ralph Butler with Audrey I. Richards and Beatrice Hooke (London: Oxford University Press, 1965); originally published in French as *Dieu d'eau* (Paris: Éditions du Chêne, 1948). The complex systems of symbols of the Dogon studied by Griaule and other ethnologists from 1931 onward have been worked out in detail. See, among others, Geneviève Calame-Griaule, *Ethnologie et langage: La parole chez les Dogon*, 2nd ed. (Paris: Institut de l'Ethnologie, 1987), and Marcel Griaule and Germaine Dieterlen, *Le renard pâle*, 2nd ed. (Paris: Institut d'Ethnologie, 1991).

9. Griaule, *Conversations with Ogotemmêli*, 27–28.

10. Griaule, *Conversations with Ogotemmêli*, 28. On this very specifically, see Calame-Griaule, *Ethnologie et langage*, 81–82, and on the relationship of gestures, technique, and language in general, see André Leroi-Gourhan, *Gesture and Speech*, trans. Anna Bostock Berger (Cambridge, MA: MIT Press, 1993).

11. On the material culture of string games among the Dogon, see Marcel Griaule, *Jeux Dogons* (Paris: Institut d'Ethnologie, 1938), 71–72.

12. Griaule, *Conversations with Ogotemmêli*, 65.

13. Griaule, *Conversations with Ogotemmêli*, 43–44, 72–74.

14. See Bruno Latour, "Technology Is Society Made Durable," in *A Sociology of Monsters: Essays on Power, Technology and Domination*, ed. John Law (London: Routledge, 1991), 103–131.

15. See André Leroi-Gourhan, *Milieu et techniques* (Paris: Éditions Albin Michel, 1945).

16. See Paul Radin, "Zur Netztechnik der südamerikanischen Indianer," *Zeitschrift für Ethnologie* 36 (1906): 926–938; Isidor Scheftelowitz, *Das Schlingen- und Netzmotiv im Glauben und Brauch der Völker* (Giessen: Alfred Töpelmann, 1912); Julius Lips, *Fallensysteme der Naturvölker* (Leipzig: Ernst Wiegandt Verlagsbuchhandlung, 1927); Sture Lagercrantz, "Beiträge zur Jagdfallensystematik," *Ethnos* 6 (1937): 361–366 (in a dispute with Lips over the status of the net as trap or as a "trap-like method of catching"); and Kurt Lindner, *Geschichte des deutschen Weidwerks*, 2 vols. (Berlin: De Gruyter, 1937–1940).

17. See Herbert Marshall McLuhan, *The Mechanical Bride: Folklore of Industrial Man* (New York: Vanguard, 1951).

18. On this, see Erhard Schüttpelz, "Die ältesten in den neuesten Medien. Folklore und Massenkommunikation um 1950," *Navigationen* 6, no. 1 (2006), 33–46; Erich Hörl, *Sacred Channels: The Archaic Illusion of Communication*, trans. Nils F. Schott (Amsterdam: Amsterdam University Press, 2018); and Lorenz Engell, Bernhard Siegert, and Joseph Vogl, eds., *1950: Archiv für Mediengeschichte* (Weimar: Universitätsverlag, 2004).

19. This is all the more true, given that the period from 1933 to 1969 was the formational period for sociological network research. On that, see section 6.4 of the present volume.

20. This is impressively demonstrated in Inge Baxmann, "Der Körper als Gedächtnisort: Bewegungswissen und die Dynamisierung der Wissenskulturen im frühen 20. Jahrhundert," *Deutungsräume: Bewegungswissen als kulturelles Archiv der Moderne*, ed. Inge Baxmann and Franz Anton Cramer (Munich: K. Kieser, 2005), 15–35. Marcel Griaule also employed the entire arsenal of recording media on his African expeditions.

21. This comes out even more in Leroi-Gourhan's later books. See Leroi-Gourhan, *Gesture and Speech*, 269–270.

22. Leroi-Gourhan, *Milieu et techniques*, 92.

23. Geoffrey Winthrop-Young, "Cultural Techniques: Preliminary Remarks," *Theory, Culture & Society* 30, no. 6 (2013): 3–19.

24. Sybille Krämer and Horst Bredekamp, "Culture, Technology, Cultural Techniques: Moving Beyond Text," trans. Michael Wutz, *Theory, Culture & Society* 30, no. 6 (2013): 20–29, esp. 27 (translation modified).

25. See Thomas Macho, "Körper der Zukunft: Vom Vor- und Nachleben der Bilder," in *Bilderfragen: Die Bildwissenschaften im Aufbruch*, ed. Hans Belting (Paderborn: Wilhelm Fink, 2007), 181–194, esp. 181–182.

26. See also Thomas Macho, "Second-Order Animals: Cultural Techniques of Identity and Identification," trans. Michael Wutz, *Theory, Culture & Society* 30, no. 6 (2013): 30–47.

27. Bruno Latour, *We Have Never Been Modern*, trans. Catherine Porter (Cambridge, MA: Harvard University Press, 1993).

28. The German-language positions are collected in a differentiated way in *Zeitschrift für Medien- und Kulturforschung* 1 (2010). See especially Harun Maye, "Was ist eine Kulturtechnik?" *Zeitschrift für Medien- und Kulturforschung* 1, no. 1 (2010): 121–135. The debate is summarized in an integrative way in Christian Kassung and Thomas Macho, "Einleitung," in *Kulturtechniken der Synchronisation*, ed. Christian Kassung and Thomas Macho (Paderborn: Wilhelm Fink, 2013), 9–21. More recent impulses from a legal perspective can be found in Fabian Steinhauer, *Vom Scheiden: Geschichte und Theorie einer juristischen Kulturtechnik* (Berlin: Duncker & Humblodt, 2014); from an ethnographic perspective in Thomas Widlok, "Kulturtechniken: Ethnographisch fremd und anthropologisch fremd; Eine Kritik an ökologisch-phänomenologischen und kognitiv-modularisierenden Ansätzen," in *Universitätsforschungen zur prähistorischen Archäologie 264: Fremdheit—Perspektiven auf das Andere*, ed. Tobias L. Kienlin (Bonn: Dr. Rudolf Habelt, 2015), 41–59; from the perspective of literary studies in Jörg Dünne et al., eds. *Cultural Techniques* (Berlin: De Gruyter, 2020); and from the perspective of cultural studies in Christian Kassung, *Fleisch: Die Geschichte einer Industrialisierung* (Leiden: Brill | Schöningh, 2020), esp. 174–175; Alwin J. Cubasch, Vanessa Engelmann, and Christian Kassung, "Theorie des Filterns: Zur Programmatik eines Experimentalsystems," *Zenodo*, April 30, 2021, https://doi.org/10.5281/zenodo.4731045.

29. See Geoffrey Winthrop-Young, Ilinca Iurascu, and Jussi Parikka, eds., special issue "Cultural Techniques," *Theory, Culture & Society* 30, no. 6 (2013).

30. See Bernhard Siegert, *Cultural Techniques: Grids, Filters, Doors, and Other Articulations of the Real*, trans. Geoffrey Winthrop-Young (New York: Fordham University Press, 2015), 1–2.

31. Hartmut Böhme, "Netzwerke: Zur Theorie und Geschichte einer Konstruktion," in *Netzwerke: Eine Kulturtechnik der Moderne*, ed. Jürgen Barkhoff, Hartmut Böhme, and Jeanne Riou (Cologne: Böhlau, 2004), 17–36; Erhard Schüttpelz, "Die medienanthropologische Kehre der Kulturtechniken," in *Archiv für Mediengeschichte: Kulturgeschichte als Mediengeschichte (oder vice versa)?*, ed. Lorenz Engell, Joseph Vogl, and Bernhard Siegert (Weimar: Universitätsverlag, 2006), 87–110; Erhard Schüttpelz,

"Elemente einer Akteur-Medien-Theorie," in *Akteur-Medien-Theorie*, ed. Tristan Thielmann and Erhard Schüttpelz (Bielefeld: transcript, 2013), 9–67.

32. "Certainly, there is no such thing as 'German Media Theory,' whether old or new." Eva Horn, "Editor's Introduction: 'There Are No Media,'" *Grey Room*, no. 29 (2007): 6–13.

33. Friedrich Kittler, "Aufschreibesysteme 1800/1900. Vorwort," *Zeitschrift für Medienwissenschaft* 6 (2012): 117–126, esp. 126.

34. Tristan Thielmann and Erhard Schüttpelz, eds., *Akteur-Medien-Theorie* (Bielefeld: transcript, 2013).

35. Cornelia Vismann, "Cultural Techniques and Sovereignty," trans. Ilinca Iurascu, *Theory, Culture & Society* 30, no. 6 (2013): 83–93, esp. 83.

36. Christoph Borbach and Tristan Thielmann, "Chains of Co-operation in the 1940s: Working on the Air Situation Picture," in *Materiality of Cooperation*, ed. Sebastian Giessmann et al. (Wiesbaden: Springer VS, 2023), 91–130.

37. Erhard Schüttpelz, "The Irreducibility of Technical Skills: Before and After Science," in *Materiality of Cooperation*, ed. Sebastian Giessmann et al. (Wiesbaden: Springer VS, 2023), 307–325.

38. Michel Serres, "Introduction," in *A History of Scientific Thought: Elements of a History of Science*, ed. Michel Serres (Oxford: Blackwell, 1995), 1–16; Latour, *We Have Never Been Modern*.

39. Cf. Latour's reflection on terminology and his argument for "worknets" or "action nets" in Bruno Latour, *Reassembling the Social: An Introduction to Actor-Network-Theory* (Oxford: Oxford University Press, 2005), 132.

40. On the history of the concept, see Christian Jürgen Emden, "Netz," in *Lexikon der philosophischen Metaphern*, ed. Ralf Konersmann (Darmstadt: Wissenschaftliche Buchgesellschaft, 2007), 248–260; and Alexander Friedrich, *Metaphorologie der Vernetzung: Zur Theorie kultureller Leitmetaphern* (Paderborn: Wilhelm Fink, 2015).

41. Samuel Johnson, "Network," in *A Dictionary of the English Language: In Which the Words Are Deduced from Their Originals and Illustrated in Their Different Significations by Examples from the Best Writers [. . .]*, vol. 2, (London: Printed by W. Strahan for J. and P. Knapton et al., 1755), n.p.

42. Heinrich Gustav Flörke, "Netz," in *Oekonomische Encyclopädie; oder, Allgemeines System der Staats- Stadt- Haus- u. Landwirtschaft in alphabetischer Ordnung*, ed. D. Johann Georg Krünitz, vol. 102 (Berlin: Trassler, 1806), 444–473.

43. Johann Heinrich Zedler, ed., "Netz," in *Johann Heinrich Zedlers Grosses vollständiges Universal-Lexicon aller Wissenschafften und Künste*, vol. 23 (N–Net) (Halle: Zedler, 1740), cols. 1993–2006.

44. Gabriele Schabacher, "Im Zwischenraum der Lösungen: Reparaturarbeit und Workarounds," *ilinx: Berliner Beiträge zur Kulturwissenschaft* 4 (2017): xiii–xxviii.

45. See section 6.1 of the present volume.

46. Denis Diderot, "Le rêve de d'Alembert," in *Œuvres complètes*, vol. 17, *Le rêve de d'Alembert, Idées IV*, ed. Jean Varloot (Paris: Hermann, 1987), 89–209.

47. Vitaliano Donati, *Della storia naturale marina dell' Adriatico* (Venice: Francesco Storti, 1750).

48. Michel Serres, *The Parasite*, trans. Lawrence R. Schehr (Baltimore: Johns Hopkins University Press, 1982), 224–225.

49. Dirk Heinrich and Johann Tischler, "Fischerei und Fischereimethoden," in *Reallexikon der Germanischen Altertumskunde*, vol. 9, *Fidel–Friedlosigkeit*, ed. Johannes Hoops, 130–146 (Berlin: De Gruyter, 1995).

50. Susanne Möller-Wiering, "Maschenstoffe," in *Reallexikon der Germanischen Altertumskunde*, vol. 19, *Luchs–Metrum*, ed. Johannes Hoops (Berlin: De Gruyter, 2001), 377–383, esp. 379.

51. Arthur Seligo, "Die Fischerei in den Fliessen, Seen und Strandgewässern Mitteleuropas," in *Handbuch der Binnenfischerei Mitteleuropas*, vol. 5, ed. Reinhard Demoll and Hermann Nikolaus Maier (Stuttgart: E. Schweizerbart'sche Verlagsbuchhandlung, 1926), 1–422; Lindner, *Geschichte des deutschen Weidwerks*; Leroi-Gourhan, *Milieu et techniques*.

52. Leroi-Gourhan, *Gesture and Speech*, 106.

53. Alfred Gell, "Vogel's Net: Traps as Artworks and Artworks as Traps," in *The Art of Anthropology: Essays and Diagrams*, ed. Eric Hirsch (London: Athlone, 1999), 187–214, esp. 200–201.

54. Elisabeth von der Osten-Sacken, "Netz," in *Reallexikon der Assyriologie und Vorderasiatischen Archäologie*, vol. 9, ed. Otto Dietz Edzard (Berlin: De Gruyter, 2001), 235–242, esp. 235–236.

55. Louvre, Paris, inv. no. AO 50.

56. Gebhard Selz, *Untersuchungen zur Götterwelt des altsumerischen Stadtstaates von Lagaš* (Philadelphia: University of Pennsylvania Museum, 1995), 51–52.

57. Uwe Bartels, *Edfu: Die Darstellungen auf den Aussenseiten der Umfassungsmauer und auf den Pylonen. Strichzeichnungen und Photographien*, ed. Dieter Kurth, Die Inschriften des Tempels von Edfu. Abteilung II: Dokumentationen (Wiesbaden: Harrassowitz, 2009).

58. Mircea Eliade, *Images and Symbols: Studies in Religious Symbolism*, trans. Philip Mairet (New York: Sheed and Ward, 1961), 92–93; Scheftelowitz, *Das Schlingen- und Netzmotiv im Glauben und Brauch der Völker*

59. Aeschylus, "Agamemnon," in *The Oresteia: Agamemnon, Libation-Bearers, Eumenides*, ed. and trans. Alan H. Sommerstein (Cambridge, MA: Harvard University Press, 2008), lines 1115–1116.

60. Sylvie Ballestra-Puech, *Métamorphoses d'Arachné: L'artiste en araignée dans la littérature occidentale* (Geneva: Librairie Droz, 2006).

61. Katarzyna Michalski and Sergiusz Michalski, *Spider* (London: Reaktion, 2010), 54–55.

62. Diderot, "Le rêve de d'Alembert."

63. Galen, *On the Usefulness of the Parts of the Body / De Usu Partium*, ed. Margaret Tallmadge May (Ithaca, NY: Cornell University Press, 1968), 430–431.

64. Andreas Vesalius, *De Humani Corporis Fabrica Libri Septem* (Basel: Johann Oporinus, 1543), 310, 524, 642.

65. René Descartes, "L'homme," in *Œuvres philosophiques 1618–1637*, vol. 1 (Paris: Éditions Garnier, 1988), 441–443.

66. Richard Sennett, *Flesh and Stone: The Body and the City in Western Civilization* (New York: W. W. Norton, 1994), 255–270; Pierre Musso, *Télécommunications et philosophie des réseaux: La posterité paradoxale de Saint-Simon* (Paris: Presses Universitaires de France, 1997), 34–41; Giessmann, *Netze und Netzwerke*, 33–34, 71–72.

67. Thomas Becker, *Mann und Weib—schwarz und weiss: Die wissenschaftliche Konstruktion von Geschlecht und Rasse, 1650–1900* (Frankfurt am Main: Campus, 2005), 185–186.

68. Giessmann, *Netze und Netzwerke*, 33–34.

69. Fernand Braudel, *The Mediterranean and the Mediterranean World in the Age of Philipp II*, trans. Siân Reynolds, 2 vols. (London: Collins, 1972), esp. vol. 1, 276.

70. Michael Giesecke, *Der Buchdruck in der frühen Neuzeit: Eine historische Fallstudie über die Durchsetzung neuer Informations- und Kommunikationstechnologien* (Frankfurt am Main: Suhrkamp, 1994), 358.

71. Wolfgang Behringer, *Im Zeichen des Merkur: Reichspost und Kommunikationsrevolution in der Frühen Neuzeit* (Göttingen: Vandenhoeck & Ruprecht, 2003). Cf. Wolfgang Behringer, "Netzwerk," in *Enzyklopädie der Neuzeit*, vol. 9, *Naturhaushalt–Physiokratie* (Stuttgart: J. B. Metzler, 2009), 98–100.

72. Joad Raymond and Noah Moxham, eds., *News Networks in Early Modern Europe* (Leiden: Brill, 2016); Robert Darnton, *Poetry and the Police: Communication Networks in Eighteenth-Century Paris* (Cambridge, MA: Belknap Press of Harvard University Press, 2010).

73. Gabriel Lamé et al., *Vues politiques et pratiques sur les travaux publics* (Paris: Everat, 1832), 81–82.

74. Michel Chevalier, "Système de la Méditerranée," *Le Globe: Journal de la religion saint-simonienne* 8, no. 43 (February 12, 1832): 169–171.

75. Pierre-Alexandre d'Allent, "Essai de reconnaissance militaire," *Flux: Cahiers scientifiques internationaux réseaux et territoires* (1989): 15–17. Cf. André Guillerme, "Réseau: genèse d'un mot," *Cahiers de la Médiologie* 3 (1997): 7–16.

76. Raymond Génieys, *Note sur un projet de distribution générale de l'eau dans l'intérieur de Paris* (Paris: Carilian-Goeury, 1827), xi–xii.

77. Abraham Chappe and René Chappe, *Mémoire sur la télégraphie* (Paris: Béthune, 1829); Patrice Flichy, *Dynamics of Modern Communication: The Shaping and Impact of New Communication Technologies*, trans. Liz Libbrecht (Thousand Oaks, CA: Sage, 1995), 30.

78. Antoine Picon, *Les saint-simoniens: Raison, imaginaire et utopie* (Paris: Belin, 2002).

79. Prosper Enfantin, *Colonisation de l'Algérie* (Paris: P. Bertrand, 1843), 293–294, 402, 429–430.

80. Erhard Schüttpelz, "Ein absoluter Begriff: Zur Genealogie und Karriere des Netzwerkkonzepts," in *Vernetzte Steuerung: Soziale Prozesse im Zeitalter technischer Netzwerke*, ed. Stefan Kaufmann (Zurich: Chronos, 2007), 25–46.

81. Thomas Parke Hughes, *Networks of Power: Electrification in Western Society, 1880–1930* (Baltimore: Johns Hopkins University Press, 1983).

82. Georg Simmel, "Space and the Spatial Ordering of Society," in *Sociology: Inquiries into the Construction of Social Forms*, ed. and trans. Anthony J. Blasi, Anton K. Jacobs, and Mathew Kanjirathinkal (Leiden: Brill, 2009), 549–626, esp. 565; Pierre Mercklé, *Sociologie des réseaux sociaux* (Paris: La Découverte, 2004).

83. Gabriel Tarde, *The Laws of Imitation*, trans. Elsie Clews Parsons (New York: Henry Holt, 1903).

84. Jacob Levy Moreno and Helen Jennings, *Who Shall Survive? Foundations of Sociometry, Group Psychotherapy and Sociodrama*, 2nd ed. (Beacon, NY: Beacon House, 1953).

85. Cf. section 6.4 of this book.

86. Ulf Hannerz, *Exploring the City: Inquiries toward an Urban Anthropology* (New York: Columbia University Press, 1980), 163–164; Christopher McCarty and José Luis Molina, "Innovative Use of Social Network Analysis in Cultural Anthropology," in *Knoten und Kanten 2.0. Soziale Netzwerkanalyse in Medienforschung und Kulturanthropologie*, ed. Markus Gamper, Linda Reschke, and Michael Schönhuth, 217–230 (Bielefeld: transcript, 2012).

87. Alfred Reginald Radcliffe-Brown, *Structure and Function in Primitive Society: Essays and Addresses*, 6th ed. (Glencoe, IL: Free Press, 1952), 191.

88. J. Clyde Mitchell, ed., *Social Networks in Urban Situations: Analyses of Personal Relationships in Central African Towns* (Manchester: Manchester University Press, 1969); Lyn Schumaker, *Africanizing Anthropology: Fieldwork, Networks, and the Making of Cultural Knowledge in Central Africa* (Durham, NC: Duke University Press, 2001).

89. Cf. section 6.3 of this book.

90. Duncan Watts, *Six Degrees: The Science of a Connected Age* (New York: W. W. Norton, 2003); Albert-László Barabási, *Linked: The New Science of Networks*

(Cambridge, MA: Perseus, 2002); Albert-László Barabási with Márton Pósfai, *Network Science* (Cambridge: Cambridge University Press, 2016).

91. Alexander Galloway, *Protocol: How Control Exists after Decentralization* (Cambridge, MA: MIT Press, 2004), 205–229.

92. Fred Turner, *From Counterculture to Cyberculture: Stewart Brand, the Whole Earth Network and the Rise of Digital Utopianism* (Chicago: University of Chicago Press, 2006).

93. Manuel Castells, *The Rise of the Network Society*, 2nd ed. (Oxford: Blackwell, 2000).

94. Luc Boltanski and Ève Chiapello, *The New Spirit of Capitalism*, trans. Gregory Elliott (London: Verso, 2007).

95. Marilyn Strathern, "Cutting the Network," *Journal of the Royal Anthropological Institute* 2, no. 3 (1996): 517–535. Cf. chapter 11 of the present volume.

96. Urs Stäheli, "Entnetzt Euch! Praktiken und Ästhetiken der Anschlusslosigkeit," in *Kunst und Arbeit: Zum Verhältnis von Ästhetik und Arbeitsanthropologie vom 18. Jahrhundert bis zur Gegenwart*, ed. Anja Lemke and Alexander Weinstock (Paderborn: Wilhelm Fink, 2014), 77–97.

97. Schüttpelz, "Ein absoluter Begriff."

98. Urs Stäheli, *Soziologie der Entnetzung* (Berlin: Suhrkamp, 2021).

99. Marcel Mauss, "Techniques of the Body." *Economy and Society* 2, no. 1 (1973): 70–88.

100. Donna Haraway, "A Cyborg Manifesto: Science, Technology and Socialist-Feminism in the Late Twentieth Century," in *Simians, Cyborgs and Women. The Reinvention of Nature* (New York: Routledge, 1991), 149–181.

CHAPTER 3

1. See Manuel Castells, *The Rise of the Network Society*, 2nd ed. (Oxford: Blackwell, 2000), 38–39.

2. See the chapters on the history of the network diagram in the present volume.

3. One important starting point was and is Albert-László Barabási, *Linked: The New Science of Networks* (Cambridge, MA: Perseus, 2002). On the growing reference literature, see Wikipedia, s.v. "Complex Networks," https://en.wikipedia.org/wiki/Complex_networks.

4. John Guare, *Six Degrees of Separation*, 2nd ed. (New York: Vintage, 1994). For an explicit borrowing for popular science, which misunderstands Guare's clever title, see Duncan Watts, *Six Degrees: The Science of a Connected Age* (New York: W. W. Norton, 2003). By contrast, specialized research in graph theory and the specific formulations of the analysis of social networks offer more subtle arguments.

5. On this, see Erhard Schüttpelz, "Elemente einer Akteur-Medien-Theorie," in *Akteur-Medien-Theorie*, ed. Tristan Thielmann and Erhard Schüttpelz (Bielefeld: transcript, 2013), 7–67.

6. On this concept, see Gustav Rossler, "Galerie neuer Dingbegriffe: Hybriden, Quasi-Objekte, Grenzobjekte, epistemische Dinge," in *Bruno Latours Kollektive: Kontroversen zur Entgrenzung des Sozialen*, ed. Georg Kneer, Markus Schroer, and Erhard Schüttpelz (Frankfurt am Main: Suhrkamp, 2008), 76–107. In actor–network theory, the network and the quasi-object become complementary concepts. See Rossler, "Galerie neuer Dingbegriffe," 84.

7. See Michel Serres, *The Parasite*, trans. Lawrence R. Schehr (Baltimore: Johns Hopkins University Press, 1982), 224–225; and Michel Serres, *The Natural Contract*, trans. Elizabeth MacArthur and William Paulson (Ann Arbor: University of Michigan Press, 1995), 105–106.

8. See Serres, *The Parasite*, 226.

9. See Gunnar Schmidt, *Ästhetik des Fadens: Zur Medialisierung eines Materials in der Avantgardekunst* (Bielefeld: transcript, 2007); Julia Gelshorn and Tristan Weddigen, "Das Netzwerk: Zu einem Denkbild in Kunst und Wissenschaft," in *Grammatik der Kunstgeschichte: Sprachproblem und Regelwerk im Bild-Diskurs: Oskar Bätschmann zum 65. Geburtstag*, ed. Hubert Locher and Peter J. Schneemann (Zurich: Schweizerisches Institut für Kunstwissenschaft, 2008), 54–77; and many of the contributions to the Venice Biennale in 2009. The works of Tomás Saraceno in particular emphasize materialized nets as a general condition of the surroundings. See Tomás Saraceno, *Cloud Cities*, ed. Marion Ackermann et al. (Berlin: Distanz, 2011).

10. See Andréa Belliger and David J. Krieger, "Einführung in die Akteur-Netzwerk-Theorie," in *ANThology: Ein einführendes Handbuch zur Akteur-Netzwerk-Theorie*, ed. Andréa Belliger and David J. Krieger (Bielefeld: transcript, 2006), 13–50, esp. 40, 44.

11. See Alexander Friedrich, "Vernetzte Zwischenräume," in *Bewegen im Zwischenraum*, ed. Uwe Wirth (Berlin: Kadmos, 2012), 55–74.

12. See Wolfgang Hagen, "Metaxy: Eine historiosemantische Fussnote zum Medienbegriff," in *Was ist ein Medium?* ed. Stefan Münker and Alexander Roesler (Frankfurt am Main: Suhrkamp, 2008), 13–29, esp. 23.

13. See Aristotle, *De Anima*, ed. and trans. Christopher Shields (Oxford: Clarendon Press, 2016), 419a20, 421b9, 422b22, 423a16, 423b26, 434b29, 435a17. Christopher Shields translates "to metaxy" as "medium," as has been common since Thomas Aquinas. On this, see Hagen "Metaxy," and the essays in *Medien vor den Medien*, ed. Friedrich Kittler and Ana Ofak (Paderborn: Wilhelm Fink, 2007).

14. See Hartmut Böhme, "Netzwerke: Zur Theorie und Geschichte einer Konstruktion," in *Netzwerke: Eine Kulturtechnik der Moderne*, ed. Jürgen Barkhoff, Hartmut Böhme, and Jeanne Riou (Cologne: Böhlau, 2004), 17–36, esp. 21. See also Walter

Seitter, *Die Physik der Medien: Materialien, Apparate, Präsentierungen* (Weimar: Verlag und Datenbank für Geisteswissenschaften, 2002), 23, 32.

15. See Sigrid Schwenk, *Zur Terminologie des Vogelfangs im Deutschen: Eine sprachliche Untersuchung aufgrund der deutschen didaktischen Literatur des 14. bis 19. Jahrhunderts* (Clausthal-Zellerfeld: Bönecke-Druck, 1967), 69–70. See also Kurt Lindner, *Geschichte des deutschen Weidwerks*, vol. 2, *Die Jagd im frühen Mittelalter* (Berlin: De Gruyter, 1940).

16. See Warren S. McCulloch, "A Heterarchy of Values Determined by the Topology of Nervous Nets," *Bulletin of Mathematical Biophysics* 7, no. 2 (1945): 89–93, esp. 93. See also, more generally, Heinz von Foerster and Bernhard Pörksen, *Wahrheit ist die Erfindung eines Lügners: Gespräche für Skeptiker*, 4th ed. (Heidelberg: Carl Auer, 2001), 86–87.

17. François Jacob, *The Logic of Life: A History of Heredity*, trans. Betty E. Spillmann (New York: Random House 1973), 300.

18. At the same time, biology was seized by information theory—a veritable intersection, as Evelyn Fox Keller has impressively shown. See Evelyn Fox-Keller, *Refiguring Life: Metaphors of Twentieth-Century Biology* (New York: Columbia University Press, 1995), 89–90.

19. See Katie Hafner and Matthew Lyon, *Where Wizards Stay Up Late: The Origins of the Internet* (New York: Touchstone, 1998), 56–57. See also Jochen Koubek, "Vernetzung als kulturelles Paradigma" (PhD diss., Humboldt Universität zu Berlin, 2003), 59, https://edoc.hu-berlin.de/handle/18452/15629.

20. See Jörg Potthast, *Die Bodenhaftung der Netzwerkgesellschaft: Eine Ethnografie von Pannen an Grossflughäfen* (Bielefeld: transcript, 2007).

21. See "protocol, n." *OED Online*, March 2021, Oxford University Press, https://www.oed.com/view/Entry/153243?rskey=3NlvLN&result=1&isAdvanced=false, and Wolfgang Pfeifer, ed., *Etymologisches Wörterbuch des Deutschen*, 7th ed. (Munich: dtv, 2004), 1052. On the genealogy of the terms, see also chapter 9 of the present volume.

22. See Alexander Galloway, *Protocol: How Control Exists after Decentralization* (Cambridge, MA: MIT Press, 2004), 8.

23. On the changeability and dynamics of protocols, see Lisa Gitelman, *Always Already New: Media, History, and the Data of Cultures* (Cambridge, MA: MIT Press, 2006), 5–6.

24. Manuel Castells, *The Internet Galaxy: Reflections on the Internet, Business, and Society* (Oxford: Oxford University Press, 2001), 228.

25. See Castells, *The Rise of the Network Society*, 502.

26. See chapter 5 in the present volume.

27. Bruno Latour, "On Technical Mediation: Philosophy, Sociology, Genealogy," *Common Knowledge* 3, no. 2 (1994): 29–64, esp. 53.

28. See Erhard Schüttpelz, "Der Punkt des Archimedes: Einige Schwierigkeiten des Denkens in Operationsketten," in *Bruno Latours Kollektive: Kontroversen zur Entgrenzung des Sozialen*, ed. Georg Kneer, Markus Schroer, and Erhard Schüttpelz (Frankfurt am Main: Suhrkamp, 2008).

29. Michel Callon, "Techno-Economic Networks and Reversibility," in *A Sociology of Monsters?* ed. John Law (London: Routledge, 1991), 132–161, esp. 145.

30. See Callon, "Techno-Economic Networks and Reversibility," 156.

31. Michel Callon to the author, e-mail, August 28, 2012. This goes back to the point made by Michel Callon, "Some Elements of a Sociology of Translation: Domestication of the Scallops and the Fishermen of St Brieuc Bay," in *Power, Action and Belief: A New Sociology of Knowledge?* ed. John Law (London: Routledge, 1986), 196–233.

32. Ludwik Fleck, *Genesis and Development of a Scientific Fact*, ed. Thaddeus J. Trenn and Robert K. Merton, trans. Fred Bradley and Thaddeus J. Trenn (Chicago: University of Chicago Press, 1981), 79 (emphasis added).

33. See Alexander Galloway and Eugene Thacker, *The Exploit: A Theory of Networks* (Minneapolis: University of Minnesota Press, 2007), 33.

34. See Anna Munster, *An Aesthesia of Networks: Conjunctive Experience in Art and Technology* (Cambridge, MA: MIT Press, 2013).

35. Michel Serres and Bruno Latour, *Conversations on Science, Culture, and Time*, trans. Roxanne Lapidus (Ann Arbor: University of Michigan Press, 1995), 109.

36. See Callon, "Techno-Economic Networks and Reversibility," 149–152.

37. On this in general, see Niklas Luhmann, for whom synchronization binds the medium of time to forms in an attempt to find suitable forms. See Niklas Luhmann, "Gleichzeitigkeit und Synchronisation," in *Soziologische Aufklärung 5: Konstruktivistische Perspektiven*, 2nd ed. (Wiesbaden: Westdeutscher Verlag, 1993), 95–130, esp. 118.

38. See J. T. Fraser, *Time, the Familiar Stranger* (Amherst: University of Massachusetts Press, 1987), 310–311; and Peter Galison, *Einstein's Clocks, Poincarés Maps: Empires of Time* (New York: W. W. Norton, 2003).

39. See Christian Kassung, *Das Pendel: Eine Wissensgeschichte* (Paderborn: Wilhelm Fink, 2007), 318. On the relationship of simultaneity and synchrony, see Kassung, *Das Pendel*, 116–117.

40. Following Arkady Pikovsky, Michael Rosenblum, and Jürgen Kurths, *Synchronization: A Universal Concept in Nonlinear Sciences* (Cambridge: Cambridge University Press, 2001), 8. Self-synchronization is the optimal goal. See Stefan Kaufmann, "Netzwerk," in *Glossar der Gegenwart*, ed. Ulrich Bröckling, Susanne Krasmann, and Thomas Lemke (Frankfurt am Main: Suhrkamp, 2004), 182–189, esp. 187.

41. See Uwe Wirth, "Kultur als Pfropfung: Pfropfung als Kulturmodell; Prolegomena zu einer allgemeinen Greffologie (2.0)," in *Impfen, Pfropfen, Transplantieren*, ed. Uwe Wirth (Berlin: Kadmos, 2011), 9–27, esp. 15.

42. Markus Krajewski, *World Projects: Global Information before World War I*, trans. Charles Marcrum II (Minneapolis: University of Minnesota Press, 2014), 13.

43. On the metaphor of the parasite as an expert in grafting to existing relationships, see Serres, *The Parasite*, 38–39.

44. It is no coincidence that transformation theory is at home, above all, in political science. For an approach from cultural theory, see Hartmut Böhme et al., eds., *Transformation: Ein Konzept zur Erforschung kulturellen Wandels* (Paderborn: Wilhelm Fink, 2011).

45. For a more detailed typology of cultural transformation processes, see Bergemann et al., "Transformation: Ein Konzept zur Erforschung kulturellen Wandels," in Böhme et al., *Transformation*, 39–56.

46. Bruno Latour, *We Have Never Been Modern*, trans. Catherine Porter (Cambridge, MA: Harvard University Press, 1993), 48.

47. See Geert Lovink, *The Principle of Notworking: Concepts in Critical Internet Culture* (Amsterdam: Hogeschool van Amsterdam, 2005); and Tony D. Sampson and Jussi Parikka, "Learning from Network Dysfunctionality: Accidents, Enterprise, and Small Worlds of Infection," in *A Companion to New Media Dynamics*, ed. John Hartley, Jean Burgess, and Axel Bruns (Oxford: Blackwell, 2013), 450–460.

48. See Albert Kümmel and Erhard Schüttpelz, eds., *Signale der Störung* (Munich: Wilhelm Fink, 2003).

49. See Sebastian Giessmann, "Netzstörungen: Erzählungen vom Ende der Netzwerke," *Störfälle: Zeitschrift für Kulturwissenschaften* 2 (2011): 125–133.

50. See Potthast, *Die Bodenhaftung der Netzwerkgesellschaft*, 187–188.

51. See Kümmel and Schüttpelz, *Signale der Störung*, 9.

52. See Eric S. Raymond, *The New Hacker's Dictionary*, 2nd ed. (Cambridge, MA: MIT Press, 1997), 323–324.

53. See Marilyn Strathern, "Cutting the Network," *Journal of the Royal Anthropological Institute* 2, no. 3 (1996): 517–535, esp. 531–532. On this, see section 11.2 of the present volume.

54. See Jörg Sydow and Arnold Windeler, "Steuerung von und in Netzwerken: Perspektiven, Konzepte, vor allem aber offene Fragen," in *Steuerung von Netzwerken: Konzepte und Praktiken*, ed. Jörg Sydow and Arnold Windeler (Wiesbaden: Westdeutscher Verlag, 2000), 1–24.

55. See Samuel M. Weber, *Targets of Opportunity: On the Militarization of Thinking* (New York: Fordham University of Press, 2005), 90–91.

56. See Sebastian Giessmann, *Netze und Netzwerke: Archäologie einer Kulturtechnik, 1740–1840* (Bielefeld: transcript, 2006).

CHAPTER 4

1. Lucien Febvre and Marc Bloch, "Ports, fleuves, mers," *Annales d'histoire sociale* 1 (1940): 66–70, esp. 70: "La méditerranée, ce sont des routes. Des positions sur ces routes. Un moyen, plutôt qu'une fin."

2. See Fernand Braudel, *The Mediterranean and the Mediterranean World in the Age of Philip II*, trans. Siân Reynolds, 2 vols. (London: Collins, 1972), vol. 1, 276.

3. Braudel, *The Mediterranean and the Mediterranean World*, 1:277, 281.

4. Braudel, *The Mediterranean and the Mediterranean World*, 1:319–320.

5. On the history of transportation and its fluctuations, see Braudel, *The Mediterranean and the Mediterranean World*, 1:295.

6. On the inertia of action at a distance, see Braudel, *The Mediterranean and the Mediterranean World*, 1:360–361.

7. See Braudel, *The Mediterranean and the Mediterranean World*, 1:312, 355.

8. Braudel, *The Mediterranean and the Mediterranean World*, 1:164.

9. Commission Imperiale, ed., *Exposition Universelle de 1867 à Paris: Catalogue general. 6ᵉ livraison. Instruments et procédés des art usuels* (Paris: E. Dentu, 1867), 289.

10. On the history of its construction, see Nathalie Montel, "Le chantier du Canal de Suez (1859–1869): Techniques et techniciens de travaux publics du XIXème siècle" (PhD diss., Université de Lille, 1995), 610.

11. Leonz Niderberger, *Der Suezkanal: Seine Geschichte, Lage und Bedeutung für den Weltverkehr und im Weltkrieg* (Limburg an der Lahn: Steffen, 1916), 5.

12. See Philipp Felsch, "Aufsteigesysteme, 1800–1900," in *Nach Feierabend: Zürcher Jahrbuch für Wissensgeschichte* 1, Bilder der Natur—Sprachen der Technik (Zurich: diaphanes, 2005), 15–32; and Charlotte Bigg, "The Panorama, or, La Nature à Coup d'Œil," in *Observing Nature—Representing Experience: The Osmotic Dynamics of Romanticism, 1800–1850*, ed. Erna Fiorentini (Berlin: Reimer, 2007), 73–95.

13. On the panorama, see Stephan Oettermann, *The Panorama: History of a Mass Medium*, trans. Deborah Lucas Schneider (New York: Zone Books, 1997); and Jörg Brauns, *Schauplätze: Zur Architektur visueller Medien* (Berlin: Kadmos, 2007), 11–12.

14. See Jonathan Crary, *Techniques of the Observer: On Vision and Modernity in the Nineteenth Century* (Cambridge, MA: MIT Press, 1990), 24.

15. See Bernhard Siegert, *Cultural Techniques: Grids, Filters, Doors, and Other Articulations of the Real*, trans. Geoffrey Winthrop-Young (New York: Fordham University Press, 2015), chap. 5. See also Bernhard Siegert, "Repräsentationen diskursiver

Räume: Einleitung," in *Topographien der Literatur: Deutsche Literatur im transnationalen Kontext*, ed. Hartmut Böhme (Stuttgart: J. B. Metzler, 2005), 3–11, esp. 6–7; and Robert Stockhammer, *Kartierung der Erde: Macht und Lust in Karten und Literatur* (Paderborn: Wilhelm Fink, 2007).

16. On the praxeology of cartography, see Steffen Siegel and Petra Weigel, eds., *Die Werkstatt des Kartographen: Materialien und Praktiken visueller Welterzeugung* (Paderborn: Wilhelm Fink, 2011).

17. See Joseph Vogl, "Regierung und Regelkreis. Historisches Vorspiel," in *Cybernetics/Kybernetik: The Macy Conferences, 1946–1953*, ed. Claus Pias, vol. 2 (Zurich: diaphanes, 2004), 67–79.

18. Michel Foucault, *Security, Territory, Population: Lectures at the Collège de France, 1977–78*, ed. Michel Sennelart, trans. Graham Burchell (New York: Palgrave Macmillan, 2009), 1–86.

19. Foucault, *Security, Territory, Population*, 45.

20. Foucault, *Security, Territory, Population*, 21.

21. See Foucault, *Security, Territory, Population*, 18. It is about controlling the causes and effects of circulations. See Foucault, *Security, Territory, Population*, 21. Foucault's understanding of "circulation," whose origin he scarcely justifies himself, derives from the discourse on political economy. It was presumably decisively inspired by Marx's concept of acts of circulation in the process of capital. Cf. Claudia Aradau and Tobias Blanke, "Governing Circulation: A Critique of the Biopolitics of Security," in *Security and Global Governmentality: Globalization, Governance and the State*, ed. Miguel de Larrinaga and Marc G. Doucet (London: Routledge, 2010), http://oro.open.ac.uk/21481/2/security_circulationAradauBlanke.pdf, 3, 6–7.

22. On this, see the historical sketch in Sebastian Giessmann, *Netze und Netzwerke: Archäologie einer Kulturtechnik, 1740–1840* (Bielefeld: transcript, 2006). The proposed use of the concept of the network follows the historical sources and not so much the theoretical approach that Foucault reserved for the disciplinary society and the dynamic makeup of the apparatus.

23. On the Count of Saint-Simon, see Émile Durkheim, *Socialism and Saint-Simon*, ed. Alvin W. Gouldner, trans. Charlotte Sattler (Yellow Springs, OH: Antioch, 1958); Martinus Emge, *Saint-Simon: Einführung in ein Leben und Werk, eine Schule, Sekte und Wirkungsgeschichte* (Munich: Oldenbourg, 1987); and the autobiography Claude Henri de Saint-Simon, "Histoire de ma vie," in *Œuvres de Claude-Henri de Saint-Simon*, vol. I/1 (Paris: Éditions anthropos, 1966), 64–88. Pierre Musso was the first to call attention to Saint-Simon's significance for the history of global networking. See Pierre Musso, *Télécommunications et philosophie des réseaux: La posterité paradoxale de Saint-Simon* (Paris: Presses Universitaires de France, 1997). See also Giessmann, *Netze und Netzwerke*, 86–87.

24. See Nathalie Coilly and Philippe Régnier, eds., *Le siècle des saint-simoniens: Du nouveau christianisme au canal de Suez* (Paris: Bibliothèque Nationale de France, 2006), 23,

52, 147. See also Claire Goldberg Moses and Leslie Wahl Rabine, *Feminism, Socialism, and French Romanticism* (Bloomington: Indiana University Press, 1993); and Michèle Riot-Sarcey, *Histoire du feminisme* (Paris: La Decouverte, 2002), 29–30; Pamela Pilbeam, *Saint-Simonians in Nineteenth-Century France: From Free Love to Algeria* (New York: Palgrave Macmillan, 2013).

25. See Walter Benjamin, *The Arcades Project*, trans. Howard Eiland and Kevin McLaughlin (Cambridge, MA: Belknap Press of Harvard University Press, 2002) convolute U, 571–602.

26. Benjamin, *The Arcades Project*, 594.

27. See Gilles Jacoud, ed., *Political Economy and Industrialism: Banks in Saint-Simonian Economic Thought* (London: Routledge, 2010).

28. See, for example, *Le Globe* of April 2, 9, 11, and 13, 1832, written by Michel Chevalier, Stéphane Flachat, and Henri Fournel.

29. See William Harvey, *Exercitatio anatomica de motu cordis et sanguinis in animalibus* (Frankfurt am Main: Fitzer, 1628). Translated as Harvey, *On the Motion of the Heart and Blood in Animals: A New Edition of William Harvey's "Exercitatio anatomica de motu cordis et sanguinis in animalibus,"* ed. Jarrett A. Carty, trans. Robert Willis (Eugene, OR: Wipf and Stock, 2016).

30. Philippe Régnier, ed., *Le livre nouveau des Saint-Simoniens: Manuscrits d'Émile Barrault, Michel Chevalier, Charles Duveyrier, Prosper Enfantin, Charles Lambert, Léon Simon et Thomas-Ismayl Urbain, 1832–1833* (Tusson: Lérot, 1991), 225. For a summary of the *longue durée* of political metaphors of the body, see Albrecht Koschorke et al., *Der fiktive Staat: Konstruktionen des politischen Körpers in der Geschichte Europas* (Frankfurt am Main: Fischer, 2007). On the connection of the body politic and landscape, see Martin Warnke, *Political Landscape: The Art History of Nature*, trans. David McLintock (Cambridge, MA: Harvard University Press, 1995), 89–90.

31. Régnier, *Le livre nouveau des Saint-Simoniens*, 229–230.

32. John Tresch has identified this temple as a "cosmogram" in which the worldview of the Saint-Simonians finds its apt expression. See John Tresch, *The Romantic Machine: Utopian Science and Technology after Napoleon* (Chicago: University of Chicago Press, 2012), chap. 7.

33. Riot-Sarcey, *Histoire du feminisme*, 31–32; Moses and Rabine, *Feminism, Socialism, and French Romanticism*, 30–75.

34. See Coilly and Régnier, *Le siècle des Saint-Simoniens*, 52. *La femme libre* was continued, with Suzanne Voilquin as editor in chief, as *L'Apostalat des femmes*, and from 1832 to 1834 as *Tribune des femmes*. See Riot-Sarcey, *Histoire du feminisme*, 31; and Moses and Rabine, *Feminism, Socialism, and French Romanticism*, 349.

35. See Richard Sennett, *Flesh and Stone: The Body and the City in Western Civilization* (New York: W. W. Norton, 1994), 255–256.

36. Antoine Picon, *Les saint-simoniens: Raison, imaginaire et utopie* (Paris: Belin, 2002), 225.

37. Foucault, *Security, Territory, Population*, 21. Foucault does not make explicit here the obvious reference to the role of the milieu in Braudel's *géohistoire*.

38. See Sennett, *Flesh and Stone*, 324–325, 329–332, on Haussmann's three networks of different streets and their specific forms of mobilization and social controls.

39. See Foucault, *Security, Territory, Population*, 45.

40. Raymond Génieys, *Note sur un projet de distribution générale de l'eau dans l'intérieur de Paris* (Paris: Carilian-Goeury, 1827), xi–xii. On this development, see André Guillerme, "L'émergence du concept du réseau (1820–1830)," in *Réseaux territoriaux*, ed. Georges Dupuy (Caen: Paradigme, 1988), 33–49.

41. In military history, these water systems were preceded by plans in 1821 for the geographical organization of French fortification. See André Guillerme, "Réseau: genèse d'un mot," *Cahiers de la Médiologie* 3 (1997): 7–16, esp. 8. The first evidence of a human-made "réseau de communications" is also found in a military context, in the work of Pierre-Alexandre d'Allent, an engineer of the Mézières school. See Pierre-Alexandre d'Allent, "Essai de reconnaissance militaire" (1802), in "Réseaux: Anthologie, 1781–1963," special issue, *Flux: Cahiers scientifiques internationaux réseaux et territoires* (1989), ed. GDR 903 Réseaux, 15–17, esp. 16; orig. pub in *Mémorial topographique et militaire*.

42. Abraham Chappe and René Chappe, *Mémoire sur la télégraphie* (Paris: Béthune, 1829), January, Musée de la Poste, Paris, D 6812. Cf. Patrice Flichy, *Dynamics of Modern Communication: The Shaping and Impact of New Communication Technologies*, trans. Liz Libbrecht (Thousand Oaks, CA: Sage, 1995), 30.

43. For the history of science of this project, which provided fresh water and the transportation of goods by means of a canal, see Frédéric Graber, "Purity and Theory: Theoretical Tools at Ponts et Chaussées, circa 1800," *Technology and Culture* 49, no. 4 (2008): 860–883.

44. Victor Hugo, *Les Misérables*, trans. Charles E. Wilbour (New York: Modern Library, 1902), 1057.

45. See Hugo, *Les Misérables*, 1057–1059.

46. Hugo, *Les Misérables*, 1066. The mapping of subterranean Paris can be followed in an impressive folio format of 100 cm × 65 cm, in Eugène de Fourcy, *Atlas souterrain de la ville de Paris exécuté conformément au vote émis en 1853 par commission municipale et suivant les ordres de M. le Bon G.E. Haussmann . . .* (Paris: Charles de Mourgues Frères, 1859).

47. Régnier, *Le livre nouveau des Saint-Simoniens*, 251.

48. Michel Chevalier, "Fin du choléra par coup d'état," *Le Globe: Journal de la religion saint-simonienne* 8, no. 102 (April 11, 1832): 405; "Système de la Méditerranée," *Le Globe: Journal de la religion saint-simonienne* 8, no. 43 (1832): 169–171.

49. Chevalier's text does indeed have an implicit plan for infrastructure. The traditions of such programs also lead to the establishment of the governmental concept of infrastructure in France in 1875. On this, see Dirk van Laak, "Der Begriff Infrastruktur und was er vor seiner Erfindung besagte," *Archiv für Begriffsgeschichte* 41 (1999): 280–299.

50. The talk of spatial structures of the first, second, and third order becomes part of France's territorial politics in the nineteenth century. Direct links to Chevalier are made, for example, by Gabriel Lamé et al., *Vues politiques et pratiques sur les travaux publics* (Paris: Everat, 1832), 81–82.

51. See Markus Krajewski, *World Projects: Global Information Before World War I*, trans. Charles Marcrum II (Minneapolis: University of Minnesota Press, 2014).

52. On this, see, in a critical continuity with Edward Said, the essays in Michael Bernsen and Martin Neumann, eds., *Die französische Literatur des 19. Jahrhunderts und der Orientalismus* (Tübingen: Max Niemeyer, 2006).

53. André-Marie Ampère, *Essai sur la philosophie des sciences; ou, Exposition analytique d'une classification naturelle de toutes les connaissances humaines* (Paris: Bachelier, 1834), 141.

54. Prosper Enfantin, *Colonisation de l'Algérie* (Paris: P. Bertrand, 1843), 293–294, 402, and 429–430.

55. Krajewski, *World Projects*, 15–16.

56. Friedrich Ratzel, *Politische Geographie*, 3rd ed. (Munich: Oldenbourg, 1923 orig. pub. 1897), 325.

57. Ratzel, *Politische Geographie*, 380.

58. Régnier, *Le livre nouveau des Saint-Simoniens*, 251–263.

59. Régnier, *Le livre nouveau des Saint-Simoniens*, 250.

60. Gerhard Herrmann, *Der Suezkanal* (Bern: Wilhelm Goldmann, 1936), 13.

61. Prosper Enfantin to Émile Barrault, August 8, 1833, "Bibliothèque de l'Arsenal, Paris, Fonds Enfantin 7619, f.3r, quoted in Philippe Régnier, "Le mythe oriental des saint-simoniens," in *Les Saint-Simoniens et l'Orient: Vers la modernité*, ed. Magali Morsy (Aix en Provence: Édisud, 1989), 29–49, esp. 40.

62. On the Saint-Simonian imagination of the Orient as "material and feminine," see Tresch, *The Romantic Machine*, 202.

63. Clorinde Rogé, Caroline Carbonnel (who also moved in with Enfantin), and Judith Grégoire, who lived with Lambert. See Moses and Rabine, *Feminism, Socialism, and French Romanticism*, 55.

64. See Montel, "Le chantier du Canal de Suez," 405–406.

65. Henry-René d'Allemagne, *Prosper Enfantin et les grandes entreprises du XIXe siècle* (Paris: Gründ, 1935), 93. For a German-language perspective, see Alfred

Birk and Karl Hermann Müller-Hamburg, *Der Suezkanal: Seine Geschichte und seine wirtschaftspolitische Bedeutung für Europa, Indien und Ägypten* (Hamburg: Boysen & Maasch, 1925), 1–2; and, in general, Philippe Régnier, *Les Saint-Simoniens en Egypte, 1833–1851* (Cairo: Banque de l'Union Européene, 1989).

66. D'Allemagne, *Prosper Enfantin et les grandes entreprises du XIXe siècle*, 88.

67. See Carl H. Jacobs, *Die Schiffahrtsfreiheit im Suezkanal* (Göttingen: E.A. Huth, 1912), 32. On exterritoriality, see Caroline Piquet, *Histoire du canal de Suez* (Paris: Perrin, 2009), 99.

68. Countries have also been elevated to neutrality status under international law, such as Switzerland (1815), Belgium (1831–1839), and Luxembourg (1867).

69. Georg Simmel, "The Philosophy of Landscape," trans. Josef Bleicher, *Theory, Culture & Society* 24, nos. 7–8 (2007): 20–29, esp. 23.

70. I wish to sincerely thank Patrick Ramponi for this idea.

71. Claude Elwood Shannon, "A Mathematical Theory of Communication," *The Bell System Technical Journal* 27, no. 3 (1948): 379–423, esp. 381.

72. See Albert Kümmel, "Mathematische Medientheorie," in *Medientheorien: Eine Einführung*, ed. Daniela Kloock and Angela Spahr (Munich: Wilhelm Fink, 1997), 205–236, esp. 206; and Erhard Schüttpelz, "Frage nach der Frage, auf die das Medium eine Antwort ist," in *Signale der Störung*, ed. Albert Kümmel and Erhard Schüttpelz (Munich: Wilhelm Fink, 2003), 15–29, esp. 16–17.

73. Erhard Schüttpelz, "'Get the message through': From the Channel of Communication to the Message of the Medium, 1945–1960," in *Media, Culture, and Mediality: New Insights into the Current State of Research*, ed. Ludwig Jäger, Erika Linz, and Irmela Schneider (Bielefeld: transcript, 2010), 109–138.

74. Michel Chevalier, "Le Temple," in Régnier, *Le livre nouveau des Saint-Simoniens*, 238.

75. *Documents diplomatiques de la conférence télégraphique internationale de Paris* (Paris: Imprimerie Imperiale, March 4, 1865), 115.

76. See Benedict Anderson, *Imagined Communities: An Inquiry into the Origins and Spread of Nationalism*, 2nd ed. (London: Verso, 1991).

77. I am sincerely grateful to the ITU archivists Kristine Klara and Heather Heywood for making materials on the conferences available. They are now available online: http://handle.itu.int/11.1004/020.2000/s.004. On the history of the ITU in general, see George Arthur Codding and Anthony M. Rutkowski, *The International Telecommunication Union: An Experiment in International Cooperation* (Dedham, MA: Artech House, 1982); and Gabriele Balbi and Andreas Fickers, eds., *History of the International Telecommunication Union (ITU)*, vol. 1 (Berlin: De Gruyter Oldenbourg, 2020).

78. This was true of German-speaking countries in particular. On October 3, 1849, the first agreement between Austria and Prussia regulating the exchange of telegraphic messages was concluded, followed by Prussia and Saxony and then

Austria and Bavaria. On July 25, 1850, the Deutsch-Österreichischer Telegraphenverein (German-Austrian Telegraph Association) was founded as a consequence. See Codding and Rutkowski, *The International Telecommunication Union*, 13. On the relationship of state-oriented and market-oriented communication in the history of telegraphy, see Flichy, *Dynamics of Modern Communication*, 7–28.

79. On the discourse on the cable as a canal-shaped "umbilical cord," see Christian Holtorf, *Der erste Draht zur Neuen Welt: Die Verlegung des transatlantischen Telegrafenkabels* (Göttingen: Wallstein, 2013), esp. 259–260.

80. For a classic account of the relationship of state-oriented and market-oriented communication, see Flichy, *Dynamics of Modern Communication*, 7–28.

81. See *Documents diplomatiques . . . de Paris*, art. 2, 3. This passage would not remain uncontroversial in the future. The modified convention of the Vienna agreement of 1868 provided for both Morse and Hughes apparatuses. See *Documents diplomatiques de la conférence télégraphique internationale de Vienne* (Paris: Imprimerie Imperiale, 1868), art. 3. Britain, which was not invited to the first conference because its national telegraph networks were mainly private prior to 1870, first joined with the Vienna convention.

82. *Documents diplomatiques . . . de Paris*, art. 4, 7.

83. *Documents diplomatiques . . . de Paris*, art. 33.

84. *Documents diplomatiques . . . de Paris*, 37–38. The calculation of tariffs always had great potential for conflict. See Codding and Rutkowski, *The International Telecommunication Union*, 57–58.

85. *Documents diplomatiques . . . de Paris*, art. 19, 20, 59.

86. *Documents diplomatiques . . . de Vienne*, art. 61–63.

87. See *Documents diplomatiques . . . de Paris*, pp. 41–42.

88. The passionate argument of the first director of the ITU, Louis Curchod, on July 4, 1868, for a central office can be read in *Documents diplomatiques . . . de Vienne*, 381–382.

89. See Codding and Rutkowski, *The International Telecommunication Union*, 27.

90. For more detailed accounts based on other cases studies, see, for example, Peter J. Hugill, *Global Communications since 1844: Geopolitics and Technology* (Baltimore: Johns Hopkins University Press, 1999) (for England's empire and communications); Dirk van Laak, *Imperiale Infrastruktur: Deutsche Planungen für eine Erschließung Afrikas 1880 bis 1960* (Paderborn: Schöningh, 2004); Christopher Alan Bayly, "Myths and Technologies of the Modern State," in *The Birth of the Modern World, 1780–1914: Global Connections and Comparisons* (Malden, MA: Blackwell, 2004), 247–283; and Erik van der Vleuten and Arne Kaijser, eds., *Networking Europe: Transnational Infrastructures and the Shaping of Europe, 1850–2000* (Sagamore Beach, MA: Science History Publications, 2006).

91. On making good circulation possible while tolerating poor exchange in the population, see Foucault, *Security, Territory, Population*, 65.

92. See Krajewski, *World Projects*, 6–7. On the connection of the media theory of space and "geobiopolitics," see Niels Werber, *Die Geopolitik der Literatur* (Munich: Hanser, 2007).

CHAPTER 5

1. There are a few collections on analog telephony where materiality and practice come together. See https://www.sammlung-schmidt.de.

2. See Hans-Jörg Rheinberger, *Toward a History of Epistemic Things: Synthesizing Proteins in the Test Tube* (Stanford, CA: Stanford University Press, 1997).

3. See Patrice Flichy, *Dynamics of Modern Communication: The Shaping and Impact of New Communication Technologies*, trans. Liz Libbrecht (Thousand Oaks, CA: Sage, 1995), 89–93; Tim Wu, *The Master Switch: The Rise and Fall of Information Empires* (New York: Alfred A. Knopf, 2011), 45–46.

4. Franz Kafka, *The Castle*, trans. Mark Harman (New York: Schocken, 1998), 72. On the telephonic uncanny, see John Durham Peters, *Speaking into the Air: A History of the Idea of Communication* (Chicago: University of Chicago Press, 1999), 195–196.

5. Kafka, *The Castle*, 71.

6. Kafka, *The Castle*, 72.

7. See Jens Ruchatz, "Das Telefon: Ein sprechender Telegraf," in *Einführung in die Geschichte der Medien*, ed. Albert Kümmel, Leander Scholz, and Eckhard Schumacher (Paderborn: UTB | Fink, 2004), 125–149, as well as the discourse on newspapers and advertising surrounding Bell's telephone in section 5.1 of this book. On dialogical uses of the telegraph, see Flichy, *Dynamics of Modern Communication*, 85–86.

8. Kafka, *The Castle*, 4, 20.

9. Wolfgang Hagen, "Gefühlte Dinge: Bells Oralismus, die Undarstellbarkeit der Elektrizität und das Telefon," in *Telefonbuch: Beiträge zu einer Kulturgeschichte des Telefons*, ed. Stefan Münker and Alexander Roesler (Frankfurt am Main: Suhrkamp, 2000), 35–60, esp. 52.

10. Thomas Watson to Elizabeth Watson, July 18, 1877, Alexander Graham Bell Box 402, Alexander Graham Bell Family Papers (hereafter AGB FP), Manuscript Division, Library of Congress, Washington, DC.

11. See *New York Herald*, *The World*, May 12, 1877; *The Evening Post*, May 16, 1877; and *New York Times*, May 19, 1877, Box 402, AGB FP. Watson was assigned the role of speaking and singing into the telephone. See Jonathan Sterne, *The Audible Past: Cultural Origins of Sound Reproduction*, 3rd ed. (Durham, NC: Duke University Press, 2006), 250–251.

12. *The Daily Graphic*, March 6, 1877, Box 402, AGB FP.

13. Bell's competitor Elisha Gray also presented his telephone live to large audiences. See the poster in Sterne, *The Audible Past*, 253.

14. Avital Ronell is not alone in describing this constellation as a psychoanalytical primal scene of telephonic desire. See Avital Ronell, *The Telephone Book: Technology, Schizophrenia, Electric Speech* (Lincoln: University of Nebraska Press, 1989), 61, 90, 194. See also Seth Shulman, *The Telephone Gambit: Chasing Alexander Graham Bell's Secret* (New York: W. W. Norton, 2009), 63–76.

15. Alexander Graham Bell to Mabel G. Hubbard, February 13, 1877, typescript, Box 402, AGB FP.

16. See Heinrich von Stephan, *Das Telephon betreffend: Brief des Generalpostmeisters Heinrich v. Stephan an Seine Durchlaucht den Fürsten Reichskanzler Otto v. Bismarck; Vorschlag für die Einführung der ersten Telephon-Verbindungen Berlin 9.11.1877*, ed. Uwe Otto (Berlin: Berliner Handpresse, 1984).

17. See Frederick Parker Lewis, "Chronicles of Early Telephone Days in Connecticut with the Personal Stories of George W. Coy and Herrick P. Frost concerning the First Commercial Telephone Exchange at New Haven, Connecticut 1934," Southern New England Telephone Company Records (hereafter SNET). All quotations from archival material refer to the state of the SNET archive in 2007. Slight variations in archival order might occur. Archives and Special Collections, University of Connecticut Library, Box 305, Folder 1923, typescript, 20.

18. Despite this, in October 1877, Heinrich Stephan in Germany was already testing maximum distances. Whereas a working connection could be achieved across sixty-one kilometers from Berlin to Brandenburg, the 142 kilometers between Berlin and Magdeburg proved to be too much. See Stephan, *Das Telephon betreffend*; Roland Gööck, *Die grossen Erfindungen: Nachrichtentechnik und Elektronik* (Künzelsau: Sigloch, 1988), 114.

19. See Wolfgang König, "Nutzungswandel, Technikgenese und Technikdiffusion: Ein Essay zur Frühgeschichte des Telefons in den Vereinigten Staaten und Deutschland," in *Fern-Sprechen: Internationale Fernmeldegeschichte, -soziologie und -politik*, ed. Jörg Becker (Berlin: Vistas, 1994), 147–163, esp. 157.

20. See Frank Hartmann, *Globale Medienkultur: Technik, Geschichte, Theorien* (Vienna: WUV, 2006), 100–101.

21. See Reuel A. Benson, Jr., "The First Century of the Telephone in Connecticut," SNET, January 1978, Box 305, Folder 2986. See also https://collections.ctdigitalarchive.org/islandora/object/20002:860138740.

22. See Milton Mueller, "The Switchboard Problem: Scale, Signaling, and Organization in Manual Telephone Switching," *Technology and Culture* 30, no. 3 (1989): 534–560.

23. See Lewis, "Chronicles of Early Telephone Days in Connecticut," SNET, Box 305, Folder 1923, typescript, 17.

24. George W. Coy to Mr. Lewis, December 26, 1910, transcription, SNET, Box 305, Folder 3024a, 1.

25. The degree to which New England functioned as an experimental field in early telephony can be seen from the repeated battles over the myth of the "first" telephone exchange in the memoirs of those involved. Local connections were available in Boston und Hartford in 1877 and in Chicago in early 1878 as well. The New Haven company is very important as an institution because it was the first whose telephone network extended to a larger territory beyond the city. On Boston, Chicago, and the status of New Haven's telephone exchange, see Robert J. Chapuis, *100 Years of Telephone Switching, 1878–1978*, part I, *Manual and Electromechanical Switching, 1878–1960s* (Amsterdam: North-Holland, 1982), 48–49. On Boston and Hartford, see Gööck, *Die grossen Erfindungen*, 124–125.

26. On the company history, see the finding aid of the corporate archive by Cynthia McElroy and Stephen Showers (2005), SNET, https://collections.ctdigitalarchive.org/islandora/object/20002:860138740.

27. A similar story is told of Berlin, where only eight parties expressed interest in 1880. See Uwe Otto's commentary in Stephan, *Das Telephon betreffend*, n.p.

28. In Connecticut, telephone numbers were not introduced until 1889.

29. In 1893, the first commercially operated switchboard was destroyed by a fire. In the 1920s, the second one added to the collection of the Bell Telephone Labs Historical Museum. See SNET, Box 305, Folder 3044.

30. See JoAnne Yates, *Control through Communication: The Rise of System in American Management* (Baltimore: Johns Hopkins University Press, 1989) for early uses of telephone switching in companies such as Scovill and DuPont.

31. On this, see Oskar Blumtritt, *Nachrichtentechnik: Sender, Empfänger, Übertragung, Vermittlung* (Munich: Deutsches Museum, 1997), 57.

32. On the French Théâtrophone, see Carolyn Marvin, *When Old Technologies Were New: Thinking about Electric Communication in the Late Nineteenth Century* (New York: Oxford University Press, 1988), 209–210; and Sterne, *The Audible Past*, 192–193. See also Wikipedia, s.v. "Telephone Newspaper," https://en.wikipedia.org/wiki/Telephone_newspaper.

33. On everyday culture in the United States, see Marvin, *When Old Technologies Were New*; and Claude S. Fischer, *America Calling: A Social History of the Telephone to 1940* (Berkeley: University of California Press, 1992). On Canada, see Michèle Martin, *Hello Central? Gender, Technology, and Culture in the Formation of the Telephone System* (Montreal: Gill University Press, 1991), esp. 140–141.

34. See Bruno Latour, "On Technical Mediation: Philosophy, Sociology, Genealogy," *Common Knowledge* 3, no. 2 (1994): 53.

35. See Christoph Asendorf, *Ströme und Strahlen: Das langsame Verschwinden der Materie um 1900* (Giessen: Anabas, 1989), 58–59.

36. Walter Rathenau, "Die Resurrection Co.," trans. Louis Kaplan, *New German Critique*, no. 62 (1994): 63–69, esp. 65–66. Orig. pub. in Rathenau, *Gesammelte Schriften*, vol. 4, *Reflexionen und Aufsätze* (Berlin: S. Fischer, 1925), 337–349, esp. 343. Sincere thanks to Albert Kümmel-Schnur for pointing me to Rathenau.

37. See Gööck, *Die grossen Erfindungen*, 126; and Rolf Barnekow, "Wählen statt Rufen: Die automatische Vermittlung," in *Fräulein vom Amt*, ed. Helmut Gold and Annette Koch (Munich: Prestel, 1993), 86–93, esp. 86. On its development in Germany, see also Birgitta Godt, "Die Entwicklung der Handvermittlung," in Gold and Koch, *Fräulein vom Amt*, 68–85.

38. See Chapuis, *100 Years of Telephone Switching I*, 49, and section 5.1 of this book.

39. On its use, see Venus Green, *Race on the Line: Gender, Labor, and Technology in the Bell System, 1880–1980* (Durham, NC: Duke University Press, 2001), 20–21; and Southern New England Telephone Company Records (SNET), Archives and Special Collections, University of Connecticut Library, Box 305, Folder 3044.

40. See P. C. Smith, "Curious Patents in Mechanical Switching," *Bell Laboratories Record* 7 (1929): 265–269.

41. See Chapuis, *100 Years of Telephone Switching I*, 51; Mueller, "The Switchboard Problem," 542–543.

42. See Mueller, "The Switchboard Problem," 547.

43. Marcel Proust, *The Guermantes Way*, trans. Charles Kenneth Scott-Moncrieff and Dennis Joseph Enright (New York: Vintage, 1996), 147.

44. On this, see the detailed sociohistorical essays in Helmut Gold and Annette Koch, eds., *Fräulein vom Amt* (Munich: Prestel, 1993).

45. See the classic article by Kenneth Lipartito, "When Women Were Switches: Technology, Work, and Gender in the Telephone Industry, 1890–1920," *The American Historical Review* 99, no. 4 (1994): 1075–1111.

46. Herbert N. Casson, *The History of the Telephone* (Moscow: Dodo Press, 2006), 65. By contrast, there were a number of historically well-documented breakdowns. See, among others, Bernhard Siegert, "Das Amt des Gehorchens: Hysterie der Telefonistinnen; oder, Wiederkehr des Ohres, 1874–1913," in *Armaturen der Sinne: Literarische und technische Medien, 1870 bis 1920*, ed. Jochen Hörisch and Michael Wetzel (Munich: Wilhelm Fink, 1990), 83–106; and Andreas Killen, "Die Telefonzentrale," in *Orte der Moderne: Erfahrungswelten des 19. und 20. Jahrhunderts*, ed. Alexa Geisthövel and Habbo Knoch (Frankfurt am Main: Campus, 2005), 81–90.

47. Green, *Race on the Line*, 195.

48. See Green, *Race on the Line*, chap. 7 for further detail.

49. This happened circa 1892. See Gööck, *Die grossen Erfindungen*, 128. The dating is uncertain. See Chapuis, *100 Years of Telephone Switching I*, 55–56.

50. See Mueller, "The Switchboard Problem," 554.

51. See Casson, *The History of the Telephone*, 50. On telephone sound and hearing tests, see Mara Mills, "Testing Hearing with Speech," in *Testing Hearing: The Making of Modern Aurality*, ed. Viktoria Tkaczyk, Mara Mills, and Alexandra Hui (Oxford: Oxford University Press, 2020), 23–48.

52. In the United States, communication was viewed as a service, whereas in Germany, the official language of telephonic regulations was standardized in the imperative.

53. On the principle of "user transparency" in discussions of AT&T's switchboard committee, see Mueller, "The Switchboard Problem," 544.

54. In some variants, one also reads that all operators had been bribed by Strowger's competitor.

55. Almon Brown Strowger, "Automatic Telephone Exchange," United States Patent Office, Patent 447 918 (filed March 12, 1889, granted March 10, 1891), lines 15–19.

56. Strowger, "Automatic Telephone Exchange," lines 13–14. On the rhetoric of patents, see Alain Pottage and Brad Sherman, *Figures of Invention: A History of Modern Patent Law* (Oxford: Oxford University Press, 2010); Albert Kümmel-Schnur, "Patente als Agenten von Mediengeschichte," in *Bildtelegraphie: Eine Mediengeschichte in Patenten, 1840–1930*, ed. Albert Kümmel-Schnur and Christian Kassung (Bielefeld: transcript, 2012), 15–38; and Nadine Taha, "Patent in Action: Das US-amerikanische Patent aus der Perspektive der Science and Technology Studies," *Zeitschrift für Medienwissenschaft* 6 (2012): 36–47.

57. Daniel Connolly, Thomas A. Connolly, and Thomas J. McTighe, "Improvement in Automatic Telephone-Exchanges," United States Patent Office, Patent 222 458 (filed September 10, 1879), 4.

58. Ezra T. Gilliland, "Telephone-Circuit," "Metallic-Circuit Telephone System," and "Multiple-Circuit Changer," United States Patent Office, Patents 306 239–241 (patented October 7, 1884).

59. See Chapuis, *100 Years of Telephone Switching I*, 58–59.

60. Strowger, "Automatic Telephone Exchange," lines 27–28, 32–33. "By and through" is contemporaneous spiritualist vocabulary. Thanks to Irma Klerings for this reference.

61. On the copying telegraph and its transformation into telephotography or phototelegraphy, see Christian Kassung, "Isochronie und Synchronie: Zur apparativen und epistemologischen Genese des Kopiertelegraphen," in *Lebendige Zeit: Wissenskulturen im Werden*, ed. Henning Schmidgen (Berlin: Kadmos, 2005), 195–209; and Albert Kümmel-Schnur and Christian Kassung, eds., *Bildtelegraphie: Eine Mediengeschichte in Patenten, 1840–1930* (Bielefeld: transcript, 2012). Sincere thanks to Marius Hug for documenting the historical differences of machine naming.

62. "Hello, Main, Good By: Strowger's Telephone Attachment Dispenses with Girls," *Chicago Daily Tribune*, November 4, 1892, 1.

63. See "No More 'Hello' Girls: Description of an Invention for Making Telephone Connections," *Chicago Daily Tribune*, October 31, 1892, 7.

64. See "Takes the Place of 'Hello' Girls: Automatic Telephone System in Successful Operation at La Porte," in *Chicago Daily Tribune*, November 2, 1892, 1.

65. See "Hello, Main, Good By," 1.

66. See Chapuis, *100 Years of Telephone Switching I*, 62.

67. See John Patrick Barrett, *Electricity at the Columbian Exposition, including an Account of the Exhibits in the Electricity Building, the Power Plant in Machinery Hall, the Arc and Incandescent Lighting of the Grounds and Buildings etc.* (Chicago: Donnelley & Sons, 1894), 340.

68. "The Strowger Company is Fully Equipped to Install Telephone Exchanges," Advertisement, *Chicago Daily Tribune*, February 1, 1894, 4.

69. For details on the reasons, see Mueller, "The Switchboard Problem," 558–559.

70. See Chapuis, *100 Years of Telephone Switching I*, 62.

71. See R. B. Hill, "Early Work on Dial Telephone Systems," *Bell Laboratories Record* 31 (1953): 22–29, esp. 22.

72. On the first telephone exchange in Hildesheim in 1908, see Erwin Horstmann, *75 Jahre Fernsprecher in Deutschland* (Bonn: Bundesministerium für Post- und Fernmeldewesen, 1952), 238–239. On national initiatives and the process of European agreements at international conferences of telephone technicians in Budapest (1908) and Paris (1910), see Chapuis, *100 Years of Telephone Switching I*, 73–80; König, "Nutzungswandel, Technikgenese und Technikdiffusion" (see note 19); and Peter J. Hugill, *Global Communications since 1844: Geopolitics and Technology* (Baltimore: Johns Hopkins University Press, 1999), 61.

73. See Ithiel de Sola Pool, *Forecasting the Telephone: A Retrospective Technology Assessment of the Telephone* (Norwood, NJ: Ablex, 1983), 26.

74. On the post office's Dead Letter Office, see, for example, Peters, *Speaking into the Air*, 167–171.

75. Edgar Allan Poe, "The Purloined Letter," in Poe, *Poetry and Tales* (New York: Library of America, 1984), 680–698.

76. See Chapuis, *100 Years of Telephone Switching I*, 69.

77. Mueller, "The Switchboard Problem," 559.

78. Ludwik Fleck, *Genesis and Development of a Scientific Fact*, ed. Thaddeus J. Trenn and Robert K. Merton, trans. Fred Bradley and Thaddeus J. Trenn (Chicago: University of Chicago Press, 1979), 21.

79. See Killen, "Die Telefonzentrale"; Gold and Koch, *Fräulein vom Amt*; and Bernhard Siegert, "Switchboards and Sex: The Nut(t) Case," in *Inscribing Science: Scientific Texts and the Materiality of Communication*, ed. Timothy Lenoir (Stanford, CA: Stanford University Press, 1998), 78–90.

80. See Rüdiger Campe, "Pronto! Telefonate und Telefonstimmen (57322)," in *Diskursanalysen 1: Medien*, ed. Friedrich Kittler, Manfred Schneider, and Samuel Weber (Opladen: Westdeutscher Verlag, 1987), 68–93; Siegert, "Das Amt des Gehorchens"; and Hagen, "Gefühlte Dinge." Newer approaches can be found in Ulla Autenrieth et al., eds., *Dis Connecting Media: Technik, Praxis und Ästhetik des Telefons; Vom Festnetz zum Handy* (Basel: Merian, 2011).

81. On the telephone exchange in *Three Days of the Condor*, see section 10.2 of this volume.

82. See Georg Seesslen, *Die Matrix entschlüsselt* (Berlin: Bertz, 2003).

83. *The Matrix*, directed by Lilly [Andy] Wachowski and Lana [Laurence] Wachowski (1999; USA: Warner Brothers), timecode ca. 0:29:03.

84. Christian Marclay, *Telephones*, 1995, video installation, single-channel video, black-and-white and color, with sound, 7 min 30 s, dimensions variable, editing: Timothy Frank. https://vimeo.com/176259496. Sincere thanks to Hartmut Böhme for this reference.

85. On the elective affinity of film and the telephone in general, see, among others, Bernhard Debatin and Hans Jürgen Wulff, eds., *Telefon und Kultur: Das Telefon im Spielfilm* (Berlin: Spiess, 1991); Marli Feldvoss, "Bei Anruf Film: Telefon und Kino—eine ideale Partnerschaft," *Neue Zuercher Zeitung*, December 14, 2002, https://www.nzz.ch/article89SX3-ld.238194; Kirsten von Hagen, "Wahlverwandtschaften: Spielformen des Telefons im Film," in *Intermedialität Analog/Digital: Theorien, Methoden, Analysen*, ed. Joachim Paech and Jens Schröter (Paderborn: Wilhelm Fink, 2008), 333–343; and Hansmartin Siegrist, "Screening the Phone: Wie das Telefon ins Kino kam," in *Dis Connecting Media. Technik, Praxis und Ästhetik des Telefons: Vom Festnetz zum Handy*, ed. Ulla Autenrieth et al. (Basel: Merian, 2011), 11–52. On the relationship of the visual arts and the telephone, see Ingrid Severin, "Technische Vernetzungen und ihre Auswirkungen auf zeitgenössische Kunst," in *Technik ohne Grenzen*, ed. Ingo Braun and Bernward Joerges (Frankfurt am Main: Suhrkamp, 1994), 212–250, 386–409.

86. *Grand Hotel* was marketed with a spectacular advertising campaign as Hollywood's first all-star movie.

87. *Grand Hotel*, directed by Edmund Goulding (1932; USA: Turner Entertainment and Warner Brothers, 2004), timecode ca. 1:28:30–1:30:00.

88. *Grand Hotel*, timecode ca. 1:29:45.

89. See Chapuis, *100 Years of Telephone Switching I*, 5 (on the hotel), and 84 (in general). See also Gööck, *Die grossen Erfindungen*. On communications technology and electronics, see Horstmann, *75 Jahre Fernsprecher in Deutschland*.

90. In German "Fräulein vom Amt," meaning "the girl from the administrative office."

91. See sections 4.2 and 7.4 of this book on the relationship of city, canal infrastructure, and blood circulation.

92. See Chapuis, *100 Years of Telephone Switching I*, 198.

93. The Parisian schema was as follows: 1: blank; 2: A B C; 3: D E F; 4: G H I; 5: J K L; 6: M N; 7: P R S; 8: T U V; 9: W X Y; 0: O Q. It is striking that the distributor in the center of this picture is missing the letter Z, though that does not mean there is a direct connection to this schema, since twenty-five-part distributors were common, even on number switches that did not have letters. Sincere thanks to Hans-Dieter Schmidt for this information.

94. Virginie Julliard, "Une 'femme machine' au travail: la 'demoiselle du telephone,'" *Quaderni* 56 (2004): 23–32, 28. Julliard refers mainly to Proust's mythological references for telephony. Yet, she also integrates historical accounts of the French "demoiselle du telephone," emphasizing the corporeal experience of female telephone workers that felt like robots placed beside other robots.

95. Ernst Kapp, *Grundlinien einer Philosophie der Technik: Zur Entstehungsgeschichte der Cultur aus neuen Gesichtspunkten* (Braunschweig: Westermann, 1877); translated by Lauren K. Wolfe as *Elements of a Philosophy of Technology: On the Evolutionary History of Culture*, ed. Jeffrey West Kirkwood and Leif Weatherby (Minneapolis: University of Minnesota Press, 2018). On the question of media anatomies, see Annette Keck and Nicolas Pethes, eds., *Mediale Anatomien: Menschenbilder als Medienprojektionen* (Bielefeld: transcript, 2001). The primal scene of such forms of knowledge transfer in the nineteenth century was in the laboratory of Hermann von Helmholtz. On this, see Timothy Lenoir, "Helmholtz and the Materialities of Communication," *Osiris* 9 (1994): 184–207.

96. On this, see, for example, Olaf Breidbach, *Die Materialisierung des Ichs: Zur Geschichte der Hirnforschung im 19. und 20. Jahrhundert* (Frankfurt am Main: Suhrkamp, 1995); Douwe Draaisma, *Die Metaphernmaschine: Eine Geschichte des Gedächtnisses* (Darmstadt: Primus Verlag, 1999); and Christian Jürgen Emden, "Epistemische Konstellationen, 1800–1900," in *Netzwerke: Eine Kulturtechnik der Moderne*, ed. Hartmut Böhme, Jürgen Barkhoff, and Jeanne Riou (Cologne: Böhlau, 2004), 127–154.

97. See Norbert Wiener, *Cybernetics; or, Control and Communication in the Animal and the Machine*, 2nd ed. (Cambridge, MA: MIT Press, 1965), 13: "Independently, they [McCulloch and Pitts] had used the technique of mathematical logic for the discussion of what were after all switching problems."

98. See Warren S. McCulloch and Walter Pitts, "A Logical Calculus of the Ideas Immanent in Nervous Activity," *Bulletin of Mathematical Biophysics* 5, no. 4 (1943): 115–133.

99. Lecture at the general assembly of the twelfth Conference of Scientists and Physicians, Moscow, December 28, 1909. Ivan Pavlov, "Natural Science and the Brain," in

Selected Works, ed. J. Gibbons, trans. S. Belsky, 2nd ed. (Moscow: Foreign Languages Publishing House, 1955), 204–218, esp. 209. Pavlov's ideas on the conditioned reflex distinguish it from lower, unconditioned reflex.

100. Ivan Pavlov, "Natural Science and the Brain," 206.

101. Golgi progressed holistically from the neural brain as an overall network, whereas Ramón y Cajal emphasized the autonomy of every single neuron. On this, see Laura Otis, *Networking: Communicating with Bodies and Machines in the Nineteenth Century* (Ann Arbor: University of Michigan Press, 2001), 55–56; and Lorraine Daston und Peter Galison, *Objectivity*, 2nd ed. (New York: Zone Books, 2008), 115–116.

102. William James, *The Principles of Psychology*, vol. 1 (New York: Dover, 1950). Central issues of this text can already be found ten years earlier in William James, "The Association of Ideas," *Popular Science Monthly* 16, no. 5 (1880): 577–593.

103. James, *The Principles of Psychology*, 577, 580.

104. James, *The Principles of Psychology*, 569–570.

105. Ronell, *The Telephone Book*, 4.

106. See Thomas Macho, "Stimmen ohne Körper: Anmerkungen zur Technikgeschichte der Stimme," in *Stimme: Annäherung an ein Phänomen*, ed. Doris Kolesch and Sybille Krämer (Frankfurt am Main: Suhrkamp, 2006), 130–146, esp. 134–135.

107. See Ronell, *The Telephone Book*, 210, 230–231, 442 on Watson's spiritualism. On his adaption of telephones to diving suits, see Ronell, *The Telephone Book*, 274–275.

108. Sterne, *The Audible Past*, 289–333, quotation on p. 292.

109. Friedrich Nietzsche, "On the Genealogy of Morality," trans. Adrian del Caro, in *The Complete Works of Friedrich Nietzsche*, vol. 8, *Beyond Good and Evil / On the Geneaology of Morality* (Stanford, CA: Stanford University Press, 2014), 292.

110. Walter Rathenau, "Die Resurrection Co.," in *Gesammelte Schriften*, vol. 4, *Reflexionen und Aufsätze* (Berlin: S. Fischer, 1925), 337–349. English translation by Louis Kaplan as "Die Resurrection Co.," *New German Critique*, no. 62 (Spring–Summer 1994): 63–69.

111. On the relationship of medium and death, see the classic study Friedrich Kittler, *Gramophone, Film, Typewriter*, trans. Geoffrey Winthrop-Young and Michael Wutz (Stanford, CA: Stanford University Press, 1999), 10–13.

112. Ronell, *The Telephone Book*, 434.

113. See Otis, *Networking*; and Daniel Gethmann, "Das Zittern der Luft: Die Erfindung der mechanischen Stimme," in *Medien/Stimmen*, ed. Cornelia Epping-Jäger and Erika Linz (Cologne: DuMont, 2003), 211–234.

114. See König, "Nutzungswandel, Technikgenese und Technikdiffusion."

115. See M. D. Fagen, ed., *A History of Engineering and Science in the Bell System: The Early Years, 1875–1925* (New York: Bell Telephone Laboratories, 1975); and Sidney

Millman, ed., *A History of Engineering and Science in the Bell System 5: Communication Sciences, 1925–1980* (Indianapolis, IN: AT&T Bell Laboratories, 1984).

116. See Flichy, *Dynamics of Modern Communication*, 93–98; Wu, *The Master Switch*, 51–52; and Richard R. John, *Network Nation: Inventing American Telecommunications* (Cambridge, MA: Belknap Press of Harvard University Press, 2010), chaps. 9, 10.

117. See Otis, *Networking*, 10.

CHAPTER 6

1. Umberto Eco, "The Encyclopedia as a Labyrinth," in *Semiotics and the Philosophy of Language* (Bloomington: Indiana University Press, 1986), 80–84, esp. 81.

2. Eco, "The Encyclopedia as a Labyrinth," 81.

3. On this, see the ethnological analysis of the use of material traps that function as models of their creators in Alfred Gell, "Vogel's Net: Traps as Artworks and Artworks as Traps," in *The Art of Anthropology: Essays and Diagrams*, ed. Eric Hirsch (London: Athlone, 1999), 187–214, esp. 198–199.

4. Bernd Mahr, "Modellieren: Beobachtungen und Gedanken zur Geschichte des Modellbegriffs," in *Bild—Schrift—Zahl*, ed. Sybille Krämer and Horst Bredekamp (Paderborn: Wilhelm Fink, 2003), 59–86, 65. On the cognitive dimension of models, see Jürgen Renn and Peter Damerow, "Mentale Modelle als kognitive Instrumente der Transformation von technischem Wissen," in *Übersetzung und Transformation*, ed. Hartmut Böhme, Christof Rapp, and Wolfgang Rösler (Berlin: De Gruyter, 2007), 311–331.

5. For more detail on the various forms of models with reference to Max Black and Nelson Goodman, see Sebastian Giessmann, "Graphen können alles: Visuelle Modellierung und Netzwerktheorie vor 1900," in *Visuelle Modelle*, ed. Ingeborg Reichle, Steffen Siegel, and Achim Spelten (Paderborn: Wilhelm Fink, 2008), 269–284, esp. 271–272. See also Lothar Krempel, *Visualierung komplexer Strukturen: Grundlagen der Darstellung mehrdimensionaler Netzwerke* (Frankfurt am Main: Campus, 2005).

6. Denis Diderot, "D'Alembert's Dream," in *Rameau's Nephew and D'Alembert's Dream*, ed. and trans. Leonard Tancock (London: Penguin, 1966), 165–237, esp. 183.

7. The drama over the capillaries of frog lungs and the anti-Galenic description that the lungs do not consist of flesh can be followed in the published correspondence of these two men. Borelli expresses his pleasure about Malpighi's work for the first time on August 21, 1660. But Borelli also repeatedly criticizes it because he cannot reproduce in experiments his student's findings. On January 14, 1661, he proposes the injection of quicksilver. After the publication of Malpighi's first epistle in February 1661 Borelli insists on a second one. See Marcello Malpighi, *The Correspondence of Marcello Malpighi, 1658–1669*, ed. Howard B. Adelmann, vol. 1 (Ithaca, NY: Cornell University Press, 1975). See also Howard B. Adelmann, *Marcello Malpighi and*

the Evolution of Embryology, 5 vols. (Ithaca, NY: Cornell University Press, 1966), 1:171–172.

8. See William Harvey, *Exercitatio anatomica de motu cordis et sanguinis in animalibus* (Frankfurt am Main: Fitzer, 1628); translated as Harvey, *On the Motion of the Heart and Blood in Animals: A New Edition of William Harvey's "Exercitatio anatomica de motu cordis et sanguinis in animalibus,"* ed. Jarrett A. Carty, trans. Robert Willis (Eugene, OR: Wipf and Stock, 2016).

9. James Young, "Malpighi's 'De Pulmonibus,'" *Proceedings of the Royal Society of Medicine* 23, no. 1 (1929): 1–11, 2; Marcello Malpighi, "De pulmonibus," in *Opera omnia: Figuris elegantissimis in aes incisis illustrata*, 2 vols. (Hildesheim: Olms, 1975), 2:133–145, 134: "in frustulo enim extimo pulmonum lumini objecto mirabile quod dam rete videtur extensum, quo emergentes, & extuberantes diceres singulas vesicas vinciri, & colligari, quod idem etiam obscure tamen in secto interius pulmone observatur."

10. See Luigi Belloni, "Marcello Malpighi and the Founding of Anatomical Microscopy," in *Reason, Experiment and Mysticism in the Scientific Revolution*, ed. Maria Luisa Righini Bonelli and William Shea (New York: Science History, 1975), 95–110.

11. See Harvey, *Exercitatio anatomica*, chap. 7. Relatively little is known about Malpighi's instruments, which unlike other scholars he made himself. The microscope used for *De pulmonibus* had a single lens. It is likely that the later microscopes were made by Eustachio Divini of Bologna. See Adelmann, *Marcello Malpighi*, 2:828–829.

12. See René Descartes, "L'homme" (1662), in Descartes, *Œuvres philosophiques, 1618–1637*, vol. 1 (Paris: Éditions Garnier, 1988), 369–480, esp. 441–442.

13. On this, see Belloni, "Marcello Malpighi and the Founding of Anatomical Microscopy"; Ilse Jahn, "Naturphilosophie und Empirie in der Frühaufklärung," in *Geschichte der Biologie: Theorien, Methoden, Institutionen, Kurzbiographien*, ed. Ilse Jahn, 4th ed. (Jena: Gustav Fischer, 1998), 197–230, esp. 204–205; and Hartmut Böhme, "Bildevidenz, Augentäuschung und Zeugenschaft in der Wissenschaft des Unsichtbaren im 17. Jahrhundert," in *Dissimulazione onesta; oder, Die ehrliche Verstellung; Von der Weisheit der versteckten Beunruhigung in Wort, Bild und Tat. Martin Warnke zu Ehren; Ein Symposion (2003)*, ed. Horst Bredekamp et al. (Hamburg: Philo Fine Arts, 2007), 13–42.

14. Malpighi, *Opera omnia*, 2:144, caption to figure II, letter C.

15. See the description of the reticular net of the prickly pear cactus in Marcello Malpighi, "De anatome plantarum," in *Opera omnia*, 1:1–78, esp. 35–36.

16. Giovanni Borelli to Marcello Malpighi, March 24, 1661, in *The Correspondence of Marcello Malpighi, 1658–1669*, 80.

17. See section 3.1 of the present volume.

18. See Pierre Musso, *Télécommunications et philosophie des réseaux: La posterité paradoxale de Saint-Simon* (Paris: Presses Universitaires de France, 1997), 33–51; and Sebastian Giessmann, *Netze und Netzwerke: Archäologie einer Kulturtechnik, 1740–1840*

(Bielefeld: transcript, 2006), 33–34, 71–72; and sections 4.2 and 7.3 of the present volume.

19. On this, see Thomas Becker, *Mann und Weib—schwarz und weiss: Die wissenschaftliche Konstruktion von Geschlecht und Rasse, 1650–1900* (Frankfurt am Main: Campus, 2005), 185–186.

20. See Igor J. Polianski, *Die Kunst, die Natur vorzustellen: Die Ästhetisierung der Pflanzenkunde um 1800 und Goethes Gründung des botanischen Gartens zu Jena im Spannungsfeld kunsttheoretischer und botanischer Diskussionen der Zeit* (Jena: Lehrstuhl für Kunstgeschichte und Kunsthistorisches Seminar mit Kustodie, Friedrich-Schiller-Universität, 2004), 222–223; and Sebastian Giessmann, "Netze als Weltbilder: Ordnungen der Natur von Donati bis Cuvier," in *Verwandte Bilder*, ed. Ingeborg Reichle, Steffen Siegel, and Achim Spelten, 2nd ed. (Berlin: Kadmos, 2008), 243–261. On diagrams in biology, see also Georg Töpfer, "Linien, Bäume, Netze—und die Gegenstände der Biologie," in *Netzwerke: Beiträge zur 13. Jahrestagung der DGGTB in Neuburg an der Donau 2004*, ed. Michael Kaasch, Joachim Kaasch, and Volker Wissemann (Berlin: VWB, 2006), 74–94.

21. Sybille Krämer, "'Operationsraum Schrift': Über einen Perspektivenwechsel in der Betrachtung der Schrift," in *Schrift: Kulturtechnik zwischen Auge, Hand und Maschine*, ed. Gernot Grube, Werner Kogge, and Sybille Krämer (Paderborn: Wilhelm Fink, 2005), 22–57, esp. 41.

22. On this, see, crucially, Martin Heidegger, "Age of the World Picture," in *Off the Beaten Track*, ed. and trans. Julian Young and Kenneth Haynes (Cambridge, UK: Cambridge University Press, 2002), 57–85. On the boundaries of the graphic system, see Jacques Bertin, *Semiology of Graphics: Diagrams, Networks, Maps*, trans. William J. Berg (Redlands, CA: ESRI, 2011), 42. See also Gabriele Werner, ed., *Systemische Räume: Bildwelten des Wissens* (Berlin: Akademie, 2007); and Inge Hinterwaldner, *The Systemic Image: A New Theory of Interactive Real-Time Simulations*, trans. Elizabeth Tucker (Cambridge, MA: MIT Press, 2017).

23. See Bertin, *Semiology of Graphics*, 269. I do not adopt here Bertin's remarkable systematic distinction between diagrams and networks. It is justified by the logic of graphic construction on paper. According to Bertin, for example, in a diagram, a frame of meaning is predetermined by the *x* and *y* axis and then entered subsequently, whereas with a network, the relationships can be entered provisionally and then the depiction is constructed with great concision.

24. See John Law, "Note on the Theory of the Actor Network: Ordering, Strategy and Heterogeneity," *Systems Practice* 5, no. 4 (1992): 379–393; and Hartmut Böhme, "Netzwerke: Zur Theorie und Geschichte einer Konstruktion," in *Netzwerke: Eine Kulturtechnik der Moderne*, ed. Jürgen Barkhoff, Hartmut Böhme, and Jeanne Riou (Cologne: Böhlau, 2004), 17–36, esp. 19.

25. See Bruno Latour, "Visualisation and Cognition: Drawing Things Together," in *Representation in Scientific Practice*, ed. Michael Lynch and Steve Woolgar (Cambridge, MA: MIT Press, 1990), 19–68.

26. On swarm research, see Sebastian Vehlken, *Zootechnologies: A Media History of Swarm Research*, trans. Valentine A. Pakis (Amsterdam: Amsterdam University Press, 2019).

27. See Erhard Schüttpelz, "Ein absoluter Begriff: Zur Genealogie und Karriere des Netzwerkkonzepts" in *Vernetzte Steuerung: Soziale Prozesse im Zeitalter technischer Netzwerke*, ed. Stefan Kaufmann (Zurich: Chronos, 2007), 25–46, esp. 40–41.

28. For an overview of the literature in German, see Steffen Siegel, *Tabula: Figuren der Ordnung um 1600* (Berlin: Akademie, 2009), 52–57; and Matthias Bauer and Christoph Ernst, *Diagrammatik: Einführung in ein kultur- und medienwissenschaftliches Forschungsfeld* (Bielefeld: transcript, 2010), 26–27. For the international discussion see, above all, Frederik Stjernfelt, *Diagrammatology: An Investigation on the Borderlines of Phenomenology, Ontology, and Semiotics* (Dordrecht: Springer, 2007).

29. See, early on, Petra Gehring, "Paradigma einer Methode: Der Begriff des Diagramms im Strukturdenken von Michel Foucault und Michel Serres," in *Diagrammatik und Philosophie*, ed. Petra Gehring et al. (Amsterdam: Rodopi, 1988), 89–105.

30. Michel Serres, *Hermès; ou, La communication* (Paris: Les éditions de minuit, 1968), 11–12. On this, see Giessmann, *Netze und Netzwerke,* 97–98.

31. Serres, *Hermès*, 11.

32. Serres, *Hermès*, 12.

33. Serres, *Hermès*, 20.

34. Serres, *Hermès*, 20.

35. See Michel Serres, *The Parasite*, trans. Lawrence R. Schehr (Baltimore: Johns Hopkins University Press, 1982), 20.

36. Serres, *Hermès*, 15.

37. Michel Serres, *Esthétiques sur Carpaccio* (Paris: Hermann, 1975), 7.

38. Serres, *Esthétiques sur Carpaccio,* 19.

39. See Erwin Panofsky, *Perspective as Symbolic Form*, trans. Christopher S. Wood (New York: Zone Books, 1997).

40. On this, see Gehring, "Paradigma einer Methode," 99–100.

41. See also Dieter Mersch, "Visual Arguments: The Role of Images in Sciences and Mathematics," trans. Rett Rossi, in *Science Images and Popular Images of the Sciences*, ed. Bernd-Rüdiger Hüppauf and Peter Weingart (New York: Routledge, 2011), 181–198, on the relation of image and discourse.

42. Gottfried Boehm, "Die Wiederkehr der Bilder," in *Was ist ein Bild?* ed. Gottfried Boehm (Munich: Wilhelm Fink, 1994), 11–38, esp. 29–30; "Ikonische Differenz," *Rheinsprung 11: Zeitschrift für Bildkritik* 1 (2011): 170–178. https://www.rheinsprung11.unibas.ch/fileadmin/documents/Edition_PDF/Ausgabe1/glossar-boehm.pdf.

43. Bruno Latour, "Visualisation and Cognition: Drawing Things Together."

44. See Horst Bredekamp, *Image Acts: A Systematic Approach to Visual Agency*, trans. Elizabeth Clegg (Berlin: De Gruyter, 2017), esp. 77–136 on schematic image acts and 193–264 on intrinsic image acts. This is made particularly clear by Jacob Levy Moreno's network diagrams.

45. See also section 10.3 in the present volume on Mark Lombardi's political network diagrams.

46. Albert-László Barabási, *Linked: The New Science of Networks* (Cambridge, MA: Perseus, 2002), 91.

47. Charles Sanders Peirce, "Of Reasoning in General," in *The Essential Peirce: Selected Philosophical Writings*, vol. 2 (1893–1913), ed. Peirce Edition Project (Bloomington: Indiana University Press, 1998), 11–26, esp. 13. On Peirce's prominent position within diagram scholarship, see Stjernfelt, *Diagrammatology*; and Bauer and Ernst, *Diagrammatik*, 40–41.

48. Leonhard Euler adopted this formulation with recourse to Gottfried Wilhelm Leibniz from the question of the Königsberg Bridge Problem he had sent. It would correctly be called an *analysis situs*, as Leibniz had done. See Gottfried Wilhelm Leibniz, "Characteristica Geometrica: Analysis Geometrica propria. Calculus situs," in *Mathematische Schriften*, vol. 5, ed. Carl Immanuel Gerhardt (Hildesheim: Olms, 1971), 141–219, esp. 178–179, English translation as Gottfried Wilhelm Leibniz, "On Analysis Situs," in *Philosophical Papers and Letters: A Selection*, ed. and trans. Leroy E. Loemker, 2nd ed. (Dordrecht: D. Reidel, 1969), 254–258.

49. André Sainte-Laguë, "Les réseaux (ou graphes)," *Mémorial des Sciences Mathematiques* 18 (1926): 1–62.

50. Barabási, *Linked*, 12–13.

51. On the history of science of graphs, see Thomas L. Hankins, "Blood, Dirt, and Nomograms: A Particular History of Graphs," *Isis* 90 (1999): 50–80.

52. See Leonhard Euler, "Das Königsberger Brückenproblem," in *Theorie der endlichen und unendlichen Graphen: Kombinatorische Topologie der Streckenkomplexe*, ed. Dénes König (Leipzig: BSB Teubner, 1986), 290–301. For a brief summary of the background, see Giessmann, *Netze und Netzwerke*, 27–28; and Wladimir Velminski, "Nachwort," in Leonhard Euler, *Die Geburt der Graphentheorie: Ausgewählte Schriften von der Topologie zum Sudoku*, ed. Wladimir Velminski (Berlin: Kadmos, 2008), 151–206, esp. 151–152. On modeling, see Bernd Mahr, "Cargo: Zum Verhältnis von Bild und Modell," in *Visuelle Modelle*, ed. Ingeborg Reichle, Steffen Siegel, and Achim Spelten (Paderborn: Wilhem Fink, 2008), 17–40, esp. 24–25. Since the completion of the Kaiserbrücke as the eighth crossing in 1905, the topology of the city of Königsberg now permits uninterrupted strolls.

53. See Velminski, "Nachwort," 170.

54. See Leonhard Euler, "Solution d'une question curieuse qui ne paroît soumise à aucune analyse," in *Histoire de l'Académie Royale des Sciences et des Belles-Lettres de Berlin 1759* (1766): 310–337. Translated by Todd Doucet as "A solution to a curious problem which seems not subjected to any analysis," https://scholarlycommons.pacific.edu/cgi/viewcontent.cgi?filename=0&article=1308&context=euler-works&type=additional.

55. See Euler, *Die Geburt der Graphentheorie*, 88–89, 98.

56. See Théophile Vandermonde, "Remarques sur les problèmes de situation," in *Mémoires de l'Académie Royale* 1771 (1774): 566–574, esp. 566. On the concept of tacit knowledge, see Michael Polanyi, *The Tacit Dimension* (Chicago: University of Chicago Press, 2009).

57. For a detailed account of the knowledge culture of woven materials between image, fabric, and calculation, see Birgit Schneider, *Textiles Prozessieren: Eine Mediengeschichte der Lochkartenweberei* (Zurich: diaphanes, 2007).

58. Euler took a similar approach but dispensed with the criterion of symmetry.

59. On the history of knowledge of magic squares and the knight's tour problem, see Velminski, "Nachwort," 188–189.

60. The public addressee of the underlying calculations was the Royal Irish Academy, to which Hamilton presented his ideas on November 10, 1856. See William Rowan Hamilton, "On a New System of Roots of Unity," *Proceedings of the Royal Irish Academy* 6 (1858): 415–416.

61. See Max Black, *Models and Metaphors: Studies in Language and Philosophy* (Ithaca, NY: Cornell University Press, 1962), 221–222.

62. See Johan Huizinga, *Homo Ludens: A Study of the Play-Element in Culture*, trans. R. F. C. Hull (London: Routledge, 1949), 105–118.

63. See Norman L. Biggs, Edward Keith Lloyd, and Robin James Wilson, eds., *Graph Theory, 1736–1936*, 2nd ed. (Oxford: Clarendon Press, 1977), 35.

64. On Hamilton's games, see Édouard Lucas, *Récréations mathématiques: Qui perd gagne, les dominos, les marelles . . .* , vol. 2 (Paris: A. Blanchard, 1892), 208–209; and Wilhelm Ahrens, *Mathematische Unterhaltungen und Spiele* (Leipzig: Teubner, 1901), 327–328.

65. See Natascha Adamowsky, "Spiel und Wissenschaftskultur: Eine Anleitung," in *Die Vernunft ist mir noch nicht begegnet: Zum konstitutiven Verhältnis von Spiel und Erkenntnis*, ed. Natascha Adamowsky (Bielefeld: transcript, 2005), 11–30, esp. 14–15.

66. See August Wilhelm Hofmann, "On the Combining Power of Atoms," *Notices of the Proceedings at the Meetings of the Members of the Royal Institution of Britain* 4, no. 42 (1865): 401–430. On Hofmann's repeated call for "Anschaulichkeit" in academic education, see Christoph Meinel, "Molecules and Croquet Balls," in *Models: The*

Third Dimension of Science, ed. Soraya de Chadarevian and Nick Hopwood (Stanford, CA: Stanford University Press, 2004), 242–275.

67. On the notion of "glyptic formulae," see Meinel, "Molecules and Croquet Balls," 252.

68. For details on this imaginative disposition of chemistry, see Alan J. Rocke, *Image and Reality: Kekulé, Kopp, and the Scientific Imagination* (Chicago: University of Chicago Press, 2010).

69. Thomas Pynchon, *Gravity's Rainbow* (1973; London: Vintage, 2000), 489 (emphasis in original).

70. James Joseph Sylvester, "On an Application of the New Atomic Theory to the Graphical Representation of the Invariants and Covariants of Binary Quantics," *American Journal of Mathematics* 1 (1878): 64–104, esp. 64.

71. Adopting Nelson Goodman's theory of notation, Sybille Krämer has identified this as the difference between notational and pictorial iconicity. See Sybille Krämer, "'Schriftbildlichkeit'; oder, Über eine fast vergessene Dimension der Schrift," in *Bild— Schrift—Zahl*, ed. Sybille Krämer and Horst Bredekamp (Paderborn: Wilhelm Fink, 2003), 157–176, esp. 163.

72. See Lorraine Daston and Peter Galison, "The Image of Objectivity," *Representations* 40 (1992): 81–128; and Lorraine Daston and Peter Galison, *Objectivity*, 2nd ed. (New York: Zone Books, 2008), 115–116.

73. I am very grateful to Philipp Ekardt for this idea.

74. On this, see Ursula Klein, "Visualität, Ikonizität, Manipulierbarkeit: Chemische Formeln als Paper Tools," in *Schrift: Kulturtechnik zwischen Auge, Hand und Maschine*, ed. Gernot Grube, Werner Kogge, and Sybille Krämer (Paderborn: Wilhelm Fink, 2005), 237–251; and Martin Quack, "Modelle in der Chemie," in *Modelle des Denkens: Streitgespräch in der wissenschaftlichen Sitzung der Versammlung der Berlin-Brandenburgischen Akademie der Wissenschaften am 12. Dezember 2003* (Berlin: BBAW, 2005), 21–33.

75. See Biggs, Lloyd, and Wilson, *Graph Theory*, 59.

76. See Biggs, Lloyd, and Wilson, *Graph Theory*, 65–66.

77. See Sylvester, "On an Application of the New Atomic Theory," 67, 73, 90–91.

78. Sylvester, "On an Application of the New Atomic Theory," 76.

79. Sylvester, "On an Application of the New Atomic Theory," 73.

80. Wolfgang Schäffner, "Electric Graphs: Charles Sanders Peirce und die Medien," in *Electric Laokoon: Zeichen und Medien, von der Lochkarte zur Grammatologie*, ed. Michael Franz et al. (Berlin: Akademie, 2007), 313–326, esp. 322.

81. See Charles Sanders Peirce, "What Is a Sign?" in *The Essential Peirce: Selected Philosophical Writings*, vol. 2 (1893–1913), ed. Peirce Edition Project (Bloomington: Indiana University Press, 1998), 4–10, esp. 4–5.

82. On the space of flows, see Manuel Castells, *The Rise of the Network Society*, 2nd ed. (Oxford: Blackwell, 2000), esp. 407–459. For a critical discussion, see Niels Werber, "Netzwerkgesellschaft: Zur Kommunikationsgeschichte von 'technoiden'' Selbstbeschreibungsformeln," in *Kulturgeschichte als Mediengeschichte (oder vice versa?): Archiv für Mediengeschichte*, ed. Lorenz Engell, Joseph Vogl, and Bernhard Siegert (Weimar: Universitätsverlag, 2006), 179–191. On techniques of networked control, see the essays in Stefan Kaufmann, ed., *Vernetzte Steuerung: Soziale Prozesse im Zeitalter technischer Netzwerke* (Zurich: Chronos, 2007).

83. Jacob Levy Moreno and Helen Jennings, *Who Shall Survive? Foundations of Sociometry, Group Psychotherapy and Sociodrama*, 2nd ed. (Beacon, NY: Beacon House, 1953), xx.

84. See Claude Lévi-Strauss, *Structural Anthropology*, vol. 1, trans. Claire Jacobson and Brooke Grundfest Schoepf (New York: Basic Books, 1963), 67–80. For reflections on the atom of kinship, see Claude Lévi-Strauss, *Structural Anthropology*, vol. 2, trans. Monique Layton (New York: Basic Books, 1976), 82–112. On reciprocal obligations in Margaret Mead, see Lévi-Strauss, *Structural Anthropology*, esp. 107–108.

85. Such trends have been clearly criticized by theoretically advanced sociology. Talcott Parsons, for example, in his theory of social action, first published in 1937, described the atomism of the individual as part of the functionalist utilitarian system of social theory: "The theoretical action system characterized by these four features, atomism, rationality, empiricism and randomness of ends will be called in the present study the utilitarian system of social theory." Talcott Parsons, *The Structure of Social Action*, vol. 1, *Marshall, Pareto, Durkheim* (New York: Free Press, 1968), 60; see also 47–52 and 87–88. In social and economic theory, this attitude, known as "methodological individualism," has, since Joseph Schumpeter, been the subject of a recurring debate over principles concerning the human condition of *homo oeconomicus*. See Joseph Heath, "Methodological Individualism," in *The Stanford Encyclopedia of Philosophy*, ed. Edward N. Zalta, Spring 2005 ed. (Stanford, CA: Metaphysics Research Lab, Stanford University, 2005). https://plato.stanford.edu/archives/spr2005/entries/methodological-individualism. I am grateful to Jens Schröter for inspiring this excursus.

86. On the crossover of chemical knowledge and chemistry, see Joseph Vogl, "Mittler und Lenker: Goethes Wahlverwandtschaften," in *Poetologien des Wissens um 1800* (Munich: Wilhelm Fink, 1999), 145–161.

87. "Emotions Mapped by New Geography: Charts Seek to Portray the Psychological Currents of Human Relationships," *New York Times*, April 3, 1933, 17.

88. Moreno and Jennings, *Who Shall Survive?*

89. See Moreno and Jennings, *Who Shall Survive?*; and Jacob Levy Moreno and Helen Jennings, *Die Grundlagen der Soziometrie: Wege zur Neuordnung der Gesellschaft*, 2nd ed. (Cologne: Westdeutscher Verlag, 1967). First published 1954.

90. Zerka Moreno has passed on a statement by Moreno that summarizes his will to make use of his female collaborators: "I know what to do with women better than with men." *Jacob Levy Moreno: His Life and His Muses*, directed by Marco J. D. Maida (1997; USA: Psychotherapy.net, 2003), video, timecode ca. 00:09:30.

91. See Jonathan D. Moreno, *Impromptu Man: J. L. Moreno and the Origins of Psychodrama, Encounter Culture, and the Social Network* (New York: Bellevue Literary Press, 2014). On the performative aspect of Moreno's work, see Brigitte Marschall, *"Ich bin der Mythe": Von der Stegreifbühne zum Psychodrama Jakob Levy Morenos* (Vienna: Böhlau, 1988); and Brigitte Marschall, "Jakob Levy Morenos Theaterkonzept: Die Zeit-Räume des Lebens als Szenenraum der Begegnung," *Zeitschrift für Psychodrama und Soziometrie* 4, no. 2 (2005): 229–243. For a practical look at psychodrama and sociodrama, see Ferdinand Buer, ed., *Morenos therapeutische Philosophie: Die Grundlagen von Psychodrama und Soziometrie*. 3rd ed. (Opladen: Leske + Budrich, 1999).

92. On Moreno's position in the genealogy of sociophysics and the physical subject matter of his concepts, see Martin Donner, "Rekursion und Wissen: Zur Emergenz technosozialer Netze," in *Rekursionen: Von Faltungen des Wissens*, ed. Ana Ofak and Philipp von Hilgers (Paderborn: Wilhelm Fink, 2010), 77–113, esp. 92–93.

93. See Moreno and Jennings, *Who Shall Survive?*, 438.

94. Moreno and Jennings, *Who Shall Survive?*, 440.

95. Moreno was in contact with Kurt Lewin. On the influence of Lewin on the theory of social networks, see Linton C. Freeman, *The Development of Social Network Analysis: A Study in the Sociology of Science* (Vancouver: Empirical Press, 2004), 66–67.

96. Moreno, *Who Shall Survive?*, 430.

97. For additional examples of the structuralist modeling of kinship relations using diagrams, see Lévi-Strauss, *Structural Anthropology*, vol. 1, 45–46; and Lévi-Strauss, *Structural Anthropology*, vol. 2, 91–92.

98. See Pierre Mercklé, *Sociologie des réseaux sociaux* (Paris: La Découverte, 2004), 29–30.

99. See Alfred Reginald Radcliffe-Brown, *Structure and Function in Primitive Society: Essays and Addresses*, 6th ed. (Glencoe, IL: Free Press, 1952), 188–189; J. Clyde Mitchell, ed., *Social Networks in Urban Situations: Analyses of Personal Relationships in Central African Towns* (Manchester: Manchester University Press, 1969); and Lyn Schumaker, *Africanizing Anthropology: Fieldwork, Networks, and the Making of Cultural Knowledge in Central Africa* (Durham, NC: Duke University Press, 2001).

100. See John A. Barnes, "Class and Committees in a Norwegian Island Parish," *Human Relations* 7 (1954): 39–58. On the criticism of modernity under the sign of networks, see Bruno Latour, *We Have Never Been Modern*, trans. Catherine Porter (Cambridge, MA: Harvard University Press, 1993), 39–40, 117–118.

101. On this, see Schüttpelz, "Ein absoluter Begriff."

102. George A. Lundberg, "Social Attraction-Patterns in a Rural Village: A Preliminary Report," *Sociometry* 1, nos. 1–2 (1937): 77–80, esp. 80. George A. Lundberg and Mary Steele, "Social Attraction-Patterns in a Village," *Sociometry* 1, nos. 3–4 (1938): 375–419, esp. 375.

103. Lundberg and Steele, "Social Attraction-Patterns in a Village," 379.

104. Lundberg and Steele, "Social Attraction-Patterns in a Village," 376.

105. One of these was the so-called Chapin scale of socioeconomic status. See Francis Stuart Chapin, *The Measurement of Social Status by the Use of the Social Status Scale* (Minneapolis: University of Minnesota Press, 1933). The instructions for using this scale reveal several informative aspects of their background. First, it is taken for granted that everyone is diligently intent on social status (p. 1). Second, this motive is seen against the backdrop of behaviorism as "configurational stimulus" (p. 2). Social status is measured by field researchers on the level of material culture: visits to the living rooms of families count and value the existing objects and, in some cases, the general impression of the ambience. Every window with curtains earns two points, electric light eight points—two points are deducted for kerosene lamps. A piano bench earns four points, but two points are deducted for a sewing machine. Each newspaper subscription increases the evaluation by eight points, as do a telephone, a radio, and every bookshelf. Two points each are subtracted for dirt, dust, disorder, and neutral monotony. A bizarre and inharmonious atmosphere results in a deduction of four points, while good taste is rewarded with two points for attractiveness, harmony, and quiet. See Chapin, *The Measurement of Social Status*, 6–7.

106. See Lundberg and Steele, "Social Attraction-Patterns," 376–377.

107. See Alden S. Klovdahl, "A Note on Images of Networks," in *Social Networks* 3 (1981): 197–214, esp. 201–202. ORTEP dated from the 1960s. See C. K. Johnson, *ORTEP: A FORTRAN Thermal-Ellipsoid Plot Program for Crystal Structure Illustrations*, ORNL Report 3794. Techn. Rep. (Oak Ridge, TN: Oak Ridge National Laboratory, 1965).

108. On this, see, paradigmatically, the Java applet at https://www.cmu.edu/joss/content/articles/volume1/images/fig30A, which uses data sets from a study in 1941. Cf. Allison Davis, Burleigh B. Gardner, and Mary R. Gardner, *Deep South: A Social Anthropological Study of Caste and Class. Directed by W. Lloyd Warner* (Chicago: University of Chicago Press, 1941), 148–149. This survey followed fourteen informal coffee klatches of eighteen black women in Old City, Mississippi. Their proximity and distance are remodeled by the online application. The numbers above the nodes are based on people in question meeting. See Linton C. Freeman, "Visualizing Social Networks," *Journal of Social Structure* 1, no. 1 (2000), http://www.cmu.edu/joss/content/articles/volume1/Freeman.html.

109. See Klovdahl, "A Note on Images of Networks," 210–222.

110. This corresponds in turn to Latour's observations on the translation of scientific insights in the mode of "drawing things together," which here become "immutable mobiles."

111. Walter Benjamin, "Capitalism as Religion," trans. Rodney Livingstone, in *Selected Writings*, vol. 1 (1913–1926), ed. Marcus Bullock and Michael W. Jennings (Cambridge, MA: Belknap Press of Harvard University Press, 1996), 288–291, esp. 288.

112. Serres, *Esthétiques sur Carpaccio,* 142.

113. *Kinsey*, directed by Bill Condon (2004; USA: 20th Century Fox Home Entertainment, 2005), timecode ca. 01:01:00, DVD.

114. See Michel Foucault, *Security, Territory, Population: Lectures at the Collège de France, 1977–78*, ed. Michel Sennelart, trans. Graham Burchell (New York: Palgrave Macmillan, 2009), 45–46. On liberal governmentality and its spatial orders, see section 4.5 of this book.

115. Benjamin, "Capitalism as Religion," 288.

116. See, pars pro toto, the Netzwerkforschung series of the Verlag für Sozialwissenschaften in Wiesbaden—for example, Christian Stegbauer and Roger Häussling, eds., *Handbuch Netzwerkforschung* (Wiesbaden: VS Verlag für Sozialwissenschaften, 2010), 525–526, on the visualization of networks. The approach of reflecting on methods has also been further developed in numerous workshops on historical network research, which have been held regularly since 2009. See https://www.historicalnetworkresearch.org.

CHAPTER 7

1. For a detailed biography of Pick, see Christian Barman, *The Man Who Built London Transport: A Biography of Frank Pick* (Newton Abbot, UK: David & Charles, 1979). See also Oliver Green, *Underground Art: London Transport Posters, 1908 to the Present*, 2nd ed. (London: Laurence King, 2001), 8–9; *Frank Pick's London: Art, Design and the Modern City* (London: V&A Publishing, 2013).

2. See Beck's own statement in Ken Garland, *Mr Beck's Underground Map* (Harrow Weld, UK: Capital Transport, 1994), 17.

3. Henry Charles Beck to Christian Barman, Publicity Office, June 2, 1938, in Garland, *Mr Beck's Underground Map*, 32.

4. Siegfried Kracauer called his era simply and accurately the "age of transportation." See Siegfried Kracauer, *The Salaried Masses: Duty and Distraction in Weimar Germany*, trans. Quintin Hoare (London: Verso, 1998), 93 (translation modified).

5. Cf. Sigfried Giedion, *Mechanization Takes Command: A Contribution to Anonymous History* (New York: Oxford University Press, 1948).

6. Harry Beck in conversation with Ken Garland, recounted by Garland in a BBC broadcast *Design Classics* in 1994, quoted in Janin Hadlaw, "The London

Underground Map: Imagining Modern Time and Space," *Design Issues* 19, no. 1 (2003): 25–35, esp. 32.

7. Ludwig Troske, *Die Londoner Untergrundbahnen: Sonder-Abdruck aus der Zeitschrift des Vereines deutscher Ingenieure 1891 und 1892* (Berlin: Julius Springer, 1892), 63.

8. See Hadlaw, "The London Underground Map."

9. See Janet Vertesi, "Mind the Gap: The London Underground Map and Users' Representation of Urban Space," *Social Studies of Science* 38, no. 1 (2008): 7–33, following Bruno Latour, "Visualisation and Cognition: Drawing Things Together," in *Representation in Scientific Practice*, ed. Michael Lynch and Steve Woolgar (Cambridge, MA: MIT Press, 1990), 19–68.

10. See Paul Elliman, "Signal Failure," in *Else/Where: Mapping New Cartographies of Networks and Territories*, ed. Janet Abrams and Peter Hall (Minneapolis: University of Minnesota Press, 2006), 166–175, esp. 175, on the status of the Tube map as an icon of network efficiency and metaphor of the city as electrified machine.

11. This is clearest in Garland, *Mr Beck's Underground Map*. On the history of London Transport in general, see Theodore Cardwell Barker and Michael Robbins, *A History of London Transport: Passenger Travel and the Development of the Metropolis*, 2 vols. (London: George Allen & Unwin, 1974); and Desmond Felix Croome and Alan Arthur Jackson, *Rails through the Clay: A History of London's Tube Railways*, 2nd ed. (Harrow Weld, UK: Capital Transport, 1993). For a diagrammatic history of the metro network, see Douglas Rose, *The London Underground: A Diagrammatic History*, 7th ed. (Harrow Weld, UK: Capital Transport, 2000).

12. See also chapter 6 of the present volume.

13. Harry Beck described this as follows: "I have tried to imagine that I was using a convex lens or mirror, so as to present the central area on a larger scale." Quoted in Garland, *Mr Beck's Underground Map*, 17.

14. See Richard Sennett, *Flesh and Stone: The Body and the City in Western Civilization* (New York: W. W. Norton, 1994), 332–333.

15. On Man Ray, see Green, *Underground Art*, 77. On Moholy-Nagy, see Achim Borchardt-Hume, ed., *Albers and Moholy-Nagy: From Bauhaus to the New World* (London: Tate, 2006), 58, 64.

16. Sarah Johnstone and Tom Masters, *London City Guide* (Melbourne: Lonely Planet, 2006), 34.

17. Nelson Goodman, *Languages of Art: An Approach to a Theory of Symbols* (Indianapolis, IN: Bobbs-Merrill, 1968), 170–171. The relationship of diagrams and topography is captured only vaguely by such a view. For a critique of it, see Bernhard Siegert, "Repräsentationen diskursiver Räume: Einleitung," in *Topographien der Literatur: Deutsche Literatur im transnationalen Kontext*, ed. Hartmut Böhme (Stuttgart: J. B. Metzler, 2005), 3–11, esp. 4.

18. See, for example, Barker and Robbins, *A History of London Transport*, 352–353.

19. See the early report Henry Mayhew, "The Metropolitan Railway," in *London Underground: Poems and Prose about the Tube*, ed. Tobias Döring (Stuttgart: Reclam, 2003), 14–34, esp. 29–30.

20. On the institutional background, see David Leboff and Tim Demuth, *No Need to Ask! Early Maps of London's Underground Railways* (Harrow Weld, UK: Capital Transport, 1999), 44. On the iconography of the logo, see Barker and Robbins, *A History of London Transport*, 179. On the in-house magazine, see Garland, *Mr Beck's Underground Map*, 21.

21. See Garland, *Mr Beck's Underground Map*, 68–76.

22. Charles Baudelaire, "Le Cygne," in Charles Baudelaire, *Œuvres complètes*, ed. Claude Pichois, 2 vols. (Paris: Bibliothèque de la Pléïade, 1975), 1:85–87, esp. 87; Charles Baudelaire, "Le Cygne/The Swan," in *Les Fleurs du Mal: The Complete Text of the Flowers of Evil*, trans. Richard Howard (Boston: D. R. Godine, 1982), 90–92, esp. 90.

23. See Elliman, "Signal Failure," 172.

24. See "Centralised Control of Railway Traffic," *The Railway Gazette and Railway News*, January 22, 1932, 106–107, 115.

25. "Centralised Control of Railway Traffic," 107.

26. *Signalling on the Ealing and Shepherd's Bush Railway*, directed by Underground Group Signals Section (1921; UK: London Transport Museum, 2007), Film Collection, no. 121.

27. See John Pattinson Thomas, *Handling London's Underground Traffic* (London: London's Underground, 1928), 83.

28. See Norbert Wiener, *Cybernetics; or, Control and Communication in the Animal and the Machine*, 2nd ed. (Cambridge, MA: MIT Press, 1965) 98. I am sincerely grateful to Lasse Scherffig for this reference.

29. Thomas, *Handling London's Underground Traffic*, 58.

30. See Thomas, *Handling London's Underground Traffic*, 55–56. This was part of a prominent international discourse on rationalization. See Anette Schlimm, *Ordnungen des Verkehrs: Arbeit an der Moderne; Deutsche und britische Verkehrsexpertise im 20. Jahrhundert* (Bielefeld: transcript, 2011), 43–44.

31. See Thomas, *Handling London's Underground Traffic*, 82–86.

32. Thomas, *Handling London's Underground Traffic*, 74–75.

33. Thomas, *Handling London's Underground Traffic*, 73.

34. Thomas, *Handling London's Underground Traffic*, 61, 86.

35. Thomas, *Handling London's Underground Traffic*, 97.

36. On this, see Peter Galison, *Einstein's Clocks and Poincaré's Maps: Empires of Time* (New York: W. W. Norton, 2003), 128–144.

37. Marc Augé, *In the Metro*, trans. Tom Conley (Minneapolis: University of Minnesota Press, 2002), 24.

38. Thomas, *Handling London's Underground Traffic*, 101.

39. Franz Kafka, *America*, trans. Willa and Edwin Muir (Harmondsworth, UK: Penguin, 1981), 139.

40. Kafka, *America*, 247.

41. Kafka, *America*, 247.

42. See Croome and Jackson, *Rails through the Clay*, 154. An expression common today to excuse the cancellation of a train begins, "Due to vandalism. . . ."

43. *The Right and Wrong Way to Board an Electric Train* (1920; UK: London Transport Museum, 2007), Film Collection, no. 125. The film was distributed by Gaumont.

44. See Hartmut Böhme, "Netzwerke: Zur Theorie und Geschichte einer Konstruktion," in *Netzwerke: Eine Kulturtechnik der Moderne*, ed. Jürgen Barkhoff, Hartmut Böhme, and Jeanne Riou (Cologne: Böhlau, 2004), 17–36, esp. 19.

45. Thomas, *Handling London's Underground Traffic*, 31.

46. Hadlaw, "The London Underground Map," 32.

47. See Thomas, *Handling London's Underground Traffic*, 170–171.

48. On the tactics of the urban passenger, see also Stefan Höhne, *Riding the New York Subway: The Invention of the Modern Passenger* (Cambridge, MA: MIT Press, 2021).

49. On this, with reference to Siegfried Kracauer, see Johannes Roskothen, *Verkehr: Zu einer poetischen Theorie der Moderne* (Munich: Wilhelm Fink, 2003), 77.

50. See Augé, *In the Metro*, 29–30.

51. Augé, *In the Metro*, 16, 20, 29–30.

52. Roskothen, *Verkehr*, 83.

53. See "Public Notes on Railways," *The Railway Gazette and Railway News*, January 8, 1932, 44–45.

54. *Bulldog Jack*, directed by Walter Forde (1935; UK: British Picture Corporation, 1994), VHS.

55. The detective in this film from the series played by a different person who merely claims to be Bulldog Drummond but is in fact Bulldog "Jack."

56. On this, see Wolfgang Schivelbusch, *The Railway Journey: The Industrialization of Time and Space in the Nineteenth Century* (Oakland: University of California Press, 2014).

57. On the active forgetting of passage as an aesthetic phenomenon, see Paul Virilio, *The Aesthetics of Disappearance*, 2nd ed., trans. Philip Beitchman (Los Angeles: Semiotext(e), 2009), 19–22.

58. Siegert, "Repräsentationen diskursiver Räume," 10. On diverse contemporaneous practices of orientation using the Tube map, see Vertesi, "Mind the Gap," 13–14.

59. Schlimm, *Ordnungen des Verkehrs*, 162–163.

60. See London County Council, Special Committee on Electricity Supply, ed., *Report on London Electricity Supply by Merz and McLellan* (London: The Council, 1914), app. 2, 37–38; Thomas Parke Hughes, *Networks of Power: Electrification in Western Society, 1880–1930* (Baltimore: Johns Hopkins University Press, 1983), 227. For an early report on the electric Tube, see also Troske, *Die Londoner Untergrundbahnen*, 90–91.

61. "Electrical Development in England," *Electrical World* 82, no. 21 (1923): 1055–1058, esp. 1056.

62. See Hughes, *Networks of Power*, 232. See also Croome and Jackson, *Rails through the Clay*, 75. On London Transport's power plant on Lots Road, see Elliman, "Signal Failure," 167. This avoided complex problems that the engineers of the electrical grid confronted on a national level. The approach to the limits of the predictability of synchronizations within the grid presented the occasion in the United States to build an analog "network analyzer" at MIT under the direction of Vannevar Bush. See Hughes, *Networks of Power*, 376.

63. For the most extreme frustrated reaction, see the design of a penny transport in the *Railway Gazette*, March 25, 1932, 459–461.

64. The first combined map of nearly all the railway lines was printed in 1908. The participants were the Underground Railways of London (UERL), Central London, Metropolitan, Great Northern & City and the City & South London Railways. See Garland, *Mr Beck's Underground Map*, 9; and Hadlaw, "The London Underground Map," 29.

65. See the *Railway Gazette*—for example, the issues of May 8 and 15, 1931—and the weekly column "In Parliament" in *Modern Transport*. For a detailed account, see Barker und Robbins, *A History of London Transport*, 272–273.

66. On visual representations in the discourse of transportation studies, see Schlimm, *Ordnungen des Verkehrs*, 129–130.

67. The institution's first annual report was published in 1934. See London Passenger Transport Board, ed., *First Annual Report and Statement of Accounts and Statistics for the Year Ended 30 June 1934* (London: LPTB, 1934).

68. See Gilles Deleuze, "A New Cartographer (*Discipline and Punish*)," in *Foucault*, ed. and trans. Seán Hand (London: Bloomsbury, 2016), 21–38.

69. I am pursuing here a thesis from Johannes Roskothen. See Roskothen, *Verkehr*, 67. See also Michel de Certeau, *The Practice of Everyday Life*, trans. Steven Rendall (Berkeley: University of California Press, 1984), 94–95.

70. Green, *Underground Art*, 155; *Frank Pick's London*, 18. Dring's image has formerly been dated in the early 1930s or 1933. Recently, its creation date has been updated to "circa 1938" in the London Transport Museum database (inv. no. 1985/123). See

https://www.ltmuseum.co.uk/collections/collections-online/artwork/item/1985-123. Since 2020, there has been a Wikipedia article on Dring. See Wikipedia, s.v. "Lilian Dring," https://en.wikipedia.org/wiki/Lilian_Dring. See http://collections.vam.ac.uk/name/dring-lilian/A5535 for Dring's works at the Victoria and Albert Museum.

71. On this, see Sennett, *Flesh and Stone*, 255–270.

72. See chapter 2 and section 6.1 of the present volume.

73. See Transport for London, "Thank You for Getting Ahead of the Games," August 4, 2012. https://web.archive.org/web/20120808132057/http://www.getaheadofthegames.com/news/1364.html; BBC, "Games Bring Record London Underground Users," August 4, 2012. http://www.bbc.co.uk/news/uk-england-london-19125580; Michael Hirst, "London's Travel Network Clears Its First Olympic Hurdle," July 30, 2012, http://www.bbc.co.uk/news/uk-19039072.

74. See Roskothen, *Verkehr*, 77–78. On British Modernism and the Tube, see Andrew Thacker, "Uncompleted Life: The Modernist Underground," in *The Railway and Modernity: Time, Space and the Machine Ensemble*, ed. Matthew Beaumont and Michael Freeman (Bern: Peter Lang, 2007), 101–124.

75. See Mark Ovenden, *Metro Maps of the World*, 2nd ed. (Harrow Weld, UK: Capital Transport, 2004).

76. Beck also provided designs for Paris—with limited success. See Garland, *Mr Beck's Underground Map*, 67.

77. Siegfried Kracauer, "Two Planes," in Kracauer, *The Mass Ornament: Weimar Essays*, ed. and trans. Thomas Y. Levin (Cambridge, MA: Harvard University Press, 1995), 37–39.

78. See Certeau, *The Practice of Everyday Life*, 97.

79. I can only mention in passing here Ernst Cassirer, *Substance and Function and Einstein's Theory of Relativity* (Chicago: Open Court, 1923).

CHAPTER 8

1. Markus Krajewski, ed., *Projektemacher: Zur Produktion von Wissen in der Vorform des Scheiterns* (Berlin: Kadmos, 2004), 13.

2. See chapter 4 in the present volume.

3. On the drawing from the early modern period onward, see Bernhard Siegert, *Cultural Techniques: Grids, Filters, Doors, and Other Articulations of the Real*, trans. Geoffrey Winthrop-Young (New York: Fordham University Press, 2015), chap. 7; and Barbara Wittmann and Christoph Hoffmann, "Introduction," in *Science in Context* 26, no. 2 (2013): 203–213.

4. See Gustav Rossler, "Galerie neuer Dingbegriffe: Hybriden, Quasi-Objekte, Grenzobjekte, epistemische Dinge," in *Bruno Latours Kollektive: Kontroversen zur Entgrenzung*

des Sozialen, ed. Georg Kneer, Markus Schroer, and Erhard Schüttpelz (Frankfurt am Main: Suhrkamp, 2008), 76–107, esp. 101.

5. Rolph Berg et al., *Netzplantechnik: Grundlagen, Methoden, Praxis* (Zurich: Verlag industrielle Organisation, 1973), 15.

6. Kenneth Gordon McLaren and Eric Leonard Buesnel, *Network Analysis in Project Management: An Introductory Manual Based on Unilever Experience* (London: Cassell, 1969). Unilever first employed graphic network analysis internally in the United Kingdom. Courses for managers introduced the new tool, supported by the "Manual on Network Planning Methods," which had been drawn up in 1964. It was mainly applied to smaller projects—although "small" could include the building of a factory. Some other handbooks include: Friedrich Rosenkranz, *Netzwerktechnik und wirtschaftliche Anwendung: Mit besonderer Berücksichtigung der Netzplantechnik* (Meisenheim am Glam: Anton Hain, 1968); Abraham Mario Becker, *Handbuch der Dynamischen Netzplantechnik für die Planung und Überwachung von Aufträgen*, 3rd ed. (Winterthur: Gebrüder Sulzer, 1970); and Berg et al., *Netzplantechnik.*

7. On the technical image, see Horst Bredekamp, Birgit Schneider, and Vera Dünkel, eds., *The Technical Image: A History of Styles in Scientific Imagery* (Chicago: University of Chicago Press, 2015).

8. See Manuel Castells, *The Rise of the Network Society*, 2nd ed. (Oxford: Blackwell, 2000), 38–39.

9. For a more comprehensive network history of the economy that begins already in the early modern period, see the works of Fernand Braudel. See also chapter 4 in the present volume.

10. See Michel Foucault, *Security, Territory, Population: Lectures at the Collège de France, 1977–1978*, ed. Michel Senellart, trans. Graham Burchell (New York: Palgrave Macmillan, 2009), 1–86, and section 4.5 in the present volume.

11. See JoAnne Yates, *Control through Communication: The Rise of System in American Management* (Baltimore: Johns Hopkins University Press, 1989).

12. See the analysis of economic entanglements in chapter 10 in the present volume.

13. See Gabriele Schabacher, "Raum-Zeit-Regime: Logistikgeschichte als Wissenszirkulation zwischen Medien, Verkehr und Ökonomie," in *Agenten und Agenturen: Archiv für Mediengeschichte*, ed. Lorenz Engell, Joseph Vogl, and Bernhard Siegert (Weimar: Universitätsverlag, 2008), 135–148; and Alexander Klose, *The Container Principle: How a Box Changes the Way We Think*, trans. Charles Marcrum II (Cambridge, MA: MIT Press, 2015), 161–165. Monika Dommann rightly points to industrial techniques of storage and distribution of the nineteenth century as an additional source of this knowledge. See Monika Dommann, "Handling, Flowcharts, Logistik: Zur Wissensgeschichte und Materialkultur von Warenflüssen," *Nach Feierabend: Zürcher Jahrbuch für Wissensgeschichte* 7 (2011): 75–103.

14. The original form in British English is "operational research." On this and on the genealogy of OR, see Wolfgang Pircher, "Im Schatten der Kybernetik: Rückkopplung im operativen Einsatz: 'operational research,'" in *Die Transformation des Humanen: Beiträge zur Kulturgeschichte der Kybernetik*, ed. Michael Hagner and Erich Hörl (Frankfurt am Main: Suhrkamp, 2008), 348–376, esp. 349–350.

15. John Francis Magee and David Morris Boodman, *Production Planning and Inventory Control*, 2nd ed. (New York: McGraw-Hill, 1967), 385. Similar definitions that place more emphasis on the role of software and simulation can also be found in current OR handbooks. See, for example, Wolfgang Domschke and Andreas Drexl, *Einführung in Operations Research*, 7th ed. (Berlin: Springer, 2007), 1. On mathematical models, see also section 6.3 in the present volume.

16. Girolamo Cardano, *Liber de ludo aleae*, ed. Massimo Tamborine (Milan: Angeli, 2006; orig. pub. 1564).

17. John Graunt, *Natural and Political Observations Mentioned in a following Index, and made upon the Bills of Mortality* (London: Martin, Allestry, and Dicas, 1662).

18. See Agner Krarup Erlang, "Sandsynlighedsregning og Telefonsamtaler," *Nyt tidsskrift for Matematik* 20 (1909): 33–39.

19. On this, with a tendency to heroize scientists, see Saul I. Gass and Arjang A. Assad, *An Annotated Timeline of Operations Research: An Informal History* (Boston: Kluwer Academic, 2005).

20. See Gass and Assad, *An Annotated Timeline of Operations Research*, 45. A more detailed account can be found in Pircher, "Im Schatten der Kybernetik," 351–352.

21. In times of peace, Wiener had a considerably broader scope. See Norbert Wiener Papers, MC22, Box 25D, 360 (Patent data. Yuk Wing Lee and Norbert Wiener 1938, 1960); Box 36, 362, 363 (Patent data of others); Box 28A, 560 (Final report, re: design of a predicting network circa 1942). Massachusetts Institute of Technology Archives and Special Collections, Cambridge, MA. "Network" is understood here primarily in the sense of a system capable of learning, adapting, and self-correcting. On this, see Flo Conway and Jim Siegelmann, *Dark Hero of the Information Age: In Search of Norbert Wiener, the Father of Cybernetics* (New York: Basic Books, 2005), 76.

22. See Peter Galison, "The Ontology of the Enemy: Norbert Wiener and the Cybernetic Vision," *Critical Inquiry* 21, no. 1 (1994): 228–266.

23. See Paul N. Edwards, *The Closed World: Computers and the Politics of Discourse in Cold War America* (Cambridge, MA: MIT Press, 1996), 114–115, 393–394.

24. For a comprehensive overview, see Fritz Nordsieck, *Die schaubildliche Erfassung der Betriebsorganisation* (Stuttgart: Poeschel, 1932). On the organization of offices and companies, see Christine Schnaithmann, "'Fifty Ways to Cut Expenses': Praktiken der Effizienzsteigerung in amerikanischen Management-Handbüchern, 1910–1920,"

ilinx: Berliner Beiträge zur Kulturwissenschaft 3 (2013). On flowcharts, see Dommann, "Handling, Flowcharts, Logistik," 87–88.

25. More precisely, its Integrated Engineering Control Group (IEC). See James E. Kelley, Jr., and Morgan R. Walker, "Critical-Path Planning and Scheduling," in *Proceedings of the Eastern Joint Computer Conference* (New York: Eastern Joint Computer Conference, 1959), 160–173, esp. 160.

26. See, among others, Joseph John Moder and Cecil R. Philipps, *Project Management with CPM and PERT*, 4th ed. (New York: Reinhold, 1967), 2–3; and Graham K. Rand and Luis Valadares Tavares, "Network Planning," in *Encyclopedia of Operations Research and Management Science*, ed. Saul I. Gass (Boston: Kluwer, 2001), 562.

27. The corresponding notational technique was developed by the French mathematician Bernard Roy (1958) for the consulting company Société d'Économie et de Mathématiques Appliquées. It was first employed for the planning of the building of the SS *France* ocean liner. For an overview of the French-language discourse, see Bernard Roy, ed., *Les problèmes d'ordonnancement* (Paris: Dunod, 1964). The parallel emergence of French and American methods was grasped quickly in the European reception. See Rosenkranz, *Netzwerktechnik und wirtschaftliche Anwendung*, 9.

28. See section 8.2 in the present volume.

29. See Moder and Philipps, *Project Management with CPM and PERT*, 5; and McLaren and Buesnel, *Network Analysis in Project Management*, 26.

30. Kelley and Walker, "Critical-Path Planning and Scheduling," 160.

31. On this, see also Roy, *Les problèmes d'ordonnancement*, 2–3.

32. See McLaren and Buesnel, *Network Analysis in Project Management*, 2, 45. On "hand methods," see also Moder and Philipps, *Project Management with CPM and PERT*, 12.

33. A narration of this form of contingency is found in the management novel *Critical Chain*, whose main character, Rob, a university teacher with an uncertain future, discovers the pleasures of applied project management. It states explicitly. "Uncertainty is what typifies projects. It's the nature of the beast." Eliyahu M. Goldratt, *Critical Chain: A Business Novel* (Great Barrington, MA: North River Press, 1997), 40.

34. McLaren and Buesnel, *Network Analysis in Project Management*, viii.

35. See Rosenkranz, *Netzwerktechnik und wirtschaftliche Anwendung*, 13–14.

36. McLaren and Buesnel, *Network Analysis in Project Management*, 45.

37. They can be taken over by separate departments or external partners.

38. See McLaren and Buesnel, *Network Analysis in Project Management*, 8.

39. Operations research therefore tends to follow the program of Frank Gilbreth more than that of Frederick Taylor. This is particularly true of the just-in-time production model of Toyota, which radicalized the Fordist vision of a constantly flowing assembly line and transformed it into the switching of flows of information and

materials. See Taiichi Ohno, *Toyota Production System: Beyond Large-Scale Production* (Boca Raton, FL: CRC, 1988; orig. pub. in Japanese in 1978), 33–34.

40. On the limits of synchronization in organizations, see Niklas Luhmann, "Gleichzeitigkeit und Synchronisation," in *Soziologische Aufklärung 5: Konstruktivistische Perspektiven*, 2nd ed. (Wiesbaden: Westdeutscher Verlag, 1993), 95–130, esp. 125–126.

41. See Kelley and Walker, "Critical-Path Planning and Scheduling," 462.

42. "Uncertainty exists in every project situation." McLaren and Buesnel, *Network Analysis in Project Management*, 65. For a detailed account of this, see Magee and Boodman, *Production Planning and Inventory Control*, 81–82, and their informative summary: "The field of forecasting has hardly reached the handbook engineering stage." Magee and Boodman, *Production Planning and Inventory Control*, 97.

43. "It provides a clear picture of the scope of a project that can easily be read and understood." Kelley and Walker, "Critical-Path Planning and Scheduling," 462. On the "great advantage" of the clarity of "network representations," see also Rosenkranz, *Netzwerktechnik und wirtschaftliche Anwendung*, 2. The modernist tradition of this style of drawing dates back to the prewar era. For example, as early as 1932, with Bauhaus consistency, Fritz Nordsieck made "brevity, transparency, clarity, and vividness, exactness and planarity" the measure of corporate "Schaubilder" (diagrams). Nordsieck, *Die schaubildliche Erfassung der Betriebsorganisation*, 3.

44. See McLaren and Buesnel, *Network Analysis in Project Management*, 45.

45. See McLaren and Buesnel, *Network Analysis in Project Management*, viii.

46. On the history of the transatlantic liner planned using the Metra potential method, see Wikipedia, s.v. "SS *France* 1960," https://en.wikipedia.org/wiki/SS_France_(1960).

47. Strange hybrid forms resulted from this: the civilian tradition of British OR had a controversial proximity to the planned economy. See Pircher, "Im Schatten der Kybernetik," 366.

48. See Claude Elwood Shannon, "A Mathematical Theory of Communication," *The Bell System Technical Journal* 27, no. 3 (1948): 379–423.

49. On flow as a central principle, with reference to Fritz Nordsieck as a classic of internal corporate organization, see Wolf-Rüdiger Bretzke, *Logistische Netzwerke*, 2nd ed. (Heidelberg: Springer, 2010), 7–8.

50. McLaren and Buesnel, *Network Analysis in Project Management*, 34.

51. Michel Serres and Bruno Latour, *Conversations on Science, Culture, and Time*, trans. Roxanne Lapidius (Ann Arbor: Michigan University Press, 1995), 105.

52. See Castells, *The Rise of the Network Society*, 188–189.

53. Castells, *The Rise of the Network Society*, 199.

54. See Saskia Sassen, *Territory, Authority, Rights: From Medieval to Global Assemblages*, 4th ed. (Princeton, NJ: Princeton University Press, 2008), 348–349.

55. See Castells, *The Rise of the Network Society*, 208. As a programmatic text on network corporations, see Robert B. Reich, *The Work of Nations: Preparing Ourselves for 21st-Century Capitalism* (New York: Alfred A. Knopf, 1991), esp. 87–88.

56. See Castells, *The Rise of the Network Society*, 180–181.

57. Colin Cherry, *On Human Communication: A Review, a Survey, and a Criticism*, 2nd ed. (Cambridge, MA: MIT Press, 1966), 27–28.

58. See section 6.4 and chapter 7 in the present volume.

59. On network protocol, see Alexander Galloway, *Protocol: How Control Exists after Decentralization* (Cambridge, MA: MIT Press, 2004), 7–20, and section 3.3 and chapter 9 in the present volume.

60. On corporations as agents of societies of control, see Gilles Deleuze, "Postscript on the Societies of Control," *October* 59 (1992): 3–7. On the relationship of machinic forms and diagrams with reference to Deleuze and Foucault, see Galloway, *Protocol*.

61. Y. Sugimoro et al., "Toyota Production System and *Kanban* System of Materialization of Just-in-Time and Respect-for-Human System," *International Journal of Production Research* 15, no. 6 (1977): 553–564; Ohno, *Toyota Production System*.

62. Toyoda Kiichiro (1894–1952) was the first director of the Toyota Motor Co., which was formed as a spin-off of the Toyoda Spinning and Weaving Co.

63. On this binary juxtaposition within Japan's "second opening," see Volker Grassmuck, *Geschlossene Gesellschaft: Mediale und diskursive Aspekte der "drei Öffnungen" Japans* (Munich: iudicium, 2002), 310–311. On the *longue durée* of Japanese technology transfer, see Erich Pauer, "Der Technologietransfer nach Japan: Strukturen und Strategien," in *Technologietransfer Deutschland–Japan*, ed. Erich Pauer (Munich: iudicium, 1992), 48–72.

64. Castells, *The Rise of the Network Society*, 188.

65. On the history of the Japanese auto industry in the global context and especially on Toyota, see Michael A. Cusumano, *The Japanese Automobile Industry: Technology and Management at Nissan and Toyota* (Cambridge, MA: Harvard University Press, 1985); Haruhito Shiomi and Kazuo Wada, eds., *Fordism Transformed: The Development of Production Methods in the Automobile Industry* (Oxford: Oxford University Press, 1995); Charles J. McMillan, *The Japanese Industrial System*, 3rd ed. (Berlin: De Gruyter, 1996); Takahiro Fujimoto, *The Evolution of the Manufacturing System at Toyota* (New York: Oxford University Press, 1999); and Christoph Neubert, "Onto-Logistik: Kommunikation und Steuerung im Internet der Dinge," in *Archiv für Mediengeschichte: Agenten und Agenturen*, ed. Lorenz Engell, Joseph Vogl, and Bernhard Siegert, 119–133 (Weimar: Universitätsverlag, 2008), esp. 121–122.

66. See Cusumano, *The Japanese Automobile Industry*, 277–278. The story about the newspaper article is not told in Ohno, *Toyota Production System*. Fujimoto, *The*

Evolution of the Manufacturing System at Toyota, 327, refers to research by H. Shioji, which dates inspiration for a "supermarket system" at the American company Lockheed to 1954.

67. See McMillan, *The Japanese Industrial System*, 273.

68. See Ohno, *Toyota Production System*, 3–4.

69. Ohno, *Toyota Production System*, 25. Sport and work integrate both practice and training as a way to acquire an embodied knowledge.

70. See Ohno, *Toyota Production System*, 5–6.

71. On the history of self-service in the supermarket, with reference to Clarence Saunders's "self-serving store" of 1916, see Christoph Neubert, "The End of the Line: Zu Theorie und Geschichte der Selbststeuerung in der modernen Logistik," in *Unsichtbare Hände: Automatismen in Medien-, Technik- und Diskursgeschichte*, ed. Hannelore Bublitz et al. (Paderborn: Wilhelm Fink, 2011), 191–214, esp. 196–197.

72. For more detail on this, see Fujimoto, *The Evolution of the Manufacturing System at Toyota*, 110–111.

73. Bruno Latour, "On Technical Mediation: Philosophy, Sociology, Genealogy," *Common Knowledge* 3, no. 2 (1994): 29–64, esp. 53.

74. See the essays in Shumpei Kumon and Henry Rosovsky, eds., *The Political Economy of Japan: Cultural and Social Dynamics*, vol. 3 (Stanford, CA: Stanford University Press, 1992), esp. Shumpei Kumon, "Japan as a Network Society," in Kumon and Rosovsky, *The Political Economy of Japan*, 109–141.

75. See Axel C. Botzenhardt, *Japan als Netzwerkgesellschaft* (Munich: iudicium, 1997), 123–166, esp. 160.

76. See Neubert, "Onto-Logistik," 123.

77. Charles McMillan traces this approach back to textile production, Toyota's original core business: "In the same way that the loom stops when wool runs out from a spindle, the line stops when a defect is discovered or an excess or shortage of parts occurs." McMillan, *The Japanese Industrial System*, 288.

78. Ohno, *Toyota Production System*, 5.

79. See Shiomi and Wada, *Fordism Transformed*, 38.

80. See McMillan, *The Japanese Industrial System*, 299–300.

81. Ohno, *Toyota Production System*, 28.

82. Ohno, *Toyota Production System*, 21, 121.

83. Hitoshi Takeda, *The Synchronized Production System: Going beyond Just-in-Time through Kaizen*, trans. Greta Konradt (London: Kogan Page, 2006), 65.

84. See jishio.org, s.v. "Kanban," https://jisho.org/search/kanban.

85. See Frank B. Gibney, "The Marks of a Japanese Merchant," in *Kanban: The Art of the Japanese Shop Sign*, ed. Dana Levy, Lea Sneider, and Frank Gibney (San Francisco: Chronicle Books, 1983), 7–29, esp. 8, 15.

86. See Dana Levy, Lea Sneider, and Frank Gibney, eds., *Kanban: The Art of the Japanese Shop Sign* (San Francisco: Chronicle Books, 1983), viii.

87. See Gibney, "The Marks of a Japanese Merchant," 17.

88. See Levy, Sneider, and Gibney, *Kanban*, 130.

89. See Gibney, "The Marks of a Japanese Merchant," 9.

90. Neubert, "Onto-Logistik," 124.

91. See Satoshi Kamata, *Japan in the Passing Lane: An Insider's Account of Life in a Japanese Auto Factory* (Boston: George Allen & Unwin, 1983). Kamata's report is also a document of the growing oil crisis. On the accident rate at Toyota, see also Cusumano, *The Japanese Automobile Industry*, 305.

92. Bretzke, *Logistische Netzwerke*, 1.

93. See Ohno, *Toyota Production System*, 47–48.

94. On its economic network logic, see Walter P. Powell, "Neither Market nor Hierarchy: Network Forms of Organization," *Research in Organizational Behavior* 12 (1990): 295–336; Castells, *The Rise of the Network Society*, 163–164.

95. This too can be understood especially well using Toyota as an example. See Ananth V. Iyer, Sridhar Seshadri, and Roy Vasher, *Toyota Supply Chain Management: A Strategic Approach to the Principles of Toyota's Renowned System* (New York: McGraw-Hill, 2009).

96. See Schabacher, "Raum-Zeit-Regime," 138.

97. On this, see sections 3.1 and 11.1 in the present volume.

98. On visibility in the logistical system, see Frank Straube, *e-Logistik: Ganzheitliches Logistikmanagement* (Berlin: Springer, 2004), 292–293. On Internet-based workflow, see Straube, *e-Logistik*, 280–281.

99. See 4Flow AG, "Logistiksoftware," https://www.4flow.de/geschaeftsbereiche/logistiksoftware.html.

100. See, for example, the photographs of container ports in Klose, *The Container Principle*, and the photographs of infrastructure in Nicole Starosielski, *The Undersea Network* (Durham, NC: Duke University Press, 2015).

101. This follows the developments in the history of graph theory. See section 6.3 in the present volume.

102. See Bretzke, *Logistische Netzwerke*, 317–318. Federal Express is considered a pioneer of hubs. In 1981, it declared the airport of Memphis, Tennessee, which had been using since 1973 as part of the hub-and-spoke system, to be the first "superhub."

103. See Bretzke, *Logistische Netzwerke*, 328.

104. Bretzke, *Logistische Netzwerke*, 8.

105. Susan Leigh Star, "This Is Not a Boundary Object: Reflections on the Origin of a Concept," *Science, Technology, & Human Values* 35, no. 5 (2010): 601–617.

CHAPTER 9

1. See Janet Abbate, *Inventing the Internet* (Cambridge, MA: MIT Press, 2000); Lisa Gitelman, *Always Already New: Media, History, and the Data of Cultures* (Cambridge, MA: MIT Press, 2006), 89–90; Andrew L. Russell, *Open Standards and the Digital Age: History, Ideology, and Networks* (Cambridge: Cambridge University Press, 2014); Ben Peters, *How Not to Network a Nation: The Uneasy History of the Soviet Internet* (Cambridge, MA: MIT Press, 2016); and Camille Paloque-Bergès and Valérie Schafer, "Arpanet (1969–2019)," *Internet Histories* 3, no. 1 (2019): 1–14. See also Jochen Koubek, "Vernetzung als kulturelles Paradigma" (PhD diss., Humboldt Universität zu Berlin, 2003), http://edoc.hu-berlin.de/18452/15629; Jens Schröter, *Das Netz und die virtuelle Realität: Zur Selbstprogrammierung der Gesellschaft durch die universelle Maschine* (Bielefeld: transcript, 2004); Rainer Fischbach, *Mythos Netz: Kommunikation jenseits von Raum und Zeit?* (Zurich: Rotpunktverlag, 2005); Hans Dieter Hellige, "Die Geschichte des Internet als Lernprozess," *artec-paper* 138 (2006): 1–39; Paul Ferdinand Siegert, *Die Geschichte der E-Mail: Erfolg und Krise eines Massenmediums* (Bielefeld: transcript, 2008); Mercedes Bunz, *Vom Speicher zum Verteiler: Die Geschichte des Internet* (Berlin: Kadmos, 2008); and Martin Warnke, *Theorien des Internet zur Einführung* (Hamburg: Junius, 2011).

2. On the networking of stock markets and airlines as cases of communication networks that result around specific uses, see Donald Watts Davies and Derek L.A. Barber, *Communication Networks for Computers* (London: John Wiley, 1973), 9 in general, and esp. 117–125 on Lloyds, and esp. 125–131 and 295–300 on flight booking systems such as BEACON and SITA.

3. Warnke, *Theorien des Internet zur Einführung*, 174. Warnke takes up the thread of Howard Rheingold and the network theory in Albert-László Barabási, *Linked: The New Science of Networks* (Cambridge, MA: Perseus, 2002), but leaves out critical views such as Alexander Galloway, *Protocol: How Control Exists after Decentralization* (Cambridge, MA: MIT Press, 2004).

4. The Advanced Research Projects Agency was established by Directive 5105.15 on February 7, 1958, in respond to the launch of Sputnik to catch up on the scientific and military edge of the Soviets. As a research institution of the Department of Defense, from March 23, 1972, with brief interruptions, it bore the name DARPA, with the D standing for defense. See www.darpa.mil.

5. Doug Dodds, "Notes on the Development of Message Technology," in *Berkeley Workshop on Distributed Data Management and Computer Networks*, ed. Lawrence

Berkeley Laboratories (Berkeley: University of California, 1976), 144–156, esp. 145. See Hellige, "Die Geschichte des Internet als Lernprozess," 17; Siegert, *Die Geschichte der E-Mail*, 212–224.

6. Joseph Carl Robnett Licklider and Robert William Taylor, "The Computer as a Communication Device," *Science and Technology* 76 (1968): 21–31. See also M. Mitchell Waldrop, *The Dream Machine: J. C. R. Licklider and the Revolution That Made Computing Personal* (New York: Viking, 2001).

7. See Licklider and Taylor, "The Computer as a Communication Device," 30–34. On libraries as information networks, see J. C. R. Licklider, *Libraries of the Future*, 5th ed. (Cambridge, MA: MIT Press, 1971); Gitelman, *Always Already New*, 98–103; and Kirsten Wagner, *Datenräume, Informationslandschaften, Wissensstädte: Zur Verräumlichung des Wissens und Denkens in der Computermoderne* (Freiburg: Rombach, 2006).

8. See Bunz, *Vom Speicher zum Verteiler*, 88.

9. Tim Wu has even called—somewhat overeffusively—the internet's architecture the first electronic information network that existed independently of the physical infrastructure over which it operates. See Tim Wu, *The Master Switch: The Rise and Fall of Information Empires* (New York: Alfred A. Knopf, 2011), 198.

10. See Jörg Pflüger, "Writing, Building, Growing: Leitvorstellungen der Programmiergeschichte," in *Geschichten der Informatik*, ed. Hans-Dieter Hellige (Berlin: Springer, 2004), 275–320.

11. See also section 3.3 of the present volume. On ceremonies at court and diplomacy, see Tobias Nanz, *Grenzverkehr: Eine Mediengeschichte der Diplomatie* (Zurich: diaphanes, 2010), 66–90. On the relationship of legal code, protocol, and computer architecture, see Cornelia Vismann and Markus Krajewski, "Computer Juridisms," *Grey Room* 29 (2008): 99–109, esp. 102–107.

12. See section 4.5 of the present volume.

13. Galloway, *Protocol*, 7–27.

14. See Alexander Galloway and Eugene Thacker, *The Exploit: A Theory of Networks* (Minneapolis: University of Minnesota Press, 2007), 55. See also Eugene Thacker, *Biomedia* (Minneapolis: University of Minnesota Press, 2004), 15–26.

15. John Newkirk et al., "RFC 55: A Prototypical Implementation of the NCP," tech. rep., Harvard and UCLA, 1970. https://www.rfc-editor.org/rfc/rfc55.txt, p. 1.

16. D. F. Jones, *Colossus* (London: Pan, 1968), 91.

17. The historical allusion to Colossus Mark I, the tube computer in operation in Bletchley Park (Buckinghamshire, United Kingdom) from 1944 onward, remains vague.

18. *Colossus: The Forbin Project*, directed by Joseph Sargent (1970; USA: Universal Pictures, 2004), DVD.

19. On things that talk, see Lorraine Daston, ed., *Things That Talk: Object Lessons from Arts and Sciences* (New York: Zone Books, 2004), esp. 11–24.

20. See Jacques Derrida, *Of Grammatology*, trans. Gayatri Chakravorty Spivak (Baltimore: Johns Hopkins University Press, 2016); and Jacques Derrida, *Writing and Difference*, trans. Alan Bass (Chicago: University of Chicago Press, 1978).

21. The hybridization of linguistic theory and media history in the 1960s has thus far been the subject of only a few critical studies in the history of knowledge. On the potential of such historical analysis, see Oliver Lerone Schultz, "Marshall McLuhan: Medien als Infrastrukturen und Archetypen," in *Medientheorien: Eine philosophische Einführung*, ed. Alice Lagaay (Frankfurt am Main: Campus, 2004), 31–68; Erhard Schüttpelz, "'Get the message through': From the Channel of Communication to the Message of the Medium, 1945–1960," in *Media, Culture, and Mediality: New Insights into the Current State of Research*, ed. Jäger Ludwig, Erika Linz, and Irmela Schneider (Bielefeld: transcript, 2010), 109–138; and Wagner, *Datenräume, Informationslandschaften, Wissensstädte*, 127–184.

22. Thomas Haigh and Paul E. Ceruzzi, *A New History of Modern Computing* (Cambridge, MA: MIT Press, 2021), 113–127.

23. Joseph Carl Robnett Licklider Papers, MC 499.3, Correspondence, 1958–1969, Massachusetts Institute of Technology. Institute Archives and Special Collections, Cambridge, MA, typescript, p. 2.

24. Joseph Carl Robnett Licklider Papers, MC499.7, "Man-Computer Communication" drafts and correspondence, c. 1965. Massachusetts Institute of Technology. Institute Archives and Special Collections, Cambridge, MA, typescript, p. 3.

25. On communication with host computers, see Bernhard Siegert, "Die trinität des gastgebers," in *Cyberspace: Gemeinschaften, virtuelle Kolonien, Öffentlichkeiten*, ed. Manfred Fassler and Wulf R. Halbach (Munich: Wilhelm Fink, 1994), 281–294.

26. Jones, *Colossus*, 40.

27. Jones, *Colossus*, 54. "KC/S" is a unit of measurement in *Colossus*: kilocycles per second.

28. Hans Blumenberg, *Die Lesbarkeit der Welt*, 2nd ed. (Frankfurt am Main: Suhrkamp, 1989), 26.

29. See Hellige, "Die Geschichte des Internet als Lernprozess," 10.

30. This selection should not obscure the fact that the actors in the United States and Europe formed as a multiplicity for which a few names can only serve as examples. For a detailed account, see Abbate, *Inventing the Internet*; Hellige, "Die Geschichte des Internet als Lernprozess"; and Siegert, *Die Geschichte der E-Mail*, 81–187.

31. "Research and Development." Project RAND was a "civilian" think thank funded largely by Air Force contracts, founded in 1946 primarily to study innovative military strategies and techniques. See Claus Pias and Sebastian Vehlken, "Einleitung,"

in *Think Tanks: Die Beratung der Gesellschaft*, ed. Thomas Brandstetter, Claus Pias, and Sebastian Vehlken (Zurich: diaphanes, 2010), 7–15, esp. 8–11.

32. Contemporaneous assessments by colleagues were made after some time had passed by, among others, James Martin, *Future Developments in Telecommunications* (Englewood Cliffs, NJ: Prentice-Hall, 1971), 124; Davies and Barber, *Communication Networks for Computers*, 443–447. The dissemination of Baran's concepts, however, should not be underestimated, since from 1964 at the latest they were accessible to a large audience in the field. See Abbate, *Inventing the Internet*, 21; and Morten Bay, "Hot Potatoes and Postmen: How Packet Switching Became ARPANET's Greatest Legacy," *Internet Histories* 3, no. 1 (2019): 15–30. Actors such as Lawrence Roberts have denied the relevance of both Baran and Donald Davies for the realization of the ARPANET. See Schröter, *Das Netz und die virtuelle Realität*, 46.

33. Paul Baran, "Memorandum RM-3097-PR: On Distributed Communications: V. History, Alternatives Approaches, and Comparisons," tech. rep., RAND Corporation, 1964. http://www.rand.org/content/dam/rand/pubs/research_memoranda/2008/RM3097.pdf, p. 1.

34. Baran, "On Distributed Communications V," 10.

35. One should also mention works by Frank Collbohm, who developed at RAND in 1955 a concept for emergency messages transmitted manually from station to station. See Baran, "On Distributed Communications V," 19.

36. "Communications" is being used here in the military sense.

37. Paul Baran, "Memorandum RM-3767-PR: On Distributed Communications: XI. Summary Overview," tech. rep., RAND Corporation, 1964. http://www.rand.org/content/dam/rand/pubs/research_memoranda/2006/RM3767.pdf, p. 1. On this, see Abbate, *Inventing the Internet*, 18.

38. Christoph Engemann, "Verteiltes Überleben: Paul Barans Antwort auf die atomare Bedrohung," in *Überleben: Historische und aktuelle Konstellationen*, ed. Falko Schmieder (Paderborn: Wilhelm Fink, 2011), 381–393, esp. 385.

39. Claude Elwood Shannon Papers, Box 8, Folder 1, Manuscript Division, Library of Congress, Washington, DC (Articles and Scientific Papers: Alphabetical File N.D. U–Z. 8.1; "Von Neumann's Contribution to Automata Theory"), typescript, 2–5. See also Edward F. Moore and Claude Elwood Shannon, *Reliable Circuits Using Crummy Relays*, tech. rep., Bell Telephone Laboratories, 1954.

40. Paul Baran, "Memorandum RM-3420-PR: On Distributed Communications: I. Introduction to Distributed Communications Networks," tech. rep., RAND Corporation, 1964, http://www.rand.org/publications/RM/RM3420, pp. 7, 13.

41. Baran, "On Distributed Communications I," 23–24.

42. This concept by Baran also became part of the ARPANET architecture. See Frank E. Heart et al., "The Interface Message Processor for the ARPA Computer

Network," in *AFIPS Conference Proceedings, Spring Joint Computers Conference, May 5–7, 1970, Atlantic City, New Jersey*, vol. 36 (New York: ACM, 1970), 551–567, esp. 555.

43. On the religious-economic tradition of "heavenly books" and worldly bookkeeping, see Blumenberg, *Die Lesbarkeit der Welt*, 17–18.

44. Baran, "On Distributed Communications I," 24. Baran referred to postal and telegraphic service but does not mention the far more advanced telex network. Manual message switching in telegraphy offices had been practiced since the 1930s. In the 1960s, the first computers were used for storage and forwarding. See Abbate, *Inventing the Internet*, 13; and Martin Campbell-Kelly et al., *Computer: A History of the Information Machine*, 3rd. ed. (Boulder, CO: Westview, 2014), 281–282. Davies and Barber trace the store-and-forward principle even further back to the Baudot telegraphs of 1874 that could employ a connection via "time sharing" of as many as six telegraphists using prepunched paper messages. Messages were passed on by hand as "tore tapes" from the arriving to the departing connection. See Davies and Barber, *Communication Networks for Computers*, 272–277.

45. The internal communication system of the SITA (Société Internationale de Télécommunications Aéronautiques) airline association also used, after a redesign in 1964, "message blocks" of 256 characters. See Davies and Barber, *Communication Networks for Computers*, 297.

46. Baran, "On Distributed Communications I," 24.

47. Paul Baran, "On Distributed Communications Networks," *IEEE Transactions on Communications* 12 (1964): 1–9, esp. 6.

48. Baran, "On Distributed Communications I," 20, 22.

49. Abbate, *Inventing the Internet*, 19.

50. One can object that Baran's network was never built. See Schröter, *Das Netz und die virtuelle Realität*, 44. But from the perspective of the history of knowledge, the project designs discussed here belong to the same archive: of network models and simulations of the 1960s.

51. See Leonard Kleinrock, *Communication Nets: Stochastic Message Flow and Delay* (New York: McGraw-Hill, 1964), 131. The published text differs from the PhD dissertation that Kleinrock submitted to MIT in December 1962. See https://dspace.mit.edu/handle/1721.1/11562.

52. Kleinrock, *Communication Nets*, 18. Analogous arguments include, for example: "An efficient topological design provides a high throughput for a given cost." Howard Frank, Robert Elliott Kahn, and Leonard Kleinrock, "Computer Communication Network Design: Experience with Theory and Practice," in *AFIPS Conference Proceedings: Spring Joint Computers Conference, May 16–17, 1972, Atlantic City, New Jersey*, vol. 40 (Montvale, NJ: AFIPS Press, 1972), 255–270, esp. 261.

53. See Kleinrock, *Communication Nets*, 61–69.

54. Kleinrock, *Communication Nets*, 64–67.

55. This is in keeping with the central assumption of the network engineer that digital data traffic is largely characterized by the transmission of messages that tend to be short in a short time. See Davies and Barber, *Communication Networks for Computers*, 290.

56. On the American–Japanese cultural transfer in just-in-time production, see section 8.2 of this book.

57. Kleinrock, *Communication Nets*, 133–134. On the disposition of media agents as service agents or servants, see Markus Krajewski, *The Server: A Media History from the Present to the Baroque*, trans. Ilinca Iurascu (New Haven, CT: Yale University Press, 2018).

58. Kleinrock, *Communication Nets*, 190.

59. Kleinrock recommends precisely this. See Kleinrock, *Communication Nets*, 132.

60. Campbell-Kelly et al., *Computer: A History of the Information Machine*, 282; Bay, "Hot Potatoes and Postmen," 20–21.

61. See James Gillies and Robert Cailliau, *How the Web Was Born: The Story of the World Wide Web* (Oxford: Oxford University Press, 2000), 21.

62. Schröter, *Das Netz und die virtuelle Realität*, 45.

63. See Abbate, *Inventing the Internet*, 21–35; Koubek, "Vernetzung als kulturelles Paradigma," 61–65; and Schröter, *Das Netz und die virtuelle Realität*, 41–45. Davies remained faithful in this even later, for example, when he and Derek Barber derive the need for a worldwide data network with diverse applications from the demands of global meteorology, air travel, and the passing of the nuclear nonproliferation treaty. See Davies and Barber, *Communication Networks for Computers*, 13.

64. Abbate, *Inventing the Internet*, 23. To be even more precise, Baran's concept optimized for "survivability" versus "efficiency under normal circumstances," while Davies's was more about "interactive computing" versus "batch computing." I owe this comment to Thomas Haigh.

65. Donald Watts Davies, "Proposal for a Digital Communication Network," tech. rep., National Physical Laboratory, United Kingdom, June 1966, p. 4 (emphasis added). "Critical time" corresponds to reaction time.

66. Davies, "Proposal for a Digital Communication Network," 5.

67. Davies, "Proposal for a Digital Communication Network," 10.

68. Davies and Barber, *Communication Networks for Computers*, 407.

69. Heinz von Foerster, "Memory without Record," in *Learning, Remembering, and Forgetting*, vol. 1, *The Anatomy of Memory*, ed. Daniel P. Kimble (Palo Alto, CA: Stanford University Press, 1965), 388–433, esp. 392. The lecture was held on October 2, 1963, at the First Conference on Learning, Remembering and Forgetting in Princeton, NJ.

70. Heinz von Foerster, "Memory without Record," 392.

71. See Katie Hafner and Matthew Lyon, *Where Wizards Stay Up Late: The Origins of the Internet* (New York: Touchstone, 1998), 56–57; Stewart Brand, "Founding Father: Interview with Paul Baran," *Wired* 9, no. 3 (2001), https://www.wired.com/wired/archive/9.03/baran_pr.html; and Morten Bay, "Conversations with a Pioneer: Paul Baran in His Own Words," *Internet Histories* 1, no. 3 (2017): 273–282, 278.

72. Blumenberg, *Die Lesbarkeit der Welt*, 23.

73. See Claus Pias, "Elektronenhirn und verbotene Zone: Zur kybernetischen Ökonomie des Digitalen," in *Analog/Digital: Opposition oder Kontinuum?* ed. Jens Schröter and Alexander Böhnke (Bielefeld: transcript, 2004), 295–309.

74. On this, see the detailed account in Wagner, *Datenräume, Informationslandschaften, Wissensstädte*, 127–184.

75. Summaries can be found in Vinton Gray Cerf, "RFC 635: An Assessment of ARPANET Protocols," tech. rep., Stanford, 1974. https://www.rfc-editor.org/rfc/rfc635.txt; Jon Postel, "RFC 694. Protocol Information," tech. rep., Stanford Research Institute, 1975. https://www.rfc-editor.org/rfc/rfc694.txt; and Lawrence G. Roberts, "The Evolution of Packet Switching," *Proceedings of the IEEE* 66, no. 1 (1978): 1307–1313.

76. Davies and Barber, *Communication Networks for Computers*, 12.

77. See chapter 4 and section 6.4 of the present volume.

78. Licklider and Taylor, "The Computer as a Communication Device," 38.

79. Sargent, *Colossus: The Forbin Project*, timecode ca. 0:35:40. The slashes indicate changes in the scrolling LED display.

80. Wu, *The Master Switch*, 196–203.

81. See Heart, "The Interface Message Processor," 551. Mercedes Bunz has rightly noted that the network order of the ARPANET still followed the logic of the hardware. Bunz underestimates, however, that networks never follow purely symbolic orders but necessarily have a material foundation. See Bunz, *Vom Speicher zum Verteiler*, 88. It is also known that when in doubt, the team developing IMPs preferred software-oriented solutions. See Hafner and Lyon, *Where Wizards Stay Up Late*, 109. The first expert criticism of the hardware-oriented design of the Americans was Louis Pouzin, who was also able to choose a more software-oriented approach for Cyclades thanks to experiences with the ARPANET. See Gillies and Cailliau, *How the Web Was Born*, 36–38.

82. The authorship of this idea has often been discussed. Usually, the initial idea is credited to Wesley Clark and traced back to a meeting of the principal investigators involved in the ARPANET held in April 1967 at the University of Michigan in Ann Arbor. See Salus, *Casting the Net*, 21.

83. On BBN's role as system integrator, see Hafner and Lyon, *Where Wizards Stay Up Late*, 82–136. The standard handbook for IMPs is BBN Report 1822. See Bolt, Beranek

and Newman Inc., "Report No. 1822: Interface Message Processor; Specifications for the Interconnection of a Host and an IMP," techn. rep., 1976. http://www.bitsavers.org/pdf/bbn/imp/BBN1822_Jan1976.pdf. For an insider's view, see David C. Walden, "Looking Back at the ARPANET Effort Decades Later," in *The ARPANET Sourcebook: The Unpublished Foundations of the Internet*, ed. Peter H. Salus (Charlottesville, VA: Peer-to-Peer-Communications, 2008), 360–362. On the details of the hardware, see Walden, "Looking Back," 369.

84. See Heart, "The Interface Message Processor," 552.

85. See Krajewski, *The Server*, 297–337, esp. on the development of client–server architecture at Xerox Park.

86. Krajewski, *The Server*, 303.

87. The networking experts of the 1960s traced the emergence of the concept of daemon back to the time-sharing community at MIT and the development of the Multics operating system. See Fernando J. Corbato and Richard Steinberg, "Origin of the Word *Daemon*," *Take Our Word for It*, no. 146 (2002): 4, http://www.takeourword.com/TOW146/page4.html.

88. Hafner and Lyon, *Where Wizards Stay Up Late*, 123.

89. See Heart, "The Interface Message Processor," 554, 557–559. "Years later, some people would say that the IMP was nothing but a bit I/O [Input/Output] device and actually very simple. To the user, the IMP was to be as simple as an electrical outlet or a wall switch that does its job *without calling attention to itself*." Hafner and Lyon, *Where Wizards Stay Up Late*, 108 (emphasis added). On blackboxing in general, see Bruno Latour, *Pandora's Hope: Essays on the Reality of Science Studies* (Cambridge, MA: Harvard University Press, 1999), 304.

90. Arnold van Gennep, *The Rites of Passage*, trans. Monika B. Vizedom and Gabrielle L. Caffee (Chicago: University of Chicago Press, 1960), 19. On IMPs as boundary objects, see Fenwick McKelvey and Kevin Driscoll, "ARPANET and Its Boundary Devices: Modems, IMPs, and the Inter-Structuralism of Infrastructures," *Internet Histories* 3, no. 1 (2019): 31–50.

91. van Gennep, *The Rites of Passage*, 15.

92. Hafner and Lyon, *Where Wizards Stay Up Late*, 121. Heart was expressing this in retrospect as a memory.

93. For a critical view of Kleinrock's heroic stories, see Warnke, *Theorien des Internet zur Einführung*, 33. See also *Lo and Behold: Reveries of the Connected World*, directed by Werner Herzog (Magnolia Pictures, 2016).

94. Leonard Kleinrock "Personal History/Biography: The Birth of the Internet, "University of California at Los Angeles, Computer Science Department, 2010, https://web.archive.org/web/20190905082135/https://www.lk.cs.ucla.edu/personal_history.html. See also Salus, *The ARPANET Sourcebook*, 95.

95. A different account is found, for example, in Hafner and Lyon, *Where Wizards Stay Up Late*, 150. See also Ingrid Burrington, "How We Misremember the Internet's Origins," *The New Republic*, October 29, 2019, https://newrepublic.com/article/155532/misremember-internets-origins.

96. The geometric conventions for IMPs and computers are also used by Robert E. Kahn (BBN) in a documentary film from 1972 in which he demonstrated on a chalkboard various forms of net topology. See *Computer Networks: The Heralds of Resource Sharing*, directed by Peter Chvany (1972; International Conference on Computer Communications/archive.org, 2021), 3:40–6:25, http://archive.org/details/ComputerNetworks_TheHeraldsOfResourceSharing. BBN had a board with moving magnets so that they always had a current diagram of the structure of the net to examine. See Hafner and Lyon, *Where Wizards Stay Up Late*, 168. Positioning the IMPs in the center was typical at first and was also used in other drawings. See Salus, *The ARPANET* Sourcebook, 256.

97. See Hafner and Lyon, *Where Wizards Stay Up Late*, 153.

98. See the photograph documenting this at https://secure.flickr.com/photos/fastlizard4/6293913865.

99. See Hafner and Lyon, *Where Wizards Stay Up Late*, 154.

100. See Abbate, *Inventing the Internet*, esp. the overview on p. 60; Hellige, "Die Geschichte des Internet als Lernprozess"; Gitelman, *Always Already New*, 107–117; and Siegert, *Die Geschichte der E-Mail*.

101. Hellige, "Die Geschichte des Internet als Lernprozess," 19.

102. On these problems, see Bernhard Siegert, *Cultural Techniques: Grids, Filters, Doors, and Other Articulations of the Real*, trans. Geoffrey Winthrop Young (New York: Fordham University Press, 2015), chap. 10.

103. On the significance of TIPs for the ARPANET's increasing orientation around users, see Hafner and Lyon, *Where Wizards Stay Up Late*, 171–186.

104. Heart, "The Interface Message Processor," 556.

105. See section 6.4 and chapter 10 of the present volume.

106. See Galloway, *Protocol*, 121–143. Joyce K. Reynolds and Jon Postel note in RFC 1,000 that the protocols for and the creation of RFCs was "largely accidental." Joyce K. Reynolds and Jon Postel, "RFC 1000: The Request for Comments Reference Guide," tech. rep., ISI, 1987, https://www.rfc-editor. org/rfc/rfc1000.txt, p. 1. See also Abbate, *Inventing the Internet*, 58–60.

107. See Hafner and Lyon, *Where Wizards Stay Up Late*, 118. As head of the French Cyclades project, Louis Pouzin also relied on information exchange among developers and users and hired a special traveling messenger: "It was assumed that information around the network would circulate much better through thousands of informal channels, than it would through a rigid, formal structure." Louis Pouzin,

ed., *The Cyclades Computer Network: Towards Layered Network Architecture* (Amsterdam: North Holland, 1982), 3.

108. On the characteristics of RFC as protocols of informal dialogue, see Douglas E. Comer, *Internetworking with TCP/IP*, vol. 1, *Principles, Protocols, and Architecture*, 6th ed. (Boston: Pearson, 2014), 8–9. See also Galloway, *Protocol*, 135–143; and Gitelman, *Always Already New*, 110–117.

109. See Steve Crocker, "RFC 1: Host Software; Network Working Group Request for Comments," tech. rep., UCLA, 1969, http://www.rfc-editor.org/rfc/rfc1.txt, p. 2. RFC 1,000 notes that the first meeting was chaired by Elmer Shapiro (SRI). See Reynolds and Postel, "RFC 1000," 1.

110. Steve Crocker, "RFC 3: Documentation Conventions; Network Working Group RFC-3," tech. rep., UCLA, 1969. http://www.rfc-editor.org/rfc/rfc3.txt, n.p.

111. See Abbate, *Inventing the Internet*, 73–81; and Davies and Barber, *Communication Networks for Computers*, 306–307.

112. Its mode is bottom-up. See Gillies and Cailliau, *How the Web Was Born*, 64.

113. On this, see the detailed account in Andrew L. Russell, "Rough Consensus and Running Code and the Internet-OSI Standards War," in *IEEE Annals of the History of Computing* 28, no. 3 (2006): 48–61.

114. On this, see especially the introduction to Hendrik Blumentrath et al., eds., *Techniken der Übereinkunft: Zur Medialität des Politischen* (Berlin: Kadmos, 2009), which refers to Walter Benjamin and the "conference, considered as a technique of civil agreement." On this, see Walter Benjamin, "Critique of Violence," trans. Edmund Jephcott, in *Selected Writings*, vol. 1 (1913–1926), ed. Marcus Bullock and Michael W. Jennings (Cambridge, MA: Belknap Press of Harvard University Press, 1996), 236–252, esp. 244.

115. Christoph Engemann, *Electronic Government—vom User zum Bürger: Zur kritischen Theorie des Internets* (Bielefeld: transcript, 2003), 26.

116. See Russell, "Rough Consensus and Running Code and the Internet-OSI Standards War," 56.

117. Donald Davies, who initially still assumed a maximum critical time of ten seconds, also considered this acceleration necessary for interaction.

118. See Crocker, "RFC 1," 5–9. The concept of the protocol in the context of developing the ARPANET came from an article by Lawrence Roberts and Thomas Marill that offered the following definition: "uniform agreed-upon manner of exchanging messages between two computers in the network." Thomas Marill and Lawrence G. Roberts, "Toward a Cooperative Network of Time-Shared Computers," in *Proceedings of AFIPS Fall Joint Computer Conference, November 7–10, San Francisco, California, Fall Joint Computer Conference 29* (Washington, DC: Spartan Books, 1966), 425–431, esp. 428.

119. See RFCs 33, 36, 37, and 38.

120. Edwin E. Meyer Jr., "RFC 46. ARPA Network Protocol Notes," tech. rep., MIT, 1970. https://www.rfc-editor.org/rfc/rfc46.txt, p. 1. Similar arguments are made in Jon Postel and Steve Crocker, "RFC 48: A Possible Protocol Plateau," tech. rep., UCLA, 1970. https://www.rfc-editor.org/rfc/rfc48.txt, p. 5.

121. See Gillies and Cailliau, *How the Web Was Born*, 68.

122. See Edwin W. Meyer Jr., "RFC 49. Conversations with Steve Crocker (UCLA)," tech. rep., MIT Project MAC, 1970. https://www.rfc-editor.org/rfc/rfc49.txt, p. 8.

123. See Abbate, *Inventing the Internet*, 63.

124. See Stephen Crocker, Stephen Carr, and Vinton Gray Cerf, "RFC 33: New HOST-HOST Protocol," tech. rep., UCLA and University of Utah, 1970, http://www.rfc-editor.org/rfc/rfc33.txt, 2.

125. Crocker, Carr, and Cerf, "RFC 33," 17.

126. Edwin W. Meyer Jr., "RFC 82: Network Meeting Notes," tech. rep., MIT Project MAC, 1970, https://www.rfc-editor.org/rfc/rfc82.txt, p. 4.

127. See Lawrence G. Roberts and Barry D. Wessler, "Computer Network Development to Achieve Resource Sharing," *AFIPS Conference Proceedings: Spring Joint Computers Conference. May 5–7, 1970. Atlantic City, New Jersey*, vol. 36 (New York: ACM Press, 1970), 543–549, esp. 543.

128. The call was continually repeated from then on. See RFCs 212, 231, 462, and 584.

129. See section 9.1 of the present volume.

130. Jon Postel, "RFC 77: Network Meeting Report," tech. rep., UCLA, 1970. https://www.rfc-editor.org/rfc/rfc77.txt, 2.

131. Meyer, "RFC 82," 6.

132. Elizabeth Feinler, "The Network Information Center and its Archives," *IEEE Annals of the History of Computing* 32, no. 3 (2010): 83–89, esp. 85.

133. On what follows, see Hafner and Lyon, *Where Wizards Stay Up Late*, 175–186; Abbate, *Inventing the Internet*, 78–81; and Gillies and Cailliau, *How the Web Was Born*, 33.

134. See Alex McKenzie, "RFC 413: Traffic Statistics (October 1972)," tech. rep., BBN, 1972. https://www.rfc-editor.org/rfc/rfc413.txt; and Alex McKenzie, "RFC 422: Traffic Statistics (November 1972)," tech. rep., BBN, 1972. https://www.rfc-editor.org/rfc/rfc422.txt.

135. In 1974, this working group became part of the International Federation for Information Processing. See Kenneth Owen, "RFC 828: Data Communications: IFIP's International Network of Experts," tech. rep., IFIP, 1982. https://www.rfc-editor.org/rfc/rfc828.txt, p. 8.

136. On this, see Pouzin, *The Cyclades Computer Network*, 340.

137. See Salus, *Casting the Net*, 99.

138. See Salus, *Casting the Net*, 304.

139. See Stephen C. Carr, Stephen D. Crocker, and Vinton Gray Cerf, "HOST-HOST Communication Protocol in the ARPA network," *AFIPS Conference Proceedings: Spring Joint Computers Conference. May 5–7, 1970. Atlantic City, New Jersey*, vol. 36 (New York: ACM Press, 1970), 589–597, esp. 591. See also Salus, *The ARPANET Sourcebook*, 373.

140. See Galloway, *Protocol*, 40–53.

141. See Heart, "The Interface Message Processor," 554. The still popular Internet acronym "Full ACK" as signal for agreement probably goes back to this mode of understanding.

142. On the relationship of "call" and protocol, see Davies and Barber, *Communication Networks for Computers*, 384–388. On the handshake, see Davies and Barber, *Communication Networks for Computers*, 178–180.

143. For details on this, see Vinton Gray Cerf, Yogen Dalal, and Carl Sunshine, "RFC 675: Specification of Internet Transmission Control Program," tech. rep., 1974. https://www.rfc-editor.org/rfc/rfc675.txt, p. 22–23. See also University of Southern California Information Sciences Institute, "RFC 793: Transmission Control Protocol; DARPA Internet Program Protocol Specification," tech. rep., DARPA, 1981, https://www.rfc-editor.org/rfc/rfc793.txt, pp. 12, 30–39.

144. See Gillies and Cailliau, *How the Web Was Born*, 40. In addition, this fundamental change was influenced in part by experiences from developing the radio-based ALOHAnet at the University of Hawaii and the early design for the Ethernet.

145. On their lack, see Salus, *Casting the Net*, 101.

146. Davies and Barber, *Communication Networks for Computers*, 462.

147. On the conflicts that have come to be known as the protocol wars, between the "Internet camp" that relied on the TCP/IP protocol and the still largely public telecommunication companies that relied on the X.25 and the protocol architecture of the Open System Interconnection (OSI), see Abbate, *Inventing the Internet*, 147–179; Gillies and Cailliau, *How the Web Was Born*, 64–90; Russell: "Rough Consensus and Running Code and the Internet-OSI Standards War"; Bunz, *Vom Speicher zum Verteiler*, 96–106; and Warnke, *Theorien des Internet zur Einführung*, 46–50; JoAnne Yates and Craig Murphy, *Engineering Rules: Global Standard Setting since 1880* (Baltimore: Johns Hopkins University Press, 2019), chap. 7.

148. On gateways, see Vinton Gray Cerf and Robert Elliott Kahn, "A Protocol for Packet Network Intercommunication," *IEEE Transactions on Communications* 22, no. 5 (1974): 637–648, esp. 639.

149. See Severo M. Ornstein et al., "The Terminal IMP for the ARPA Computer Network," *AFIPS Conference Proceedings: Spring Joint Computer Conference, May 16–18,*

1972. Atlantic City, New Jersey, vol. 40 (Montvale, NJ: AFIPS Press, 1972), 243–254, esp. 244.

150. See Cerf and Kahn, "A Protocol for Packet Network Intercommunication," 637.

151. See Abbate, *Inventing the Internet*, 127. In France, intercommunication was integrated into the network design from the outset in the conceptual early phase of Cyclades. See Pouzin, *The Cyclades Computer Network*, 1, 231–256. Cerf and Kahn also refer explicitly to Cyclades. See Cerf and Kahn, "A Protocol for Packet Network Intercommunication," 642.

152. Cerf and Kahn, "A Protocol for Packet Network Intercommunication," 645.

153. See chapter 5 of the present volume.

154. See section 9.2 of the present volume.

155. Latour, *Pandora's Hope*, 304.

156. "Path control procedures establish a virtual communications medium between the communicating entities; this medium has certain desirable characteristics which may not be possessed by the physical medium (e.g. provision for addressing, synchronization, error control and flow control." David C. Walden and Alexander A. McKenzie, "The Evolution of Host-to-Host Protocol Technology," *Computer* 12, no. 9 (1979): 29–38, esp. 29.

157. See Cerf and Kahn, "A Protocol for Packet Network Intercommunication," 647.

158. Cerf, Dalal, and Sunshine, "RFC 675," 1.

159. See Cerf, Dalal, and Sunshine, "RFC 675," 18.

160. Cerf, Dalal, and Sunshine, "RFC 675," 1. See also Information Sciences Institute, "RFC 793," 7.

161. For an outstanding account of this, see Siegert, *Die Geschichte der E-Mail.*

162. Warnke, *Theorien des Internet zur Einführung*, 43.

163. See Cerf, Dalal, and Sunshine, "RFC 675," 2.

164. Vint Cerf, "The Future of Security and Open Standards on the Internet," lecture in the "Gespräche zur Netzpolitik" series, Heinrich-Böll-Stiftung, Akademie der Konrad-Adenauer-Stiftung, Berlin, May 26, 2011.

165. The military influence is especially clear in the central RFCs 791 and 793, both in the dedication to the Department of Defense and in the famous principle of robustness, which in turn follows more academic standards: "Be conservative in what you do, be liberal in what you accept from others." Information Sciences Institute, "RFC 793," 13.

166. See Abbate, *Inventing the Internet*, 131; and Vinton Gray Cerf, "Oral History Interview by Judy E. O'Neill" (OH 191), April 24, 1990. Reston, Virginia, Charles Babbage Institute, University of Minnesota, Minneapolis, 1990, https://hdl.handle.net/11299/107214, 28–30.

167. On what follows, see Salus, *Casting the Net*, 89–91; Gillies and Cailliau, *How the Web Was Born*, 56–60; and Mariann Unterluggauer, "Wie Europa beinahe das Internet erfand," *MIT Technology Review* 1 (2012): 52–55. https://www.heise.de/hintergrund/Wie-Europa-beinahe-das-Internet-erfand-1398231.html.

168. "Agreement on the Establishment of a European Informatics Network (with Annex), Concluded at Brussels on 23 November 1971," in *United Nations Treaty Series*, vol. 1262, I-20750 (New York: United Nations, 1982), 311–342, esp. 312. The treaty was passed at the Conference of Ministers in Brussels on November 22 and 23, 1971.

169. See "Agreement on the Establishment of a European Informatics Network," 341.

170. See Unterluggauer, "Wie Europa beinahe das Internet erfand." For its status in 1978, see the map in Salus, *Casting the Net*, 90.

171. See section 4.5 of the present volume.

172. Neal Stephenson, *Snow Crash* (New York: Bantam, 1992), 193.

173. Michel Callon, "Techno-Economic Networks and Irreversibility," in *A Sociology of Monsters? Essays on Power, Technology and Domination*, ed. John Law (London: Routledge, 1991), 132–161, esp. 148.

174. See Network Working Group, "RFC 1958: Architectural Principles of the Internet," tech. rep., Internet Architecture Board, 1996, http://www.rfc-editor.org/rfc/rfc1958.txt, p. 2. This analogy is a structural factor in the history of the computing, especially in the 1980s and 1990s, as Kirsten Wagner has shown in detail. See Wagner, *Datenräume, Informationslandschaften, Wissensstädte*.

175. Network Working Group, "RFC 1958," 2.

176. See Jerome H. Saltzer, David Patrick Reed, and David Dana Clark, "End-To-End Arguments in System Design," in *ACM Transactions on Computer Systems* 2, no. 4 (1984): 277–288, esp. 287. On the concept of architecture in computer science, see Barbara van Schewick, *Internet Architecture and Innovation* (Cambridge, MA: MIT Press, 2010), 19–33. On the genesis of end-to-end architecture, see also Matthias Bärwolff, "End-to-End Arguments in the Internet: Principles, Practices, and Theory," PhD diss., Technische Universität Berlin, 2010. http://dx.doi.org/10.14279/depositonce-2631.

177. Conversely, this is also true of the architecture, which responds to the computerizing of the city with its own network designs. See Mark Wigley, "Network Fever," *Grey Room* 4 (2001): 82–122; Hartmut Böhme, "Von der Vernetzung zur Virtualisierung der Städte: Ende der Philosophie—Beginn des Neuen Jerusalem?" in *Constructing Utopia: Konstruktionen künstlicher Welten*, ed. Annett Zinsmeister (Zurich: Diaphanes, 2005), 141–155; and Mark Wigley, "The Architectural Brain," in *Network Practices: New Strategies in Architecture and Design*, ed. Anthony Burke and Therese Tierney (New York: Princeton Architectural Press, 2007), 30–53.

178. "There are no connections or logical circuits (virtual or otherwise)." University of Southern California Information Sciences Institute, "RFC 791. Internet Protocol. DARPA Internet Program Control Specification," tech. rep., DARPA, 1981. https://www.rfc-editor.org/rfc/rfc791.txt, p. 2. By contrast, connections are made by the Transmission Control Program, see Information Sciences Institute, "RFC 793," 5, 30–39.

179. Lorenz Engell, *Ausfahrt nach Babylon: Essais und Vorträge zur Kritik der Medienkultur* (Weimar: Verlag und Datenbank für Geisteswissenschaften, 2000).

180. See Galloway, *Protocol*, 46.

181. For a linguistic theory of network formation by computers, with recourse to Ferdinand de Saussure, see Hartmut Winkler, *Docuverse: Zur Medientheorie der Computer* (Munich: Boer, 1997), 14–53.

CHAPTER 10

1. See Luc Boltanski and Ève Chiapello, *The New Spirit of Capitalism*, trans. Gregory Elliott (London: Verso, 2007), 138–139; Pierre Mercklé, *Sociologie des réseaux sociaux* (Paris: La Découverte, 2004), 6–7; and Christian Stegbauer and Roger Häussling, *Handbuch Netzwerkforschung* (Wiesbaden: VS Verlag für Sozialwissenschaften, 2010), 19–20, 281–282, and 859–860.

2. Élisabeth Roudinesco and Michel Plon, eds., *Dictionnaire de la psychanalyse* (Paris: Fayard, 1997), 764.

3. Eve Kosofsky Sedgwick, "Paranoid Reading and Reparative Reading; or, You're so Paranoid, You Probably Think This Essay Is about You," in *Touching Feeling: Affect, Pedagogy, Performativity* (Durham, NC: Duke University Press, 2003), 123–151, esp. 130.

4. Timothy Melley, *Empire of Conspiracy: The Culture of Paranoia in Postwar America* (Ithaca, NY: Cornell University Press, 2000), vii. On the explorability and narratability of conspiracies from an outsider's position, see Eva Horn and Anson Rabinbach, "Introduction," *New German Critique*, no. 103 (2008): 1–8, esp. 2.

5. See part 6.4 of the present volume.

6. See chapter 8 of the present volume.

7. See Dieter Groh, "The Temptation of Conspiracy Theory; or, Why Do Bad Things Happen to Good People? Part I: Preliminary Draft of a Theory of Conspiracy Theories," in *Changing Conceptions of Conspiracy*, ed. Carl F. Graumann and Serge Moscovici (New York: Springer, 1987), 1–13.

8. See Fredric Jameson, *The Geopolitical Aesthetic: Cinema and Space in the World System* (Bloomington: Indiana University Press, 1992). For a critique of Jameson, see Sedgwick, "Paranoid Reading and Reparative Reading."

9. According to a thesis by Erhard Schüttpelz. See Erhard Schüttpelz, "Ein absoluter Begriff: Zur Genealogie und Karriere des Netzwerkkonzepts," in *Vernetzte Steuerung: Soziale Prozesse im Zeitalter technischer Netzwerke*, ed. Stefan Kaufmann (Zurich: Chronos, 2007), 25–46, esp. 43. It is not exclusive. See the categorization of TCP/IP as the medium of the Internet in Martin Warnke, *Theorien des Internet zur Einführung* (Hamburg: Junius, 2011), 43.

10. See Samuel M. Weber, *Targets of Opportunity: On the Militarization of Thinking* (New York: Fordham University Press, 2005), 102. Weber adopts this view from John Arquilla and David F. Ronfeldt, *Networks and Netwars: The Future of Terror, Crime, and Militancy* (Santa Monica, CA: RAND, 2001), 328–329. The think-tank intellectuals Arquilla and Ronfeldt cite, in turn, Manuel Castells's remarks on the cultural identity of networks. Sociological network theory usually more soberly assumes "trust" as the basic currency of social networks. See Jörg Sydow and Arnold Windeler, "Steuerung von und in Netzwerken: Perspektiven, Konzepte, vor allem aber offene Fragen," in *Steuerung von Netzwerken: Konzepte und Praktiken*, ed. Jörg Sydow and Arnold Windeler (Wiesbaden: Westdeutscher Verlag, 2000), 1–24; and Walter P. Powell, "Neither Market nor Hierarchy: Network Forms of Organization," *Research in Organizational Behavior* 12 (1990): 295–336, esp. 326.

11. On the historical transformation of conspiracy theories, see Groh, "The Temptation of Conspiracy Theory," 15–37.

12. Pollux, *Trusts in der Schweiz? Die schweizerische Politik im Schlepptau der Hochfinanz* (Zurich: Verein für wirtschaftliche Studien, 1945), 99.

13. Niklas Luhmann, *Ecological Communication*, trans. John Bednarz, Jr. (Chicago: University of Chicago Press, 1989), 107. An opening up of systems theory to network analysis can be found in Dirk Baecker, "Handeln im Netzwerk: Zur Problemstellung der Soziologie," in *Relationale Soziologie: Zur kulturellen Wende der Netzwerkforschung*, ed. Jan Fuhse and Sophie Mützel (Wiesbaden: VS Verlag für Sozialwissenschaften, 2010), 233–256.

14. This quotation and those that follow in this paragraph are from Niklas Luhmann, "Kausalität im Süden," *Soziale Systeme* 1 (1995): 7–28, http://www.soziale-systeme.ch/leseproben/luhmann.htm.

15. Luhmann, "Kausalität im Süden."

16. David Ricardo, "Three Letters to the Morning Chronicle on the Bullion Report, 1810," in *The Works and Correspondence of David Ricardo*, vol. 3, *Pamphlets and Papers, 1809–1811*, ed. Piero Sraffa (Cambridge: Cambridge University Press, 1951), 129–153, esp. 148. For a genealogy of economics depicted in tables, see also the remarks on Quesnay's *Tableau Économique* (1758) in Bernhard Siegert, "Currents and Currency: Elektrizität, Ökonomie und Ideenumlauf um 1800," in *Netzwerke: Eine Kulturtechnik der Moderne*, ed. Hartmut Böhme, Jürgen Barkhoff, and Jeanne Riou (Cologne: Böhlau, 2004), 53–68, esp. 60–61.

17. Hobson comments with an urgent undertone: "Nowhere in the world has there ever existed so concentrated a form of capitalism as that represented by the financial power of the mining house in South Africa, and nowhere else does that power so completely realise and enforce the need of controlling politics." John A. Hobson, *The Evolution of Modern Capitalism: A Study of Machine Production*, rev. ed. (London: Walter Scott, 1906), 267.

18. On this, see Linton C. Freeman, *The Development of Social Network Analysis: A Study in the Sociology of Science* (Vancouver: Empirical Press, 2004), esp. 18–19. Hobson contributed additional works of unique visualizations of trade. See James A. Hobson, *The Industrial System: An Inquiry into Earned and Unearned Income* (London: Longman, Green & Co., 1910), 11.

19. Louis Brandeis, *Other People's Money and How the Bankers Use It* (New York: Frederick A. Stokes, 1914), 51. For important hints on the history of "interlocking directorates" and for making me aware of Pollux, I am sincerely grateful to Martin Lüpold.

20. Similar norms had existed prior to 1816, between banks for example, only to be forgotten *en passant*. See David Bunting, "Origins of the American Corporate Network," *Social Science History* 7, no. 2 (1983): 129–142, esp. 130–131.

21. See Bunting, "Origins of the American Corporate Network," 137–138.

22. See Frans N. Stokman, Rolf Ziegler and John Scott, eds., *Networks of Corporate Power: A Comparative Analysis of Ten Countries* (Cambridge: Polity Press, 1985), 1; and Bunting, "Origins of the American Corporate Network," 139. William Domhoff, author of the bestseller *Who Rules America?*, which has been repeatedly updated since 1967, signals already on the homepage of whorulesamerica.net that he distances himself from conspiracy theories of all kinds.

23. On the popularization of interlock analyses of fluid alliances in the context of "small world" phenomena, see Albert-László Barabási, *Linked: The New Science of Networks* (Cambridge, MA: Perseus, 2002), 204–205.

24. On this, see, to begin with, Richard Hofstadter, *The Paranoid Style in American Politics and Other Essays* (London: Jonathan Cape, 1966).

25. On his life and work, see Sarah Nichols, "Pollux's Spears," *Grey Room* 71 (2018): 141–155.

26. See Georg Baumann, *Atlantikpakt der Konzerne: Die internationale Kapitalverflechtung in Westdeutschland* (Berlin: Verlag die Wirtschaft, 1952); and Georg Baumann, *Eine Handvoll Konzernherren* (Berlin: Volk und Welt, 1953). For additional details on his life, see Markus Bürgi, "Bähler [Baehler], Georges," in *Historisches Lexikon der Schweiz (HLS)* (Bern: HLS, 2005), https://hls-dhs-dss.ch/de/articles/015156/2020-01-24 and the finding aid for the Baehler papers, Zentralbibliothek Zürich (manuscript department). With sincere thanks to Christoph Eggenberger for making it accessible.

27. Pollux, *Trusts in der Schweiz?*, 108 (italicized in the original). The comment by Karl Hediger refers to the Cadelplata affair, which Pollux analyzed.

28. *Gazette de Lausanne* (February 11, 1944), quoted in Pollux, *Trusts in der Schweiz?*, 109 (italicized in the original).

29. See, among others, Otto Neurath, "Bildstatistik und Wirtschaftsplan," in *Gesammelte bildpädagogische Schriften*, ed. Rudolf Haller and Robin Kinross (Vienna: Hölder-Pichler-Tempsky, 1991), 177–179.

30. Z. R.,"Wie sich Pollux die Schweiz vorstellt," *Neue Zürcher Zeitung*, evening edition no. 1280 (July 28, 1944), folio 5, quoted in Pollux, *Trusts in der Schweiz?*, 122–128, 124. The article in the *NZZ* mocks, among other things, Pollux's idea of overlapping personnel on supervisory boards—that is to say, his suspicion of interlocking directorates.

31. Pollux, *Trusts in der Schweiz?*, 126.

32. Pollux, *Trusts in der Schweiz?*, 129–130 (emphasis in bold in the original).

33. Pollux, *Trusts in der Schweiz?*, 129. In the original, this turn of phrase is printed in bold in a paragraph set off from the rest of the text.

34. Baehler's papers held in the manuscript department of the Universitätsbibliothek Zürich represent, characteristically, a monument to a mania for collecting data. See the finding aid to the papers of Georges Baehler, 17–18.

35. On the pair of terms pictural and notational iconicity, see Sybille Krämer, "Writing, Notational Iconicity, Calculus: On Writing as a Cultural Technique." *MLN* 118, no. 3 (2003): 518–537, esp. 524. See also section 6.2 of the present volume.

36. In the sense of the CIA as the "Company" and the organizations related to it, for example, the Parallax Corporation in Alan J. Pakula's *Parallax View*, directed by Alan J. Pakula (1974; Paramount Pictures, 2004), DVD.

37. See Mark S. Mizruchi, *The American Corporate Network, 1904–1974* (Beverly Hills, CA: Sage, 1982).

38. The dead-accurate name of a private organized parallel organization of contract killers that has decoupled from the state "corporations" in Don DeLillo, *Running Dog* (1978; London: Picador, 1999), esp. 153–154.

39. See Jameson, *The Geopolitical Aesthetic*, 9–10.

40. On this, see, for example, Fritz Nordsieck, *Die schaubildliche Erfassung der Betriebsorganisation* (Stuttgart: Poeschel, 1932), and the *Zeitschrift für Organisation*, published from 1927 onward.

41. See Pollux, *Trusts in der Schweiz?*, 34; Pollux, *Wer leitete Deutschland? Die eigentlichen Kriegsverbrecher: 50 Drahtzieher hinter den Kulissen* (Zurich: Verein für wirtschaftliche Studien, 1945), 18.

42. Pollux, *Trusts in der Schweiz?* 130.

43. David Gugerli and Daniel Speich have shown how much the Swiss domestic identity is interwoven with the mapping of the country in the nineteenth century. See David Gugerli and Daniel Speich, *Topografien der Nation: Politik, kartografische Ordnung und Landschaft im 19. Jahrhundert* (Zurich: Chronos, 2002).

44. On the "stronger" form versus the loose coupling of the medial substratum, see Niklas Luhmann, *Theory of Society*, vol. 1, trans. Rhodes Barrett (Stanford, CA: Stanford University Press, 2012), 119.

45. Pollux, *Wer leitete Deutschland?*, 8.

46. Pollux, *Wer leitete Deutschland?*, 15 (emphasis in bold in the original).

47. See Arno Meteling, "The Parallax View: Verschwörungstheorie zur Einführung," *Transkriptionen* 9 (2008): 15–18, esp. 15–16. On the constitutive suspicion of manipulation versus knowledge of mass media, see Niklas Luhmann, *The Reality of the Mass Media*, trans. Kathleen Cross (Stanford, CA: Stanford University Press, 2000), 1, 14. On the rise of the connection between conspiracy rhetoric and the democratic public sphere in ancient Greece, see Joseph Roisman, *The Rhetoric of Conspiracy in Ancient Athens* (Berkeley: University of California Press, 2006). For an outstanding historical analysis of the networks of the modern public sphere, see Robert Darnton, *Poetry and the Police: Communication Networks in Eighteenth-Century Paris* (Cambridge, MA: Belknap Press of Harvard University Press, 2010).

48. Eva Horn, *The Secret War: Treason, Espionage, and Modern Fiction*, trans. Geoffrey Winthrop-Young (Evanston, IL: Northwestern University Press, 2013), 98.

49. Sydney Pollack, *Three Days of the Condor*, directed by Sydney Pollack (1975; Paramount Pictures, 2001). DVD, time code ca. 01:50:00–51:30. On the ambiguity of this open ending to the film, between endless play and final end to the conspiracy, see Marcus Krause, Arno Meteling, and Markus Stauff, "Einleitung," in *The Parallax View: Zur Mediologie der Verschwörung*, ed. Arno Meteling, Markus Stauff, and Marcus Krause (Paderborn: Wilhelm Fink, 2011), 9–42, esp. 39.

50. DeLillo, *Running Dog*, 111.

51. Pollack, *Three Days of the Condor*, time code ca. 01:16:00.

52. The opening sequence prominently features a book scanner turning Chinese paper documents into computer code, while the title of the film is superimposed in a computer type as if being typed in above it. For a cybernetic reading of *Condor* as a film of switches and apparatuses, see Ute Holl, "Türen, Lektüren und das Wissen des Films: Zur Verschwörung der Medien in Sydney Pollack's *Three Days of the Condor*," *Parallax View*, 235–251.

53. Pollack, *Three Days of the Condor*, time code ca. 0:39:00.

54. On this in general, see Melley, *Empire of Conspiracy*, 13; and Horn, *The Secret War*, 85–88.

55. The distinction between *arcanum, mysterium,* and *secretum* is made in Horn, *The Secret War,* 85–88.

56. Pollack, *Three Days of the Condor,* time code ca. 01:24:00.

57. Pollack, *Three Days of the Condor,* time code ca. 0:42:00.

58. See James Grady, *Six Days of the Condor* (1974; Harpenden: No Exit Press, 2007).

59. Pollack, *Three Days of the Condor,* time code ca. 01:04:30–31.

60. *Three Days of the Condor* is a spy film about electronic apparatuses and instruments making noises and writing beyond the gadgets of the likes of James Bond.

61. Pollack employed this stylistic means himself in one of his last films: *The Interpreter* with Nicole Kidman and Sean Penn (2005)—again with clear reference to the New Hollywood narrations of the 1970s.

62. See Jameson, *The Geopolitical Aesthetic,* 13–14.

63. What is problematic, however, is the only briefly outlined iconography, which argues primarily by using a set photograph that does not appear in the film itself. See Jameson, *The Geopolitical Aesthetic,* 15. Its perspective runs counter to the spatial logic of a montage of moving images.

64. Antonin Artaud, *The Theater and Its Double,* trans. Mary Caroline Richards (New York: Grove, 1958), 12.

65. Jameson, *The Geopolitical Aesthetic,* 15.

66. In addition to Jameson and his student Alexander Galloway, this also applies to Michael Hardt and Antonio Negri's vision of the "multitude" in Michael Hardt and Antonio Negri, *Empire,* 4th ed. (Cambridge, MA: Harvard University Press, 2001). Donna Haraway's cyborg mythology also belongs to this tradition.

67. Commentaries on Pynchon are legion. So, I will merely mention here that the unfolding of a *longue durée* of the "other" postal service, which recalls Borges, originally aimed to create a continental network of two postal systems. See Thomas Pynchon, *The Crying of Lot 49* (London: Vintage, 2000), 113–127.

68. Luhmann, "Kausalität im Süden." "Medium" here is connected to Talcott Parsons's "generalized symbolic medium."

69. See Sydow and Windeler, "Steuerung von und in Netzwerken."

70. Pollack, *Three Days of the Condor,* time code ca. 01:38:30. The MacGuffin of the original novel, which was written under the impression of the Vietnam War, was still drugs. See Grady, *Six Days of the Condor,* 20. Against the backdrop of the oil crisis, the scriptwriters Lorenzo Semple, Jr., and David Rayfield shifted the focus to securing oil resources.

71. Stephen Gaghan's oil conspiracy thriller *Syriana* of 2004 presents a continuation of *Three Days of the Condor* as cinematic revenant of New Hollywood Cinema.

72. See Jameson, *The Geopolitical Aesthetic*, 9: "the cognitive function of the conspiratorial plot must be able to flicker in and out, like some secondary or subliminal after-image."

73. Darnton, *Poetry and the Police*, 130.

74. I am sincerely grateful to Isabelle Geisthardt and the students of the seminar *Was ist ein Netzwerk?* (What Is a Network?) at the Humboldt-Universität zu Berlin, summer semester 2007, for discussions of Lombardi.

75. See Robert Hobbs, "Mark Lombardi: Global Networks," in *Mark Lombardi: Global Networks*, ed. Robert Hobbs (New York: Independent Curators International, 2003), 11–49. Lombardi was, moreover, a highly reflective artist who identified his own role models in notes: Duchamp, Beuys, Haacke, Aycock, Zakanitch, Yoko, and Arakawa. Moreover, he had an explicit affinity to Jackson Pollock. It would be rewarded to compare Sol Le Witt's murals, to which Lombardi was often clearly closer than to the aforementioned artists. In general, his drawings follow the process of eliminating the boundaries between drawing and writing in the twentieth century. See Carolin Meister, "Einleitung," in *Randgänge der Zeichnung*, ed. Werner Busch, Oliver Jehle, and Carolin Meister (Paderborn: Wilhelm Fink, 2007), 7–10, esp. 8.

76. Angela Lammert, "Mark Lombardi: Denken als Mustererkennung," in *Räume der Zeichnung*, ed. Angela Lammert et al. (Nuremberg: Verlag für moderne Kunst, 2007), 57–71, esp. 67.

77. See Boris Groys, *Under Suspicion: A Phenomenology of Media*, trans. Carsten Strathausen (New York: Columbia University Press, 2012), 48.

78. Michel Serres, *Hermès; ou, La communication* (Paris: Éditions de minuit, 1968), 20.

79. See Robert Tolksdorf, "Lombardi Networks: Towards the Complete Mark Lombardi Digital," https://www.lombardinetworks.net. On the modeling based on graph theory, see also Christian A. Duncan et al., "Lombardi Drawings of Graphs," *Journal of Graph Algorithms and Applications* 16, no. 1 (2012): 85–108, https://doi.org/10.7155/jgaa.00251.

80. On this figure of thought so typical of Michel Serres, see part 6.2 of the present volume. The more recent positions in art theory since Umberto Eco's thoughts on the "open work" are always realized by Lombardi *en passant* as well.

81. See the speculations at http://web.archive.org/web/20080319120819/www.pierogi2000.com/memorial/lombardm.html. See also the documentary film *Mark Lombardi: Kunst und Konspiration* by Mareike Wegener, Germany, 2012.

82. Hannah Arendt, *The Origins of Totalitarianism*, 2nd ed. (New York: Meridian, 1958).

83. Herbert Marcuse, *Eros and Civilization: A Philosophical Inquiry into Freud* (Boston: Beacon Press, 1974).

84. The list is found at http://web.archive.org/web/20080309152509/www.pierogi2000.com/flatfile/lombardibibliography.html.

85. See Robert Hobbs, ed., *Mark Lombardi: Global Networks* (New York: Independent Curators International, 2003), 51.

86. See *Chicago Outfit and Satellite Regimes, ca. 1931–83; Meyer Lansky's Financial Network, ca. 1960–78*. See also Hobbs, *Mark Lombardi*, 54–55.

87. See, for example, the report by the police prefect Cesare Mori on the network of assistance of the Grisafi band, which formed in 1917, in Anton Blok, *The Mafia of a Sicilian Village, 1860–1960: A Study of Violent Peasant Entrepreneurs* (New York: Harper Torchbooks, 1974), 134–135.

88. See Melley, *Empire of Conspiracy*, vii.

89. Following Horn, *The Secret War*, 93.

90. For a biography, see Hobbs, *Mark Lombardi*, 15–16.

91. It is less surprising to find Hannah Arendt's study of totalitarianism as another central philosophical text in Lombardi's library.

92. Marcuse, *Eros and Civilization*, 16 (emphasis in the original).

93. Lombardi thus belonged to the tradition of Jacob Levy Moreno's sociograms, which primarily record emotional flows. See part 6.4 of the present volume.

94. This was compensated for by a reader on Lombardi that includes reproductions of single index cards. See Mark Lombardi, *100 Notes—100 Thoughts, No. 071*, introduction by Carolyn Christov-Bakargiev, exh. cat. Kassel, documenta 13 (Ostfildern: Hatje Cantz, 2012).

95. See Edward Tufte, *Design of Causal Diagrams: Barr Art Chart, Lombardi Diagrams, Evolutionary Trees, Feynman Diagrams, Timelines*, 2003, http://www.edwardtufte.com/bboard/q-and-a-fetch-msg?msg_id=0000yO&topic_id=1&topic=.

96. This is not a general characterization of Lombardi's drawings, which depending on their overall form also suggest other points of entry.

97. The drawing notes 1996 as the year of his inauguration, although his term started on January 17, 1995.

98. See Hobbs, *Mark Lombardi*, 47.

99. In Foucault's sense of "archive." See Michel Foucault, *Archaeology of Knowledge*, trans. A. M. Sheridan Smith (London: Routledge, 2002), 145–146.

100. On this, see Lammert, "Mark Lombardi," 68–69.

101. See Groys, *Under Suspicion*, 19–20.

102. It also has an elective affinity to Immanuel Wallerstein's works on the modern World System. See Hobbs, *Mark Lombardi*, 23.

103. The best example is the ominous BCCI-Bank, to which Lombardi dedicated a series of drawings. A similar connection is offered by the investment banker Jackson

Stephens, who had at least as much influence on Bill Clinton's presidency as he had on the career of George W. Bush.

104. See Roberta Smith, "Mark Lombardi, 48, an Artist Who Was Inspired by Scandals," *New York Times*, March 25, 2000, http://web.archive.org/web/20070927191929/https://www.pierogi2000.com/memorial/lombardiobit.html.

105. On this, see also Tan Lin, "Two or Three Sentences about Mark Lombardi," in *Mark Lombardi: Preparatory Drawings, 1994–2000* (New York: Pierogi, 2003), 14.

CHAPTER 11

1. See chapter 2 of this book. On the nineteenth and twentieth centuries, see also chapters 1–3 in Grant Bollmer, *Inhuman Networks: Social Media and the Archaeology of Connection* (New York: Bloomsbury, 2016).

2. See Matthew 13:47–48.

3. See Matthew 4:18–22; Mark 1:16–20; Luke 5:1–11.

4. Aleš Šteger, *Buch der Dinge*, trans. Urška P. Černe and Matthias Göritz (Frankfurt am Main: Suhrkamp, 2006), 27.

5. Michel Serres, *The Parasite*, trans. Lawrence R. Schehr (Baltimore: Johns Hopkins University Press, 1982), 225.

6. Nicole Loraux, "Le lien de la division," *Le Cahier (Collège international de philosophie)*, no. 4 (1987): 101–124.

7. See Michel Serres, *The Natural Contract*, trans. Elizabeth MacArthur and William Paulson (Ann Arbor: University of Michigan Press, 1995), 105–106.

8. See Marcel Mauss, *A General Theory of Magic*, trans. Robert Brain (London: Routledge, 2005), 76; and Mircea Eliade, *Images and Symbols: Studies in Religious Symbolism*, trans. Philip Mairet (New York: Sheed and Ward, 1961), 92–93.

9. On the sociology of this, see Mark S. Granovetter, "The Strength of Weak Ties," *American Journal of Sociology* 78, no. 6 (1973): 1360–1380. On the history, see Natalie Binczek and Georg Stanitzek, eds., *Strong Ties / Weak Ties: Freundschaftssemantik und Netzwerktheorie* (Heidelberg: Winter, 2010).

10. See Anna Munster, *An Aesthesia of Networks: Conjunctive Experience in Art and Technology* (Cambridge, MA: MIT Press, 2013); and Isabell Otto, *Prozess und Zeitordnung: Temporalität unter den Bedingungen digitaler Vernetzung* (Konstanz: Konstanz University Press, 2020).

11. The use of the network diagram in science and art is much closer to trapping.

12. Tero Karppi, *Disconnect: Facebook's Affective Bonds* (Minneapolis: University of Minnesota Press, 2018); Urs Stäheli, *Soziologie der Entnetzung* (Berlin: Suhrkamp, 2021).

13. See Serres, *The Parasite*, 224–225.

14. Serres, *The Parasite*, 248.

15. Patrick McGrath, *Spider* (New York: Vintage, 1990), 31.

16. Michel Serres, *La naissance de la physique dans le texte de Lucrèce: Fleuves et turbulences* (Paris: Éditions de minuit, 1977), 163.

17. Astrid Deuber-Mankowsky's describing Donna Haraway's method. Astrid Deuber-Mankowsky, *Praktiken der Illusion: Kant, Nietzsche, Cohen, Benjamin bis Donna J. Haraway* (Berlin: Vorwerk 8, 2007), 300.

18. See Eugene Thacker, "Networks, Swarms, Multitudes. Part I," *ctheory.net*, May 18, 2004, https://journals.uvic.ca/index.php/ctheory/article/view/14542/5389.

19. See Sebastian Giessmann, "Netzwerkprotokolle und Schwarm-Intelligenz: Zur Konstruktion von Komplexität und Selbstorganisation," in *Schwärme: Kollektive ohne Zentrum*, ed. Eva Horn and Lucas Marco Gisi (Bielefeld: transcript, 2009), 163–182.

20. See Sebastian Vehlken, *Zootechnologies: A Media History of Swarm Research*, trans. Valentine A. Pakis (Amsterdam: Amsterdam University Press, 2019).

21. See Adrian Mackenzie, *Wirelessness: Radical Empiricism in Network Cultures* (Cambridge, MA: MIT Press, 2010); and Regine Buschauer, *Mobile Räume: Medien- und diskursgeschichtliche Studien zur Tele-Kommunikation* (Bielefeld: transcript, 2010), 304–305.

22. Bruno Latour, *Aramis, or, The Love of Technology*, trans. Catherine Porter (Cambridge, MA: Harvard University Press, 1996).

23. See Sebastian Giessmann, "Netzstörungen: Erzählungen vom Ende der Netzwerke." *Störfälle: Zeitschrift für Kulturwissenschaften* 2 (2011): 125–133.

24. On this, see Tero Karppi et al., *Undoing Networks* (Lüneburg: Meson Press, 2021).

25. Marilyn Strathern, "Cutting the Network," *Journal of the Royal Anthropological Institute* 2, no. 3 (1996): 517–535, esp. 523.

26. "Actually, there is no story for which the question 'How does it continue?' would not be legitimate." Walter Benjamin, "The Storyteller: Observations on the Works of Nikolai Leskov," trans. Harry Zohn, in *Selected Writings*, vol. 3 (1935–1938), ed. Howard Eiland and Michael W. Jennings (Cambridge, MA: Belknap Press of Harvard University Press, 2002), 143–162, esp. 155.

27. See Strathern, "Cutting the Network," 524.

28. Strathern, "Cutting the Network," 527–528.

29. Marilyn Strathern, *Partial Connections*, rev. ed. (Walnut Creek, CA: Altamira, 2004).

30. Thomas S. Mullaney et al., eds., *Your Computer Is on Fire* (Cambridge, MA: MIT Press, 2021).

31. Serres, *The Parasite*, 11.

REFERENCES

ARCHIVES

Bibliothèque de l'Arsenal, Paris

International Telecommunication Union Library and Archives, Geneva

Library of Congress Manuscript Collection, Washington DC London Transport Museum, London

London Transport Museum, London

Massachusetts Institute of Technology, Institute Archives and Special Collections, Cambridge, Massachusetts

Musée de la Poste, Paris

Request for Comments Editor, https://www.rfc-editor.org

United States Patent Office, Alexandria, Virginia

University of Connecticut Libraries, Archives & Special Collections, Storrs, Connecticut

University of Minnesota Digital Conservancy, http://conservancy.umn.edu

Zentralbibliothek Zürich, Handschriftenabteilung

BIBLIOGRAPHY

Abbate, Janet. *Inventing the Internet*. Cambridge, MA: MIT Press, 2000.

Adamowsky, Natascha. "Spiel und Wissenschaftskultur: Eine Anleitung." In *Die Vernunft ist mir noch nicht begegnet: Zum konstitutiven Verhältnis von Spiel und Erkenntnis*, edited by Natascha Adamowsky, 11–30. Bielefeld: transcript, 2005.

Adelmann, Howard B. *Marcello Malpighi and the Evolution of Embryology*. 5 vols. Ithaca, NY: Cornell University Press, 1966.

Aelian. *On the Characteristics of Animals (De natura animalium), Books I–V*. Edited by Alwyn Faber Scholfied. Vol. 1. Cambridge, MA: Harvard University Press, 1958.

Aelian. *On the Characteristics of Animals (De natura animalium), Books VI–XI*. Edited by Alwyn Faber Scholfied. Vol. 2. Cambridge, MA: Harvard University Press, 1959.

Aeschylus. *The Oresteia: Agamemnon, Libation-Bearers, Eumenides*. Edited and translated by Alan H. Sommerstein. Cambridge, MA: Harvard University Press, 2008.

"Agreement on the Establishment of a European Informatics Network (with annex). Concluded at Brussels on 23 November 1971." In *United Nations Treaty Series*. Vol. 1262. I-20750, 311–342. New York: United Nations, 1982. https://treaties.un.org/doc/Publication/UNTS/Volume%201262/v1262.pdf.

Ahrens, Wilhelm. *Mathematische Unterhaltungen und Spiele*. Leipzig: Teubner, 1901.

Ampère, André-Marie. *Essai sur la philosophie des sciences; ou, Exposition analytique d'une classification naturelle de toutes les connaissances humaines*. Paris: Bachelier, 1834.

Anderson, Benedict. *Imagined Communities: An Inquiry into the Origins and Spread of Nationalism*. 2nd ed. London: Verso, 1991.

Aradau, Claudia, and Tobias Blanke. "Governing Circulation: A Critique of the Biopolitics of Security." In *Security and Global Governmentality: Globalization, Governance and the State*, edited by Miguel de Larrinaga and Marc G. Doucet. London: Routledge, 2010. http://oro.open.ac.uk/21481/2/security_circulationAradauBlanke.pdf.

Arendt, Hannah. *The Origins of Totalitarianism*. 2nd ed. New York: Meridian, 1958.

Aristotle. *De Anima*. Edited and translated by Christopher Shields. Oxford: Clarendon Press, 2016.

Arquilla, John, and David F. Ronfeldt. *Networks and Netwars: The Future of Terror, Crime, and Militancy*. Santa Monica, CA: RAND, 2001.

Artaud, Antonin. *The Theater and Its Double*. Translated by Mary Caroline Richards. New York: Grove, 1958.

Asendorf, Christoph. *Ströme und Strahlen: Das langsame Verschwinden der Materie um 1900*. Giessen: Anabas, 1989.

Augé, Marc. *In the Metro*. Translated by Tom Conley. Minneapolis: University of Minnesota Press, 2002.

Autenrieth, Ulla, Andreas Blättler, Regine Buschauer, and Doris Gassert, eds. *Dis Connecting Media: Technik, Praxis und Ästhetik des Telefons: Vom Festnetz zum Handy*. Basel: Merian, 2011.

Baecker, Dirk. "Handeln im Netzwerk: Zur Problemstellung der Soziologie." In *Relationale Soziologie: Zur kulturellen Wende der Netzwerkforschung*, edited by Jan Fuhse and Sophie Mützel, 233–256. Wiesbaden: VS Verlag für Sozialwissenschaften, 2010.

Balbi, Gabriele, and Andreas Fickers, eds. *History of the International Telecommunication Union (ITU)*. Vol. 1. Berlin: De Gruyter Oldenbourg, 2020.

Ballestra-Puech, Sylvie. *Métamorphoses d'Arachné: L'artiste en araignée dans la litterature occidentale*. Geneva: Librairie Droz, 2006.

Barabási, Albert-László. *Linked: The New Science of Networks*. Cambridge, MA: Perseus, 2002.

Barabási, Albert-László, and Márton Pósfai. *Network Science*. Cambridge: Cambridge University Press, 2016.

Baran, Paul. "Memorandum RM-3420-PR. On Distributed Communications." Tech. rep. RAND Corporation, 1964. https://www.rand.org/publications/RM/RM3420.

Baran, Paul. "Memorandum RM-3097-PR. On Distributed Communications: V. History, Alternatives Approaches, and Comparions." Tech. rep. RAND Corporation, 1964. https://www.rand.org/content/dam/rand/pubs/research_memoranda/2008/RM3097.pdf.

Baran, Paul. "Memorandum RM-3767-PR. On Distributed Communications: XI. Summary Overview." Tech. rep. RAND Corporation, 1964. https://www.rand.org/content/dam/rand/pubs/research_memoranda/2006/RM3767.pdf.

Baran, Paul. "On Distributed Communications Networks." In *IEEE Transactions on Communications* 12 (1964): 1–9.

Barker, Theodore Cardwell, and Michael Robbins. *A History of London Transport: Passenger Travel and the Development of the Metropolis.* 2 vols. London: George Allen & Unwin, 1974.

Barkhoff, Jürgen, Hartmut Böhme, and Jeanne Riou, eds. *Netzwerke: Eine Kulturtechnik der Moderne.* Cologne: Böhlau, 2004.

Barman, Christian. *The Man Who Built London Transport: A Biography of Frank Pick.* Newton Abbot, UK: David & Charles, 1979.

Barnekow, Rolf. "Wählen statt Rufen: Die automatische Vermittlung." In *Fräulein vom Amt*, edited by Helmut Gold and Annette Koch, 86–93. Munich: Prestel, 1993.

Barnes, John A. "Class and Committees in a Norwegian Island Parish." In *Human Relations* 7 (1954): 39–58.

Barrett, John Patrick. *Electricity at the Columbian Exposition. Including an Account of the Exhibits in the Electricity Building, the Power Plant in Machinery Hall, the Arc and Incandescent Lighting of the Grounds and Buildings etc.* Chicago: Donnelley & Sons, 1894.

Bartels, Uwe. *Edfu: Die Darstellungen auf den Aussenseiten der Umfassungsmauer und auf dem Pylonen; Strichzeichnungen und Photographien.* Edited by Dieter Kurth. Die Inschriften des Tempels von Edfu. Abteilung II: Dokumentationen. Wiesbaden: Harrassowitz, 2009.

Barthes, Roland. *The Pleasure of the Text.* Translated by Richard Miller. New York: Hill and Wang, 1975.

Bartholini, Thomas, and Marcello Malpighi. *Thomae Bartolini De Pulmonum Substantia & Motu Diatribe: Accedunt Cl.V. Marcelli Malphigij de Pulmonibus Observationes Anatomicae.* Hafniae: Typis Henrici Gödiani R. & Ac.Typ. Prostant apud P. Hauboldum, 1663.

Bärwolff, Matthias. "End-to-End Arguments in the Internet: Principles, Practices, and Theory." PhD diss., Technische Universität Berlin, 2010. http://dx.doi.org/10.14279/depositonce-2631.

Baudelaire, Charles. "Le Cygne." In *Œuvres complètes*, edited by Claude Pichois. 2 vols., 1:85–87. Paris: Bibliothèque de la Pléïade, 1975.

Baudelaire, Charles. "The Swan." In Charles Baudelaire, *Les Fleurs du Mal: The Complete Text of the Flowers of Evil.* Translated by Richard Howard, 90–92. Boston: D. R. Godine, 1982.

Bauer, Matthias, and Christoph Ernst. *Diagrammatik: Einführung in ein kultur- und medienwissenschaftliches Forschungsfeld.* Bielefeld: transcript, 2010.

Baumann, Georg [Georges Baehler]. *Atlantikpakt der Konzerne: Die internationale Kapitalverflechtung in Westdeutschland.* Berlin: Verlag die Wirtschaft, 1952.

Baumann, Georg [Georges Baehler]. *Eine Handvoll Konzernherren.* Berlin: Volk und Welt, 1953.

Baxmann, Inge. "Der Körper als Gedächtnisort: Bewegungswissen und die Dynamisierung der Wissenskulturen im frühen 20. Jahrhundert." In *Deutungsräume. Bewegungswissen als kulturelles Archiv der Moderne,* edited by Inge Baxmann and Franz Anton Cramer, 15–35. Munich: K. Kieser, 2005.

Bay, Morten. "Conversations with a Pioneer: Paul Baran in His Own Words." *Internet Histories* 1, no. 3 (2017): 273–282.

Bay, Morten. "Hot Potatoes and Postmen: How Packet Switching Became ARPANET's Greatest Legacy." *Internet Histories* 3, no. 1 (2019): 15–30.

Bayly, Christopher Alan. "Myths and Technologies of the Modern State." In *The Birth of the Modern World, 1780–1914: Global Connections and Comparisons,* 247–283. Malden, MA: Blackwell, 2004.

Bayly, Christopher Alan. *The Birth of the Modern World, 1780–1914: Global Connections and Comparisons.* Malden, MA: Blackwell, 2004.

BBC. "Games Bring Record London Underground Users." *BBC,* August 4, 2012. https://www.bbc.com/news/uk-england-london-19125580.

Becker, Abraham Mario. *Handbuch der Dynamischen Netzplantechnik für die Planung und Überwachung von Aufträgen.* 3rd ed. Winterthur: Gebrüder Sulzer, 1970.

Becker, Thomas. *Mann und Weib—schwarz und weiss: Die wissenschaftliche Konstruktion von Geschlecht und Rasse, 1650–1900.* Frankfurt am Main: Campus, 2005.

Behringer, Wolfgang. *Im Zeichen des Merkur: Reichspost und Kommunikationsrevolution in der Frühen Neuzeit.* Göttingen: Vandenhoeck & Ruprecht, 2003.

Behringer, Wolfgang. "Netzwerk." In *Enzyklopädie der Neuzeit.* Vol. 9, *Naturhaushalt–Physiokratie,* edited by Friedrich Jaeger, 98–100. Stuttgart: J. B. Metzler, 2009.

Behrman House, ed. *Pirke Avot: A Modern Commentary on Jewish Ethics.* Edited and translated by Leonard Kravitz and Kerry M. Olitzky. New York: UAHC Press, 1993.

Belliger, Andréa, and David J. Krieger. "Einführung in die Akteur-Netzwerk-Theorie." In *ANThology: Ein einführendes Handbuch zur Akteur-Netzwerk-Theorie,* edited by Andréa Belliger and David J. Krieger, 13–50. Bielefeld: transcript, 2006.

Belloni, Luigi. "Marcello Malpighi and the Founding of Anatomical Microscopy." In *Reason, Experiment and Mysticism in the Scientific Revolution,* edited by Maria Luisa Righini Bonelli and William Shea, 95–110. New York: Science History, 1975.

Benjamin, Walter. "Capitalism as Religion." Translated by Rodney Livingstone. In *Selected Writings*. Vol. 1 (1913–1926), edited by Marcus Bullock and Michael W. Jennings, 288–291. Cambridge, MA: Belknap Press of Harvard University Press, 1996.

Benjamin, Walter. "Critique of Violence." Translated by Edmund Jephcott. In *Selected Writings*. Vol. 1 (1913–1926), edited by Marcus Bullock and Michael W. Jennings, 236–252. Cambridge, MA: Belknap Press of Harvard University Press, 1996.

Benjamin, Walter. *Origins of the German Trauerspiel*. Translated by Howard Eiland. Cambridge, MA: Harvard University Press, 2019.

Benjamin, Walter. *The Arcades Project*. Translated by Howard Eiland and Kevin McLaughlin. Cambridge, MA: Belknap Press of Harvard University Press, 2002.

Benjamin, Walter. "The Storyteller: Observations on the Works of Nikolai Leskov." Translated by Harry Zohn. In *Selected Writings*. Vol. 3 (1935–1938), edited by Howard Eiland and Michael W. Jennings, 143–162. Cambridge, MA: Belknap Press of Harvard University Press, 2002.

Berg, Rolph, Arved Meyer, Max Müller, and Andreas Zogg. *Netzplantechnik: Grundlagen, Methoden, Praxis*. Zurich: Verlag industrielle Organisation, 1973.

Bergemann, Lutz, Martin Dönike, Albert Schirrmeister, Georg Toepfer, Marco Walter, and Julia Weitbrecht. "Transformation: Ein Konzept zur Erforschung kulturellen Wandels." In *Transformation: Ein Konzept zur Erforschung kulturellen Wandels*, edited by Hartmut Böhme, Lutz Bergemann, Martin Dönike, Albert Schirrmeister, Georg Toepfer, Marco Walter, and Julia Weitbrecht, 39–56. Paderborn: Wilhelm Fink, 2011.

Bernsen, Michael, and Martin Neumann, eds. *Die französische Literatur des 19. Jahrhunderts und der Orientalismus*. Tübingen: Max Niemeyer, 2006.

Bertin, Jacques. *Semiology of Graphics: Diagrams, Networks, Maps*. Translated by William J. Berg. Redlands, CA: ESRI, 2011.

Bigg, Charlotte. "The Panorama, or, La Nature à Coup d'Œil." In *Observing Nature—Representing Experience: The Osmotic Dynamics of Romanticism, 1800–1850*, edited by Erna Fiorentini, 73–95. Berlin: Reimer, 2007.

Biggs, Norman L., Edward Keith Lloyd, and Robin James Wilson, eds. *Graph Theory, 1736–1936*. 2nd ed. Oxford: Clarendon Press, 1977.

Binczek, Natalie, and Georg Stanitzek, eds. *Strong Ties / Weak Ties: Freundschaftssemantik und Netzwerktheorie*. Heidelberg: Winter, 2010.

Birk, Alfred, and Karl Hermann Müller-Hamburg. *Der Suezkanal: Seine Geschichte und seine wirtschaftspolitische Bedeutung für Europa, Indien und Ägypten*. Hamburg: Boysen & Maasch, 1925.

Black, Max. *Models and Metaphors: Studies in Language and Philosophy*. Ithaca, NY: Cornell University Press, 1962.

Blair, Ann, Paul Duguid, Anja-Silvia Goeing, and Anthony Grafton, eds. *Information: A Historical Companion*. Princeton, NJ: Princeton University Press, 2021.

Blok, Anton. *The Mafia of a Sicilian Village, 1860–1960: A Study of Violent Peasant Entrepreneurs*. New York: Harper Torchbooks, 1974.

Blumenberg, Hans. *Die Lesbarkeit der Welt.* 2nd ed. Frankfurt am Main: Suhrkamp, 1989.

Blumentrath, Hendrik, Katja Rothe, Sven Werkmeister, Michaela Wünsch, and Barbara Wurm, eds. *Techniken der Übereinkunft: Zur Medialität des Politischen.* Berlin: Kadmos, 2009.

Blumtritt, Oskar. *Nachrichtentechnik: Sender, Empfänger, Übertragung, Vermittlung.* Munich: Deutsches Museum, 1997.

Boehm, Gottfried. "Die Wiederkehr der Bilder." In *Was ist ein Bild?* edited by Gottfried Boehm, 11–38. Munich: Wilhelm Fink, 1994.

Boehm, Gottfried. "Ikonische Differenz." *Rheinsprung 11: Zeitschrift für Bildkritik* 1 (2011): 170–178. https://www.rheinsprung11.unibas.ch/fileadmin/documents/Edition_PDF/Ausgabe1/glossar-boehm.pdf.

Boenke, Michaela. *Körper, Spiritus, Geist: Psychologie vor Descartes.* Paderborn: Wilhelm Fink, 2005.

Böhme, Gernot, and Hartmut Böhme. *Feuer, Wasser, Erde, Luft: Eine Kulturgeschichte der Elemente.* Munich: Beck, 1996.

Böhme, Hartmut. "Bildevidenz, Augentäuschung und Zeugenschaft in der Wissenschaft des Unsichtbaren im 17. Jahrhundert." In *Dissimulazione onesta oder Die ehrliche Verstellung. Von der Weisheit der versteckten Beunruhigung in Wort, Bild und Tat. Martin Warnke zu Ehren. Ein Symposion (2003)*, edited by Horst Bredekamp, Michael Diers, Ruth Tesmar, and Franz-Joachim Verspohl, 13–42. Hamburg: Philo Fine Arts, 2007, 13–42.

Böhme, Hartmut. "Einladung zur Transformation." In *Transformation: Ein Konzept zur Erforschung kulturellen Wandels*, edited by Hartmut Böhme, Lutz Bergemann, Martin Dönike, Albert Schirrmeister, Georg Toepfer, Marco Walter, and Julia Weitbrecht, 7–37. Paderborn: Wilhelm Fink, 2011.

Böhme, Hartmut. "Netzwerke: Zur Theorie und Geschichte einer Konstruktion." In *Netzwerke: Eine Kulturtechnik der Moderne*, edited by Jürgen Barkhoff, Hartmut Böhme, and Jeanne Riou, 17–36. Cologne: Böhlau, 2004.

Böhme, Hartmut, ed. *Topographien der Literatur: Deutsche Literatur im transnationalen Kontext.* Stuttgart: J. B. Metzler, 2005.

Böhme, Hartmut. "Von der Vernetzung zur Virtualisierung der Städte: Ende der Philosophie—Beginn des Neuen Jerusalem?" In *Constructing Utopia: Konstruktionen künstlicher Welten*, edited by Annett Zinsmeister, 141–155. Zurich: diaphanes, 2005.

Böhme, Hartmut. "Vorwort." In *Übersetzung und Transformation*, edited by Hartmut Böhme, Christof Rapp, and Wolfgang Rösler, V–XIII. Berlin: De Gruyter, 2007.

Böhme, Hartmut, Lutz Bergemann, Martin Dönike, Albert Schirrmeister, Georg Toepfer, Marco Walter, and Julia Weitbrecht, eds. *Transformation: Ein Konzept zur Erforschung kulturellen Wandels.* Paderborn: Wilhelm Fink, 2011.

Bollmer, Grant. *Inhuman Networks: Social Media and the Archaeology of Connection.* New York: Bloomsbury, 2016.

Bolt, Beranek, and Newman, Inc. *Report No. 1822: Interface Message Processor; Specifications for the Interconnection of a Host and an IMP*. Tech. rep., 1976. http://www.bitsavers.org/pdf/bbn/imp/BBN1822_Jan1976.pdf.

Boltanski, Luc, and Ève Chiapello. *The New Spirit of Capitalism*. Translated by Gregory Elliott. London: Verso, 2007.

Borbach, Christoph, and Tristan Thielmann. "Chains of Co-operation in the 1940s: Working on the Air Situation Picture." In *Materiality of Cooperation*, edited by Sebastian Giessmann, Tobias Röhl, Ronja Trischler, and Martin Zillinger, 91–130. Wiesbaden: Springer VS, 2023.

Borchardt-Hume, Achim, ed. *Albers and Moholy-Nagy: From Bauhaus to the New World*. London: Tate, 2006.

Botzenhardt, Axel C. *Japan als Netzwerkgesellschaft*. Munich: iudicium, 1997.

Bowker, Geoffrey C. "How To Be Universal: Some Cybernetic Strategies, 1943–70." *Social Studies of Science* 23, no. 1 (1993): 107–127.

Brand, Stewart. "Founding Father." Interview with Paul Baran. *Wired* 9, no. 3 (2001). https://www.wired.com/wired/archive/9.03/baran_pr.html.

Brandeis, Louis. *Other People's Money and How the Bankers Use it*. New York: Frederick A. Stokes, 1914.

Brändle, Rainer. "Abenteuer der Dialektik: Ein Gemälde Veroneses in Sempruns Roman *Der weisse Berg*." In *Überschreitungen: Dialoge zwischen Literatur- und Theaterwissenschaft, Architektur und Bildender Kunst; Festschrift für Leonhard M. Fiedler*, edited by Jörg Sader and Anette Wörner, 329–338. Würzburg: Königshausen & Neumann, 2002.

Braudel, Fernand. "History and the Social Sciences: The *Longue Durée*." Translated by Immanuel Wallerstein. *Review (Fernand Braudel Center)* 32, no. 2 (2009): 171–203.

Braudel, Fernand. *The Mediterranean and the Mediterranean World in the Age of Philip II*. Translated by Siân Reynolds. 2 vols. London: Collins, 1972.

Brauns, Jörg. *Schauplätze: Zur Architektur visueller Medien*. Berlin: Kadmos, 2007.

Bredekamp, Horst. *Image Acts: A Systematic Approach to Visual Agency*. Translated by Elizabeth Clegg. Berlin: De Gruyter, 2017.

Bredekamp, Horst, Birgit Schneider, and Vera Dünkel, eds. *The Technical Image: A History of Styles in Scientific Imagery*. Chicago: University of Chicago Press, 2015.

Breidbach, Olaf. *Die Materialisierung des Ichs: Zur Geschichte der Hirnforschung im 19. und 20. Jahrhundert*. Frankfurt am Main: Suhrkamp, 1995.

Bretzke, Wolf-Rüdiger. *Logistische Netzwerke*. 2nd ed. Heidelberg: Springer, 2010.

Brewer, John. "Networks." In *Information: A Historical Companion*, edited by Ann Blair, Paul Duguid, Anja-Silvia Goeing, and Anthony Grafton, 620–627. Princeton, NJ: Princeton University Press, 2021.

Buer, Ferdinand, ed. *Morenos therapeutische Philosophie: Die Grundlagen von Psychodrama und Soziometrie*. 3rd ed. Opladen: Leske + Budrich, 1999.

Bunting, David. "Origins of the American Corporate Network." *Social Science History* 7, no. 2 (1983): 129–142.

Bunz, Mercedes. *Vom Speicher zum Verteiler: Die Geschichte des Internet.* Berlin: Kadmos, 2008.

Bürgi, Markus. "Bähler [Baehler], Georges." In *Historisches Lexikon der Schweiz (HLS).* Bern: HLS, 2005. https://hls-dhs-dss.ch/de/articles/015156/2020-01-24.

Burrington, Ingrid. "How We Misremember the Internet's Origins." *The New Republic,* October 29, 2019. https://newrepublic.com/article/155532/misremember-internets-origins.

Buschauer, Regine. *Mobile Räume: Medien- und diskursgeschichtliche Studien zur Tele-Kommunikation.* Bielefeld: transcript, 2010.

Calame-Griaule, Geneviève. *Ethnologie et langage: La parole chez les Dogon.* 2nd ed. Paris: Institut de l'Ethnologie, 1987.

Callon, Michel. "Some Elements of a Sociology of Translation: Domestication of the Scallops and the Fishermen of St Brieuc Bay." In *Power, Action and Belief: A New Sociology of Knowledge?* edited by John Law, 196–223. London: Routledge, 1986.

Callon, Michel. "Techno-Economic Networks and Reversibility." In *A Sociology of Monsters? Essays on Power, Technology and Domination,* edited by John Law, 132–161. London: Routledge, 1991.

Campbell-Kelly, Martin, William Aspray, Nathan Ensmenger, and Jeffrey R. Yost. *Computer: A History of the Information Machine.* 3rd ed. Boulder, CO: Westview, 2014.

Campe, Rüdiger. "Pronto! Telefonate und Telefonstimmen (57322)." In *Diskursanalysen 1: Medien,* edited by Friedrich Kittler, Manfred Schneider, and Samuel Weber, 68–93. Opladen: Westdeutscher Verlag, 1987.

Cardano, Girolamo. *Liber de ludo aleae,* edited by Massimo Tamborine. Milan: Angeli, 2006. First published 1564.

Carpenter, Brian, ed. *RFC 1958: Architectural Principles of the Internet.* Tech. rep. Internet Architecture Board. Los Angeles, CA: Network Working Group, 1996. https://www.rfc-editor.org/rfc/rfc1958.txt.

Carr, Stephen C., Stephen D. Crocker, and Vinton Gray Cerf. "HOST-HOST Communication Protocol in the ARPA network." In *AFIPS Conference Proceedings.* Vol. 36, *Spring Joint Computers Conference. May 5–7, 1970. Atlantic City, New Jersey,* 589–597. New York: ACM Press, 1970.

Cassirer, Ernst. *Substance and Function and Einstein's Theory of Relativity.* Chicago: Open Court, 1923.

Casson, Herbert N. *The History of the Telephone.* Moscow: Dodo Press, 2006. First published 1910 by A. C. McClurg (Chicago).

Castells, Manuel. *The Internet Galaxy: Reflections on the Internet, Business, and Society.* Oxford: Oxford University Press, 2001.

Castells, Manuel. *The Rise of the Network Society.* 2nd ed. Oxford: Blackwell, 2000.

Cerf, Vinton Gray. *An Interview with Vinton G. Cerf (OH 191)*. Conducted by Judy E. O'Neill. April 24, 1990, Reston, Virginia. Minneapolis: Charles Babbage Institute, University of Minnesota, 1990. https://hdl.handle.net/11299/107214.

Cerf, Vinton Gray. *RFC 635: An Assessment of ARPANET Protocols*. Tech. rep. Stanford, CA: Network Working Group, 1974. https://www.rfc-editor.org/rfc/rfc635.txt.

Cerf, Vinton Gray, Yogen Dalal, and Carl Sunshine. *RFC 675: Specification of Internet Transmission Control Program*. Tech. rep. 1974. https://www.rfc-editor.org/rfc/rfc675.txt.

Cerf, Vinton Gray, and Robert Elliot Kahn. "A Protocol for Packet Network Intercommunication." In *IEEE Transactions on Communications* 22, no. 5 (1974): 637–648.

Chambaz, Bernard. *La dialectique Véronèse*. Crest: La Sétérée, 1989.

Chapin, Francis Stuart. *The Measurement of Social Status by the Use of the Social Status Scale*. Minnesota: University of Minnesota Press, 1933.

Chappe, Abraham, and René Chappe. *Mémoire sur la télégraphie*. Paris: Béthune, 1829.

Chapuis, Robert J. *100 Years of Telephone Switching (1878–1978). Part I: Manual and Electromechanical Switching (1878–1960s)*. Amsterdam: North-Holland, 1982.

Chastel, André. *A Chronicle of Italian Renaissance Painting 1280–1580*. Ithaca, NY: Cornell University Press, 1984.

Cheek, Chris, Nicole Starosielski, and Braxton Soderman, eds. "Network Archaeologies." Special issue. *Amodern* 2 (October 2013). https://amodern.net/issues/amodern-2-network-archaeology.

Cherry, Colin. *On Human Communication: A Review, a Survey, and a Criticism*. 2nd ed. Cambridge, MA: MIT Press, 1966.

Chevalier, Michel. "Fin du choléra par coup d'état." *Le Globe. Journal de la religion saint-simonienne* 8, no. 102 (April 11, 1832): 405.

Chevalier, Michel. "Système de la Méditerranée." *Le Globe. Journal de la religion saint-simonienne* 8, no. 43 (February 12, 1832): 169–171.

Chicago Daily Tribune. "Hello, Main, Good By: Strowger's Telephone Attachment Dispenses with Girls." *Chicago Daily Tribune*, November 4, 1892, 1.

Chicago Daily Tribune. "No More 'Hello' Girls: Description of an Invention for Making Telephone Connections." *Chicago Daily Tribune*, October 31, 1892, 7.

Chicago Daily Tribune. "Takes the Place of 'Hello' Girls. Automatic Telephone System in Successful Operation at La Porte." *Chicago Daily Tribune*, November 2, 1892, 1.

Chicago Daily Tribune. "The Strowger Company is Fully Equipped to Install Telephone Exchanges," Advertisement *Chicago Daily Tribune*, February 1, 1894, 4.

Chun, Wendy Hui Kyong. "Networks NOW: Belated Too Early." In *Postdigital Aesthetics: Art, Computation and Design*, edited by David M. Berry and Michael Dieter, 289–315. London: Palgrave Macmillan, 2015.

Codding, George Arthur, and Anthony M. Rutkowski. *The International Telecommunication Union: An Experiment in International Cooperation*. Dedham, MA: Artech House, 1982.

Coilly, Nathalie, and Philippe Régnier, eds. *Le siècle des saint-simoniens: Du nouveau christianisme au canal de Suez*. Paris: Bibliothèque Nationale de France, 2006.

Comer, Douglas E. *Internetworking with TCP/IP*. Vol. 1, *Principles, Protocols, and Architecture*. 6th ed. Boston: Pearson, 2014.

Commission Imperiale, ed. *Exposition Universelle de 1867 à Paris. Catalogue général. 6e livraison. Instruments et procédés des art usuels*. Paris: E. Dentu, 1867.

Connolly, Daniel, Thomas A. Connolly, and Thomas J. McTighe. "Improvement in Automatic Telephone-Exchanges." US Patent 222 458, filed September 10, 1879.

Conway, Flo, and Jim Siegelmann. *Dark Hero of the Information Age: In Search of Norbert Wiener, the Father of Cybernetics*. New York: Basic Books, 2005.

Corbato, Fernando J. and Richard Steinberg. "Origin of the Word *Daemon*." *Take Our Word for It* 146 (January 16, 2002): 4. http://www.takeourword.com/TOW146/page4.html.

Crary, Jonathan. *Techniques of the Observer: On Vision and Modernity in the Nineteenth Century*. Cambridge, MA: MIT Press, 1990.

Crocker, Stephen, Stephen Carr, and Vinton Gray Cerf. *RFC 33: New HOST-HOST Protocol*. Tech. rep. UCLA und University of Utah, 1970. https://www.rfc-editor.org/rfc/rfc33.txt.

Crocker, Steve. *RFC 1: Host Software. Network Working Group Request for Comments*. Tech. rep. UCLA, 1969. https://www.rfc-editor.org/rfc/rfc1.txt.

Crocker, Steve. *RFC 3: Documentation Conventions. Network Working Group RFC-3*. Tech. rep. UCLA, 1969. http://www.rfc-editor.org/rfc/rfc3.txt.

Croome, Desmond Felix, and Alan Arthur Jackson. *Rails through the Clay: A History of London's Tube Railways*. 2nd ed. Harrow Weld, UK: Capital Transport, 1993.

Cubasch, Alwin J., Vanessa Engelmann, and Christian Kassung. "Theorie des Filterns: Zur Programmatik eines Experimentalsystems." *Zenodo*, April 30, 2021. https://doi.org/10.5281/zenodo.4731045.

Cusumano, Michael A. *The Japanese Automobile Industry: Technology and Management at Nissan and Toyota*. Cambridge, MA: Harvard University Press, 1985.

d'Allemagne, Henry-René. *Prosper Enfantin et les grandes entreprises du XIXe siècle*. Paris: Gründ, 1935.

d'Allent, Pierre-Alexandre. "Essai de reconnaissance militaire." *Flux: Cahiers scientifiques internationaux réseaux et territoires* (1989): 15–17.

Darnton, Robert. *Poetry and the Police: Communication Networks in Eighteenth-Century Paris*. Cambridge, MA: Belknap Press of Harvard University Press, 2010.

Daston, Lorraine, ed. *Things That Talk: Object Lessons from Arts and Sciences*. New York: Zone Books, 2004.

Daston, Lorraine, and Peter Galison. *Objectivity*. 2nd ed. New York: Zone Books, 2008.

Daston, Lorraine, and Peter Galison. "The Image of Objectivity." *Representations* 40 (1992): 81–128.

Davis, Allison, Burleigh B. Gardner, and Mary R. Gardner. *Deep South: A Social Anthropological Study of Caste and Class; Directed by W. Lloyd Warner*. Chicago: University of Chicago Press, 1941.

Davies, Donald Watts. *Proposal for a Digital Communication Network*. Tech. rep. National Physical Laboratory, Great Britain, June 1966.

Davies, Donald Watts, and Derek L. A. Barber. *Communication Networks for Computers*. London: John Wiley, 1973.

Debatin, Bernhard, and Hans Jürgen Wulff, eds. *Telefon und Kultur: Das Telefon im Spielfilm*. Berlin: Spiess, 1991.

de Certeau, Michel. *The Practice of Everyday Life*. Translated by Steven Rendall. Berkeley: University of California Press, 1984.

Deleuze, Gilles. "Postscript on the Societies of Control." *October* 59 (1992): 3–7.

Deleuze, Gilles, and Michel Foucault. "A New Cartographer (*Discipline and Punish*)." In *Foucault*. 10th ed. Edited and translated by Seán Hand, 21–38. London: Bloomsbury, 2016.

DeLillo, Don. *Running Dog*. London: Picador, 1999. First published 1978 by Vintage (New York).

Derrida, Jacques. *Of Grammatology*. Translated by Gayatri Chakravorty Spivak. Baltimore: Johns Hopkins University Press, 1976.

Derrida, Jacques. *Writing and Difference*. Translated by Alan Bass. Chicago: University of Chicago Press, 1978.

de Fourcy, Eugène. *Atlas souterrain de la ville de Paris exécuté conformément au vote émis en 1853 par commission municipale et suivant les ordres de M. le Bon G.E. Haussmann . . .* Paris: Charles de Mourgues Frères, 1859.

de Saint-Simon, Claude Henri. "Histoire de ma vie." In *Œuvres de Claude-Henri de Saint-Simon*. Vol. I/1, 64–88. Paris: Éditions anthropos, 1966.

Descartes, René. "L'homme." In *Œuvres philosophiques, 1618–1637*. Vol. 1: 369–480. Paris: Éditions Garnier, 1988.

de Sola Pool, Ithiel. *Forecasting the Telephone. A Retrospective Technology Assessment of the Telephone*. Norwood, NJ: Ablex, 1983.

Deuber-Mankowsky, Astrid. *Praktiken der Illusion: Kant, Nietzsche, Cohen, Benjamin bis Donna J. Haraway*. Berlin: Vorwerk 8, 2007.

Diderot, Denis. "D'Alembert's Dream." In *Rameau's Nephew and D'Alembert's Dream*. Translated by Leonard Tancock, 165–237. London: Penguin, 1966.

Diderot, Denis. "Le Rêve de d'Alembert." In *Œuvres completes*. Vol. 17, *Le Rêve de d'Alembert. Idées IV*. Edited by Jean Varloot. Paris: Hermann, 1987, 89–209.

Dirmoser, Gerhard. *Eine subjektive Chronologie der Netzentwicklungen (Vom Nutzen schematischer Zeichnungen XXXII)*. 2004. http://www.servus.at/kontext/diagramm/32_Netzhype.htm.

Dittrich, Sigrid und Lothar Dittrich. *Lexikon der Tiersymbole: Tiere als Sinnbilder in der Malerei des 14.–17. Jahrhunderts*. Petersberg: Michael Imhof Verlag, 2004.

Documents diplomatiques de la conférence télégraphique internationale de Paris. Paris: Imprimerie Imperiale, 1865.

Documents diplomatiques de la conférence télégraphique internationale de Vienne. Paris: Imprimerie Imperiale, 1868.

Dodds, Doug. "Notes on the Development of Message Technology." In *Berkeley Workshop on Distributed Data Management and Computer Networks*, edited by Lawrence Berkeley Laboratories, 144–156. Berkeley: University of California, 1976.

Dommann, Monika. "Handling, Flowcharts, Logistik: Zur Wissensgeschichte und Materialkultur von Warenflüssen." *Nach Feierabend: Zürcher Jahrbuch für Wissensgeschichte* 7 (2011): 75–103.

Domschke, Wolfgang, and Andreas Drexl. *Einführung in Operations Research*. 7th ed. Berlin: Springer, 2007.

Donati, Vitaliano. *Della storia naturale marina dell' Adriatico*. Venice: Francesco Storti, 1750.

Donner, Martin. "Rekursion und Wissen: Zur Emergenz technosozialer Netze." In *Rekursionen: Von Faltungen des Wissens*, edited by Ana Ofak and Philipp von Hilgers, 77–113. Paderborn: Wilhelm Fink, 2010.

Döring, Tobias, ed. *London Underground: Poems and Prose about the Tube*. Stuttgart: Reclam, 2003.

Draaisma, Douwe. *Die Metaphernmaschine: Eine Geschichte des Gedächtnisses*. Darmstadt: Primus Verlag, 1999.

Duncan, Christian A., David Eppstein, Michael T. Goodrich, Stephen G. Kobourov, and Martin Nöllenburg. "Lombardi Drawings of Graphs." *Journal of Graph Algorithms and Applications* 16, no. 1 (2012): 85–108. https://doi.org/10.7155/jgaa.00251.

Dundas, Judith. "Arachne's Web: Emblem into Art." *Emblematica: An Interdisciplinary Journal for Emblem Studies* 2, no. 1 (1987): 109–137.

Dünne, Jörg, Kathrin Fehringer, Kristina Kuhn, and Wolfgang Struck, eds. *Cultural Techniques*. Berlin: De Gruyter, 2020.

Durkheim, Émile. *Socialism and Saint-Simon*. Edited by Alvin W. Gouldner. Translated by Charlotte Sattler. Yellow Springs, OH: Antioch, 1958.

Eco, Umberto. "The Encyclopedia as a Labyrinth." In *Semiotics and the Philosophy of Language*, 80–84. Bloomington: Indiana University Press, 1986.

Edwards, Paul N. "Hyper Text and Hypertension: Post-Structuralist Critical Theory, Social Studies of Science and Software." *Social Studies of Science* 24 (1994): 229–278.

Edwards, Paul N. *The Closed World: Computers and the Politics of Discourse in Cold War America*. Cambridge, MA: MIT Press, 1996.

Edwards, Paul N., Stephen S. Jackson, Geoffrey C. Bowker, and Cory P. Knobel. *Understanding Infrastructure: Dynamics, Tensions, and Design: Report of a Workshop on "History and Theory of Infrastructure: Lessons for New Scientific Cyberinfrastructures."* Alexandria, VA: National Science Foundation, 2007. https://deepblue.lib.umich.edu/bitstream/handle/2027.42/49353/UnderstandingInfrastructure2007.pdf.

Electrical World. "Electrical Development in England." *Electrical World* 82, no. 21 (1923): 1055–1058.

Eliade, Mircea. *Images and Symbols: Studies in Religious Symbolism.* Translated by Philip Mairet. New York: Sheed and Ward, 1961.

Elliman, Paul. "Signal Failure." In *Else/Where: Mapping New Cartographies of Networks and Territories*, edited by Janet Abrams and Peter Hall, 166–175. Minneapolis: University of Minnesota Press, 2006.

Emden, Christian Jürgen. "Epistemische Konstellationen 1800–1900." In *Netzwerke: Eine Kulturtechnik der Moderne*, edited by Hartmut Böhme, Jürgen Barkhoff, and Jeanne Riou, 127–154. Cologne: Böhlau, 2004.

Emden, Christian Jürgen. "Netz." In *Lexikon der philosophischen Metaphern*, edited by Ralf Konersmann, 248–260. Darmstadt: Wissenschaftliche Buchgesellschaft, 2007.

Emerson, Lori. "Category: Other Networks." *loriemerson.net* (blog). https://loriemerson.net/category/other-networks.

Emerson, Lori. "What's Wrong with the Internet and How We Can Fix It: Interview with Internet Pioneer John Day." *loriemerson.net* (blog). July 23, 2015, https://loriemerson.net/2015/07/23/whats-wrong-with-the-internet-and-how-we-can-fix-it-interview-with-internet-pioneer-john-day.

Emge, Martinus. *Saint-Simon: Einführung in ein Leben und Werk, eine Schule, Sekte und Wirkungsgeschichte.* Munich: Oldenbourg, 1987.

Enfantin, Prosper. *Colonisation de l'Algérie.* Paris: P. Bertrand, 1843.

Engell, Lorenz. *Ausfahrt nach Babylon: Essais und Vorträge zur Kritik der Medienkultur.* Weimar: Verlag und Datenbank für Geisteswissenschaften, 2000.

Engell, Lorenz, Bernhard Siegert, and Joseph Vogl, eds. *1950: Archiv für Mediengeschichte.* Weimar: Universitätsverlag, 2004.

Engemann, Christoph. *Electronic Government: Vom User zum Bürger; Zur kritischen Theorie des Internets.* Bielefeld: transcript, 2003.

Engemann, Christoph. "Verteiltes Überleben: Paul Barans Antwort auf die atomare Bedrohung." In *Überleben: Historische und aktuelle Konstellationen*, edited by Falko Schmieder, 381–393. Paderborn: Wilhelm Fink, 2011.

Erlang, Agner Krarup. "Sandsynlighedsregning og Telefonsamtaler." *Nyt tidsskrift for Matematik* 20 (1909): 33–39.

Euler, Leonhard. "Das Königsberger Brückenproblem." In *Theorie der endlichen und unendlichen Graphen: Kombinatorische Topologie der Streckenkomplexe*, edited by Dénes König, 290–301. Leipzig: BSB Teubner, 1986.

Euler, Leonhard. *Die Geburt der Graphentheorie: Ausgewählte Schriften von der Topologie zum Sudoku*. Edited by Wladimir Velminski. Berlin: Kadmos, 2008.

Euler, Leonhard. "Solution d'une question curieuse qui ne paroît soumise à aucune analyse." In *Histoire de l'Académie Royale des Sciences et des Belles-Lettres de Berlin 1759* (1766): 310–337. Translated by Todd Doucet as "A solution to a curious problem which seems not subjected to any analysis." https://scholarlycommons.pacific.edu/cgi/viewcontent.cgi?filename=0&article=1308&context=euler-works&type=additional.

Fagen, M. D., ed. *A History of Engineering and Science in the Bell System: The Early Years (1875–1925)*. New York: Bell Telephone Laboratories, 1975.

Febvre, Lucien, and Marc Bloch. "Ports, fleuves, mers." *Annales d'histoire sociale* 1 (1940): 66–70.

Feinler, Elizabeth. "The Network Information Center and its Archives." *IEEE Annals of the History of Computing* 32, no. 3 (2010): 83–89.

Feldvoss, Marli. "Bei Anruf Film: Telefon und Kino—eine ideale Partnerschaft." *Neue Zuercher Zeitung*, December 14, 2002. https://www.nzz.ch/article89SX3-ld.238194.

Felsch, Philipp. "Aufsteigesysteme 1800–1900." In *Nach Feierabend: Zürcher Jahrbuch für Wissensgeschichte* 1, *Bilder der Natur—Sprachen der Technik*, 15–32. Zurich: diaphanes, 2005.

Fenlon, Iain. *The Ceremonial City: History, Memory and Myth in Renaissance Venice*. New Haven, CT: Yale University Press, 2008.

Ferguson, Niall. *The Square and the Tower: Networks, Hierarchies and the Struggle for Global Power*. New York: Penguin, 2017.

Fischbach, Rainer. *Mythos Netz: Kommunikation jenseits von Raum und Zeit?* Zurich: Rotpunktverlag, 2005.

Fischer, Claude S. *America Calling: A Social History of the Telephone to 1940*. Berkeley: University of California Press, 1992.

Fleck, Ludwik. *Genesis and Development of a Scientific Fact*. Edited by Thaddeus J. Trenn and Robert K. Merton. Translated by Fred Bradley and Thaddeus J. Trenn. Chicago: University of Chicago Press, 1979.

Flichy, Patrice. *Dynamics of Modern Communication: The Shaping and Impact of New Communication Technologies*. Translated by Liz Libbrecht. Thousand Oaks, CA: Sage, 1995.

Flörke, Heinrich Gustav. "Netz." In *Oekonomische Encyclopädie; oder, Allgemeines System der Staats- Stadt- Haus- u. Landwirtschaft in alphabetischer Ordnung*, edited by D. Johann Georg Krünitz, 444–473. Berlin: Trassler, 1806.

Flusser, Vilém. *Gestures*. Translated by Nancy Ann Roth. Minneapolis: University of Minnesota Press, 2014.

Foucault, Michel. "Nietzsche, la généalogie, l'histoire." In *Dits et écrits*. Vol. 2 (1970–1975), 136–156. Paris: Gallimard, 1994.

Foucault, Michel. *Security, Territory, Population: Lectures at the Collège de France, 1977–1978*. Edited by Michel Senellart. Translated by Graham Burchell. New York: Palgrave Macmillan, 2009.

Foucault, Michel. *The Archaeology of Knowledge*. Translated by A. M. Sheridan Smith. London: Routledge, 2002.

Foucault, Michel. *The Order of Things: An Archaeology of the Human Sciences*. Translated by Alan Sheridan. New York: Vintage, 1973.

Fox-Keller, Evelyn. *Refiguring Life: Metaphors of Twentieth-Century Biology*. New York: Columbia University Press, 1995.

Frank, Howard, Robert Elliott Kahn, and Leonard Kleinrock. "Computer Communication Network Design: Experience with Theory and Practice." In *AFIPS Conference Proceedings. Spring Joint Computers Conference*. Vol. 40, *May 16–17, 1972. Atlantic City, New Jersey*, 255–270. Montvale, NJ: AFIPS Press, 1972.

Fraser, J. T. *Time, the Familiar Stranger*. Amherst: University of Massachusetts Press, 1987.

Freeman, Linton C. "Visualizing Social Networks." *Journal of Social Structure* 1, no. 1 (2000). https://www.cmu.edu/joss/content/articles/volume1/Freeman.html.

Freeman, Linton C. *The Development of Social Network Analysis: A Study in the Sociology of Science*. Vancouver: Empirical Press, 2004.

Friedrich, Alexander. *Metaphorologie der Vernetzung: Zur Theorie kultureller Leitmetaphern*. Paderborn: Wilhelm Fink, 2015.

Friedrich, Alexander. "Metaphorology of Networks." In *Metaphors Shaping Culture and Theory*, edited by Herbert Grabes, Ansgar Nünning, and Sibylle Baumbach, 285–297. Tübingen: Narr, 2009.

Friedrich, Alexander. "Vernetzte Zwischenräume." In *Bewegen im Zwischenraum*, edited by Uwe Wirth, 55–74. Berlin: Kadmos, 2012.

Fujimoto, Takahiro. *The Evolution of the Manufacturing System at Toyota*. New York: Oxford University Press, 1999.

Galen. *On the Usefulness of the Parts of the Body / De Usu Partium*. Edited by Margaret Tallmadge May. Ithaca, NY: Cornell University Press, 1968.

Galison, Peter. *Einstein's Clocks and Poincaré's Maps: Empires of Time*. New York: W. W. Norton, 2003.

Galison, Peter. "The Ontology of the Enemy: Norbert Wiener and the Cybernetic Vision." *Critical Inquiry* 21, no. 1 (1994): 228–266.

Galloway, Alexander. *Protocol: How Control Exists after Decentralization*. Cambridge, MA: MIT Press, 2004.

Galloway, Alexander, and Eugene Thacker. *The Exploit: A Theory of Networks*. Minneapolis: University of Minnesota Press, 2007.

Garland, Ken. *Mr Beck's Underground Map*. Harrow Weld, UK: Capital Transport, 1994.

Gass, Saul I., and Arjang A. Assad. *An Annotated Timeline of Operations Research: An Informal History*. Boston: Kluwer Academic, 2005.

Gehring, Petra. "Paradigma einer Methode: Der Begriff des Diagramms im Strukturdenken von Michel Foucault und Michel Serres." In *Diagrammatik und Philosophie*, edited by Petra Gehring, Thomas Keutner, Jörg F. Maas, and Wolfgang Maria Ueding, 89–105. Amsterdam: Rodopi, 1988.

Gell, Alfred. "Vogel's Net: Traps as Artworks and Artworks as Traps." In *The Art of Anthropology: Essays and Diagrams*, edited by Eric Hirsch, 187–214. London: Athlone, 1999.

Gelshorn, Julia, and Tristan Weddigen. "Das Netzwerk: Zu einem Denkbild in Kunst und Wissenschaft." In *Grammatik der Kunstgeschichte: Sprachproblem und Regelwerk im Bild-Diskurs. Oskar Bätschmann zum 65. Geburtstag*, edited by Hubert Locher and Peter J. Schneemann, 54–77. Zurich: Schweizerisches Institut für Kunstwissenschaft, 2008.

Génieys, Raymond. *Essai sur les moyens de conduire, d'élever et de distribuer les eaux*. Paris: Carilian-Goeury, 1829.

Génieys, Raymond. *Note sur un projet de distribution générale de l'eau dans l'intérieur de Paris*. Paris: Carilian-Goeury, 1827.

Gethmann, Daniel. "Das Zittern der Luft: Die Erfindung der mechanischen Stimme." In *Medien/Stimmen*, edited by Cornelia Epping-Jäger and Erika Linz, 211–234. Cologne: DuMont, 2003.

Gibney, Frank B. "The Marks of a Japanese Merchant." In *Kanban: The Art of the Japanese Shop Sign*, edited by Dana Levy, Lea Sneider, and Frank B. Gibney, 7–29. San Francisco: Chronicle Books, 1983.

Giedion, Sigfried. *Mechanization Takes Command: A Contribution to Anonymous History*. New York: Oxford University Press, 1948.

Giesecke, Michael. *Der Buchdruck in der frühen Neuzeit: Eine historische Fallstudie über die Durchsetzung neuer Informations- und Kommunikationstechnologien*. Frankfurt am Main: Suhrkamp, 1994.

Giessmann, Sebastian. "Graphen können alles: Visuelle Modellierung und Netzwerktheorie vor 1900." In *Visuelle Modelle*, edited by Ingeborg Reichle, Steffen Siegel, and Achim Spelten, 269–284. Paderborn: Wilhelm Fink, 2008.

Giessmann, Sebastian. "Netze als Weltbilder: Ordnungen der Natur von Donati bis Cuvier." In *Verwandte Bilder*, edited by Ingeborg Reichle, Steffen Siegel, and Achim Spelten. 2nd ed., 243–261. Berlin: Kadmos, 2008.

Giessmann, Sebastian. *Netze und Netzwerke: Archäologie einer Kulturtechnik, 1740–1840*. Bielefeld: transcript, 2006.

Giessmann, Sebastian. "Netzstörungen: Erzählungen vom Ende der Netzwerke." *Störfälle: Zeitschrift für Kulturwissenschaften* 2 (2011): 125–133.

Giessmann, Sebastian. "Netzwerkprotokolle und Schwarm-Intelligenz: Zur Konstruktion von Komplexität und Selbstorganisation." In *Schwärme: Kollektive ohne Zentrum*, edited by Eva Horn and Lucas Marco Gisi, 163–182. Bielefeld: transcript, 2009.

Giessmann, Sebastian. "Verunreinigungsarbeit: Über den Netzwerkbegriff der Akteur-Netzwerk-Theorie." *Reinigungsarbeit: Zeitschrift für Kulturwissenschaften* 1 (2013): 149–160.

Gillies, James, and Robert Cailliau. *How the Web was Born: The Story of the World Wide Web*. Oxford: Oxford University Press, 2000.

Gilliland, Ezra T. "Telephone-Circuit," "Metallic-Circuit Telephone System," and "Multiple-Circuit Changer," US Patents 306 239–241, filed October 7, 1884.

Gitelman, Lisa. *Always Already New: Media, History, and the Data of Cultures*. Cambridge, MA: MIT Press, 2006.

Godt, Birgitta. "Die Entwicklung der Handvermittlung." In *Fräulein vom Amt*, edited by Helmut Gold and Annette Koch, 68–85. Munich: Prestel, 1993.

Gold, Helmut, and Annette Koch, eds. *Fräulein vom Amt*. Munich: Prestel, 1993. Exhibition catalog of *Fräulein vom Amt* at the Deutsche Postmuseum, Frankfurt am Main, May 4—August 15, 1993.

Goldratt, Eliyahu M. *Critical Chain: A Business Novel*. Great Barrington, MA: North River Press, 1997.

Gööck, Roland. *Die grossen Erfindungen: Nachrichtentechnik und Elektronik*. Künzelsau: Sigloch, 1988.

Goodman, Nelson. *Languages of Art: An Approach to a Theory of Symbols*. Indianapolis, IN: Bobbs-Merrill, 1968.

Graber, Frédéric. "Purity and Theory: Theoretical Tools at Ponts et Chaussées, circa 1800." *Technology and Culture* 49, no. 4 (2008): 860–883.

Grady, James. *Six Days of the Condor*. Harpenden: No Exit Press, 2007. First published 1974 by W. W. Norton.

Granovetter, Mark S. "The Strength of Weak Ties." *The American Journal of Sociology* 78, no. 6 (1973): 1360–1380.

Grassmuck, Volker. *Geschlossene Gesellschaft: Mediale und diskursive Aspekte der "drei Öffnungen" Japans*. Munich: iudicum, 2002.

Graunt, John. *Natural and Political Observations Mentioned in a following Index, and made upon the Bills of Mortality*. London: Martin, Allestry, and Dicas, 1662.

Gravelot, Hubert François, Charles Nicolas Cochin, and Charles-Étienne Gaucher. *Iconologie par Figures ou Traité complet des Allégories, Emblêmes &c. Ouvrage outile aux Artistes et Amateurs, et peuvent servir à l'éducation des jeunes personnes*. Vol. 3. Paris: Lattré Graveur, 1791.

Green, Oliver. *Frank Pick's London: Art, Design and the Modern City*. London: V&A Publishing, 2013.

Green, Oliver. *Underground Art: London Transport Posters 1908 to the Present*. 2nd ed. London: Laurence King, 2001.

Green, Venus. *Race on the Line: Gender, Labor, and Technology in the Bell System, 1880–1980*. Durham, NC: Duke University Press, 2001.

Griaule, Marcel. *Conversations with Ogotemmêli: An Introduction to Dogon Religious Ideas*. Translated by Ralph Butler with Audrey I. Richards and Beatrice Hooke. London: Oxford University Press, 1965. Originally published as *Dieu d'eau*. Paris: Éditions du Chêne, 1948.

Griaule, Marcel. *Jeux Dogons*. Paris: Institut d'Ethnologie, 1938.

Griaule, Marcel, and Germaine Dieterlen. *Le renard pâle*. 2nd ed. Paris: Institut d'Ethnologie, 1991.

Grimm, Jacob, and Wilhelm Grimm. *Deutsches Wörterbuch*. Vol. 7, *N–Quurren*, edited by Matthias von Lexer. Leipzig: S. Hirzel, 1889.

Groh, Dieter. "The Temptation of Conspiracy Theory; or, Why Do Bad Things Happen to Good People? Part I: Preliminary Draft of a Theory of Conspiracy Theories." In *Changing Conceptions of Conspiracy*, edited by Carl F. Graumann and Serge Moscovici, 1–13. New York: Springer, 1987.

Groh, Dieter. "The Temptation of Conspiracy Theory, or: Why Do Bad Things Happen to Good People? Part II: Case Studies." In *Changing Conceptions of Conspiracy*, edited by Carl F. Graumann and Serge Moscovici, 15–37. New York: Springer, 1987.

Gropp, Harald. "Hamiltonian Graphs from Kirkman to König." *Electronic Notes in Discrete Mathematics* 24 (2006): 81–88.

Groys, Boris. *Under Suspicion: A Phenomenology of Media*. Translated by Carsten Strathausen. New York: Columbia University Press, 2012.

Guare, John. *Six Degrees of Separation*. 2nd ed. New York: Vintage, 1994.

Gugerli, David, and Daniel Speich. *Topografien der Nation: Politik, kartografische Ordnung und Landschaft im 19. Jahrhundert*. Zurich: Chronos, 2002.

Guillerme, André. "L'émergence du concept du réseau (1820–1830)." In *Réseaux territoriaux*, edited by Georges Dupuy, 33–49. Caen: Paradigme, 1988.

Guillerme, André. "Réseau: genèse d'un mot." *Cahiers de la Médiologie* 3 (1997): 7–16.

Hadlaw, Janin. "The London Underground Map: Imagining Modern Time and Space." *Design Issues* 19, no. 1 (2003): 25–35.

Hafner, Katie, and Matthew Lyon. *Where Wizards Stay Up Late: The Origins of the Internet*. New York: Touchstone, 1998.

Hagen, Wolfgang. "Gefühlte Dinge: Bells Oralismus, die Undarstellbarkeit der Elektrizität und das Telefon." In *Telefonbuch: Beiträge zu einer Kulturgeschichte des Telefons*, edited by Stefan Münker and Alexander Roesler, 35–60. Frankfurt am Main: Suhrkamp, 2000.

Hagen, Wolfgang. "Metaxy: Eine historiosemantische Fußnote zum Medienbegriff." In *Was ist ein Medium?*, edited by Stefan Münker and Alexander Roesler, 13–29. Frankfurt am Main: Suhrkamp, 2008.

Haigh, Thomas, and Paul E. Ceruzzi. *A New History of Modern Computing*. Cambridge, MA: MIT Press, 2021.

Hamilton, William Rowan. "On a New System of Roots of Unity." *Proceedings of the Royal Irish Academy* 6 (1858): 415–416.

Hankins, Thomas L. "Blood, Dirt, and Nomograms: A Particular History of Graphs." *Isis* 90 (1999): 50–80.

Hannerz, Ulf. *Exploring the City: Inquiries toward an Urban Anthropology*. New York: Columbia University Press, 1980.

Haraway, Donna. "A Cyborg Manifesto: Science, Technology and Socialist-Feminism in the Late Twentieth Century." In *Simians, Cyborgs and Women: The Reinvention of Nature*, 149–181. New York: Routledge, 1991.

Haraway, Donna. "A Game of Cat's Cradle: Science Studies, Feminist Theory, Cultural Studies." *Configurations* 2, no. 1 (1994): 59–71.

Hardt, Michael, and Antonio Negri. *Empire*. 4th ed. Cambridge, MA: Harvard University Press, 2001.

Hartmann, Frank. *Globale Medienkultur: Technik, Geschichte, Theorien*. Vienna: WUV, 2006.

Hartmann, Frank, and Erwin K. Bauer, eds. *Bildersprache: Otto Neuraths Visualisierungen*. 2nd ed. Vienna: WUV, 2006.

Harvey, William. *Exercitatio anatomica de motu cordis et sanguinis in animalibus*. Frankfurt am Main: Fitzer, 1628. Translated by Robert Willis as *On the Motion of the Heart and Blood in Animals: A New Edition of William Harvey's "Exercitatio anatomica de motu cordis et sanguinis in animalibus."* Edited by Jarrett A. Carty. Eugene, OR: Wipf and Stock, 2016.

Heart, Frank E, Robert E. Kahn, Severo M. Ornstein, William R. Crowther, and David C. Walden. "The Interface Message Processor for the ARPA Computer Network." In *AFIPS Conference Proceedings*. Vol. 36, *Spring Joint Computers Conference. May 5–7, 1970. Atlantic City, New Jersey*, 551–567. New York: ACM Press, 1970.

Heath, Joseph. "Methodological Individualism." In *The Stanford Encyclopedia of Philosophy*. Edited by Edward N. Zalta. Spring 2005 Edition. Stanford, CA: Metaphysics Research Lab, Stanford University, 2005. https://plato.stanford.edu/archives/spr2005/entries/methodological-individualism.

Heidegger, Martin. "Age of the World Picture." In *Off the Beaten Track*, edited and translated by Julian Young and Kenneth Haynes, 57–85. Cambridge, UK: Cambridge University Press, 2002.

Heinrich, Dirk, and Johann Tischler. "Fischerei und Fischereimethoden." In *Reallexikon der Germanischen Altertumskunde*. Vol. 9, *Fidel–Friedlosigkeit*, edited by Johannes Hoops, 130–146. Berlin: De Gruyter, 1995.

Hellige, Hans Dieter. "Die Geschichte des Internet als Lernprozess." *artec-paper* 138 (2006): 1–39.

Herrmann, Gerhard. *Der Suezkanal*. Bern: Wilhelm Goldmann, 1936.

Hilger, Dietrich. "Industrie, Gewerbe." In *Geschichtliche Grundbegriffe. Historisches Lexikon zur politisch-sozialen Sprache in Deutschland*. Vol. 3, edited by Otto Brunner, Werner Conze, and Reinhart Koselleck, 237–304. Stuttgart: Klett-Cotta, 1982.

Hill, R. B. "Early Work on Dial Telephone Systems." *Bell Laboratories Record* 31 (1953): 22–29.

Hinterwaldner, Inge. *The Systemic Image: A New Theory of Interactive Real-Time Simulations*. Translated by Elizabeth Tucker. Cambridge, MA: MIT Press, 2017.

Hirst, Michael. "London's Travel Network Clears Its First Olympic Hurdle." *BBC*, July 30, 2012. http://www.bbc.co.uk/news/uk-19039072.

Hobbs, Robert. "Mark Lombardi: Global Networks." In *Mark Lombardi: Global Networks,* edited by Robert Hobbs, 11–49. New York: Independent Curators International, 2003.

Hobbs, Robert, ed. *Mark Lombardi: Global Networks*. New York: Independent Curators International, 2003. Exhibition catalog.

Hobson, James A. *The Industrial System: An Inquiry into Earned and Unearned Income*. London: Longman, Green & Co., 1910.

Hobson, John A. *The Evolution of Modern Capitalism: A Study of Machine Production*. Rev. ed. London: Walter Scott, 1906.

Hofmann, August Wilhelm. "On the Combining Power of Atoms." *Notices of the Proceedings at the Meetings of the Members of the Royal Institution of Britain* 4, no. 42 (1865): 401–430.

Hofstadter, Richard. *The Paranoid Style in American Politics and Other Essays*. London: Jonathan Cape, 1966.

Höhne, Stefan. *Riding the New York Subway: The Invention of the Modern Passenger*. Cambridge, MA: MIT Press, 2021.

Holl, Ute. "Türen, Lektüren und das Wissen des Films. Zur Verschwörung der Medien in Sydney Pollacks Three Days of the Condor." In *Parallax View. Zur Mediologie der Verschwörung,* edited by Marcus Krause, Arno Meteling, and Markus Stauff, 235–251. Paderborn: Wilhelm Fink, 2011.

Holtorf, Christian. *Der erste Draht zur Neuen Welt: Die Verlegung des transatlantischen Telegrafenkabels*. Göttingen: Wallstein, 2013.

Holzer, Boris. *Netzwerke*. Bielefeld: transcript, 2006.

Hörl, Erich. *Sacred Channels: The Archaic Illusion of Communication*. Translated by Nils F. Schott. Amsterdam: Amsterdam University Press, 2018.

Horn, Eva. "Editor's Introduction: 'There Are No Media.'" *Grey Room* 29 (2007): 6–13.

Horn, Eva. *The Secret War: Treason, Espionage, and Modern Fiction*. Translated by Geoffrey Winthrop-Young. Evanston, IL: Northwestern University Press, 2013. Originally published as *Der geheime Krieg: Verrat, Spionage und moderne Fiktion* (Frankfurt am Main: Fischer, 2007).

Horn, Eva, and Anson Rabinbach. "Introduction." *New German Critique*, no. 103 (2008): 1–8.

Horstmann, Erwin. *75 Jahre Fernsprecher in Deutschland*. Bonn: Bundesministerium für Post- und Fernmeldewesen, 1952.

Hughes, Thomas Parke. *Networks of Power: Electrification in Western Society, 1880–1930*. Baltimore: Johns Hopkins University Press, 1983.

Hugill, Peter J. *Global Communications since 1844: Geopolitics and Technology*. Baltimore: Johns Hopkins University Press, 1999.

Hugo, Victor. *Les Misérables*. Translated by Charles E. Wilbour. New York: Modern Library, 1902. Originally published as *Les Misérables* (Brussels: Lacroix, 1862).

Huizinga, Johan. *Homo Ludens: A Study of the Play-Element in Culture*. Translated by R. F. C. Hull. London: Routledge, 1949.

Information Sciences Institute, University of Southern California. *RFC 791: Internet Protocol. DARPA Internet Program Control Specification*. Tech. rep. Marina del Rey, CA: Information Sciences Institute, 1981. https://www.rfc-editor.org/rfc/rfc791.txt.

Information Sciences Institute, University of Southern California. *RFC 793: Transmission Control Protocol. DARPA Internet Program Protocol Specification*. Tech. rep. Marina del Rey, CA: Information Sciences Institute, 1981. https://www.rfc-editor.org/rfc/rfc793.txt.

Iyer, Ananth V., Sridhar Seshadri, and Roy Vasher. *Toyota Supply Chain Management: A Strategic Approach to the Principles of Toyota's Renowned System*. New York: McGraw-Hill, 2009.

Jacob, François. *The Logic of Life: A History of Heredity*. Translated by Betty E. Spillmann. New York: Random House, 1973.

Jacobs, Carl H. *Die Schiffahrtsfreiheit im Suezkanal*. Göttingen: E. A. Huth, 1912.

Jacoud, Gilles, ed. *Political Economy and Industrialism: Banks in Saint-Simonian Economic Thought*. London: Routledge, 2010.

Jahn, Ilse. "Naturphilosophie und Empirie in der Frühaufklärung." In *Geschichte der Biologie: Theorien, Methoden, Institutionen, Kurzbiographien*, edited by Ilse Jahn. 4th ed., 197–230. Jena: Gustav Fischer, 1998.

James, William. "The Association of Ideas." *The Popular Science Monthly* 16, no. 5 (1880): 577–593.

James, William. *The Principles of Psychology*. Vol. 1. New York: Dover, 1950. First published 1890 by Holt (New York).

Jameson, Fredric. *The Geopolitical Aesthetic: Cinema and Space in the World System*. Bloomington: Indiana University Press, 1992.

Jishio.org, s.v. "Kanban." https://jisho.org/search/Kanban.

John, Richard R. *Network Nation: Inventing American Telecommunications*. Cambridge, MA: Belknap Press of Harvard University Press, 2010.

Johnson, C. K. *ORTEP. A FORTRAN Thermal-Ellipsoid Plot Program for Crystal Structure Illustrations. ORNL Report 3794*. Tech. rep. Oak Ridge, TN: Oak Ridge National Laboratory, 1965.

Johnson, Samuel, *A Dictionary of the English Language: In Which the Words are deduced from their Originals and Illustrated in their Different Significations by Examples from the Best Writers [. . .]*. Vol. 2. London: Printed by W. Strahan, for J. and P. Knapton, T. and T. Longman, C. Hitch and L. Hawes, A. Millar, and R. and J. Dodsley, 1755.

Johnstone, Sarah, and Tom Masters. *London City Guide*. Melbourne: Lonely Planet, 2006.

Jones, D. F. *Colossus*. London: Pan, 1968. First published 1966 by G. P. Putnam's Sons (New York).

Julliard, Virginie. "Une 'femme machine' au travail: la 'demoiselle du téléphone.'" *Quaderni* 56 (2004): 23–32, 28.

Kafka, Franz. *America*. Translated by Willa and Edwin Muir. Harmondsworth, UK: Penguin, 1981. Originally published as *Amerika* (Munich: Kurt Wolff, 1927).

Kafka, Franz. *The Castle*. Translated by Mark Harman. New York: Schocken, 1998. Originally published as *Das Schloss* (Munich: Kurt Wolff, 1926).

Kamata, Satoshi. *Japan in the Passing Lane: An Insider's Account of Life in a Japanese Auto Factory*. Boston: George Allen & Unwin, 1983.

Kapp, Ernst. *Elements of a Philosophy of Technology: On the Evolutionary History of Culture*. Edited by Jeffrey West Kirkwood and Leif Weatherby. Translated by Lauren K. Wolfe. Minneapolis: University of Minnesota Press, 2018. Originally published as *Grundlinien einer Philosophie der Technik: Zur Entstehungsgeschichte der Cultur aus neuen Gesichtspunkten* (Braunschweig: Westermann, 1877).

Karppi, Tero. *Disconnect: Facebook's Affective Bonds*. Minneapolis: University of Minnesota Press, 2018.

Karppi, Tero, Urs Stäheli, Clara Wieghorst, and Lea P. Zierott. *Undoing Networks*. Lüneburg: Meson Press, 2021.

Kassung, Christian. *Das Pendel: Eine Wissensgeschichte*. Paderborn: Wilhelm Fink, 2007.

Kassung, Christian. *Fleisch: Die Geschichte einer Industrialisierung*. Leiden: Brill | Schöningh, 2020.

Kassung, Christian. "Isochronie und Synchronie: Zur apparativen und epistemologischen Genese des Kopiertelegraphen." In *Lebendige Zeit: Wissenskulturen im Werden*, edited by Henning Schmidgen, 195–209. Berlin: Kadmos, 2005.

Kassung, Christian, and Albert Kümmel-Schnur. "Wissensgeschichte als Malerarbeit? Ein Trialog über das Weißeln schwarzer Kisten." In *Bruno Latours Kollektive*, edited by Georg Kneer, Markus Schroer, and Erhard Schüttpelz, 155–179. Frankfurt am Main: Suhrkamp, 2008.

Kassung, Christian, and Thomas Macho. "Einleitung." In *Kulturtechniken der Synchronisation*, edited by Christian Kassung and Thomas Macho, 9–21. Paderborn: Wilhelm Fink, 2013.

Kaufmann, Stefan. "Netzwerk." In *Glossar der Gegenwart*, edited by Ulrich Bröckling, Susanne Krasmann, and Thomas Lemke, 182–189. Frankfurt am Main: Suhrkamp, 2004.

Kaufmann, Stefan, ed. *Vernetzte Steuerung: Soziale Prozesse im Zeitalter technischer Netzwerke*. Zurich: Chronos, 2007.

Keck, Annette, and Nicolas Pethes, eds. *Mediale Anatomien: Menschenbilder als Medienprojektionen*. Bielefeld: transcript, 2001.

Keith, Alexander E., John Erickson, and Charles G. Erickson. "Electrical Exchange," US Patent 638 249, filed December 16, 1895, and issued December 5, 1899.

Kelley, James E., Jr., and Morgan R. Walker. "Critical-Path Planning and Scheduling." In *Proceedings of the Eastern Joint Computer Conference*, 160–173. New York: Eastern Joint Computer Conference, 1959.

Killen, Andreas. "Die Telefonzentrale." In *Orte der Moderne: Erfahrungswelten des 19. und 20. Jahrhunderts*, edited by Alexa Geisthövel and Habbo Knoch, 81–90. Frankfurt am Main: Campus, 2005.

Kittler, Friedrich. "Aufschreibesysteme 1800/1900. Vorwort." *Zeitschrift für Medienwissenschaft* 6 (2012): 117–126.

Kittler, Friedrich. *Gramophone, Film, Typewriter*. Translated by Geoffrey Winthrop-Young and Michael Wutz. Stanford, CA: Stanford University Press, 1999.

Kittler, Friedrich, and Ana Ofak, eds. *Medien vor den Medien*. Paderborn: Wilhelm Fink, 2007.

Klein, Ursula. "Visualität, Ikonizität, Manipulierbarkeit: Chemische Formeln als Paper Tools." In *Schrift: Kulturtechnik zwischen Auge, Hand und Maschine*, edited by Gernot Grube, Werner Kogge, and Sybille Krämer, 237–251. Paderborn: Wilhelm Fink, 2005.

Kleinrock, Leonard. *Communication Nets: Stochastic Message Flow and Delay*. Lincoln Laboratory Publications. New York: McGraw-Hill, 1964.

Kleinrock, Leonard. "Personal History/Biography: The Birth of the Internet." University of California at Los Angeles, Computer Science Department, 2010. https://web.archive.org/web/20190905082135/https://www.lk.cs.ucla.edu/personal_history.html.

Klose, Alexander. *The Container Principle: How a Box Changes the Way We Think*. Translated by Charles Marcrum II. Cambridge, MA: MIT Press, 2015.

Klovdahl, Alden S. "A Note on Images of Networks." *Social Networks* 3, no. 3 (1981): 197–214.

Knorr Cetina, Karin. "Scopic Media and Global Coordination: The Mediatization of Face-to-Face Encounters." In *Mediatiziation of Communication*, edited by Knut Lundby, 39–62. Berlin: De Gruyter, 2014.

König, Dénes. *Theory of Finite and Infinite Graphs*. Translated by Richard McCoart. Boston: Birkhäuser, 1990.

König, Wolfgang. "Nutzungswandel, Technikgenese und Technikdiffusion: Ein Essay zur Frühgeschichte des Telefons in den Vereinigten Staaten und Deutschland." In *Fern-Sprechen: Internationale Fernmeldegeschichte, -soziologie und -politik*, edited by Jörg Becker, 147–163. Berlin: Vistas, 1994.

König, Wolfgang, and Wolfhard Weber, eds. *Propyläen-Technikgeschichte*. Vol. 4, *Netzwerke, Stahl und Strom, 1840–1914*. Berlin: Propyläen, 1990.

Koschorke, Albrecht, Susanne Lüdemann, Frank Thomas, and Ethel Matala de Mazza. *Der fiktive Staat: Konstruktionen des politischen Körpers in der Geschichte Europas*. Frankfurt am Main: Fischer, 2007.

Koubek, Jochen. "Vernetzung als kulturelles Paradigma." PhD diss., Humboldt Universität zu Berlin, 2003. https://edoc.hu-berlin.de/18452/15629.

Kracauer, Siegfried. *The Salaried Masses: Duty and Distraction in Weimar Germany*. Translated by Quintin Hoare. London: Verso, 1998.

Kracauer, Siegfried. "Two Planes." In *The Mass Ornament: Weimar Essays*, edited and translated by Thomas Y. Levin, 37–39. Cambridge, MA: Harvard University Press, 1995.

Krajewski, Markus, ed. *Projektemacher: Zur Produktion von Wissen in der Vorform des Scheiterns*. Berlin: Kadmos, 2004.

Krajewski, Markus. *The Server: A Media History from the Present to the Baroque*. Translated by Ilinca Iurascu. New Haven, CT: Yale University Press, 2018.

Krajewski, Markus. *World Projects: Global Information Before World War I*. Translated by Charles Marcrum II. Minneapolis: University of Minnesota Press, 2014.

Krämer, Sybille. "'Operationsraum Schrift': Über einen Perspektivenwechsel in der Betrachtung der Schrift." In *Schrift: Kulturtechnik zwischen Auge, Hand und Maschine*, edited by Gernot Grube, Werner Kogge, and Sybille Krämer, 22–57. Paderborn: Wilhelm Fink, 2005.

Krämer, Sybille. "'Schriftbildlichkeit'; oder, Über eine fast vergessene Dimension der Schrift." In *Bild— Schrift—Zahl*, edited by Sybille Krämer and Horst Bredekamp, 157–176. Paderborn: Wilhelm Fink, 2003.

Krämer, Sybille. "Writing, Notational Iconicity, Calculus: On Writing as a Cultural Technique." *MLN* 118, no. 3 (2003): 518–537.

Krämer, Sybille, and Horst Bredekamp. "Culture, Technology, Cultural Techniques: Moving Beyond Text." Translated by Michael Wutz. *Theory, Culture & Society* 30, no. 6 (2013): 20–29.

Krause, Marcus, Arno Meteling, and Markus Stauff. "Einleitung." In *The Parallax View: Zur Mediologie der Verschwörung*, edited by Arno Meteling, Markus Stauff, and Marcus Krause, 9–42. Paderborn: Wilhelm Fink, 2011.

Krempel, Lothar. *Visualierung komplexer Strukturen: Grundlagen der Darstellung mehrdimensionaler Netzwerke*. Frankfurt am Main: Campus, 2005.

Kümmel, Albert. "Mathematische Medientheorie." In *Medientheorien: Eine Einführung*, edited by Daniela Kloock and Angela Spahr, 205–236. Munich: Wilhelm Fink, 1997.

Kümmel, Albert, and Erhard Schüttpelz, eds. *Signale der Störung*. Munich: Wilhelm Fink, 2003.

Kümmel-Schnur, Albert. "Patente als Agenten von Mediengeschichte." In *Bildtelegraphie: Eine Mediengeschichte in Patenten (1840–1930)*, edited by Albert Kümmel-Schnur and Christian Kassung, 15–38. Bielefeld: transcript, 2012.

Kümmel-Schnur, Albert, and Christian Kassung, eds. *Bildtelegraphie: Eine Mediengeschichte in Patenten, 1840–1930*. Bielefeld: transcript, 2012.

Kumon, Shumpei. "Japan as a Network Society." In *The Political Economy of Japan: Cultural and Social Dynamics*. Vol. 3, edited by Shumpei Kumon and Henry Rosovsky, 109–141. Stanford, CA: Stanford University Press, 1992.

Kumon, Shumpei, and Henry Rosovsky, eds. *The Political Economy of Japan: Cultural and Social Dynamics*. Vol. 3. Stanford, CA: Stanford University Press, 1992.

Laertius, Diogenes. *Lives of Eminent Philosophers*. Vol. 2. Edited by R. D. Hicks. Cambridge, MA: Harvard University Press, 1972.

Lagercrantz, Sture. "Beiträge zur Jagdfallensystematik." *Ethnos* 6 (1937): 361–366.

Lamé, Gabriel, Émile Clapeyron, Stéphane Flachat, and Eugène Flachat. *Vues politiques et pratiques sur les travaux publics*. Paris: Everat, 1832.

Lammert, Angela. "Mark Lombardi: Denken als Mustererkennung." In *Räume der Zeichnung*, edited by Angela Lammert, Carolin Meister, Jan Philipp Frühsorge, and Andreas Schalhorn, 57–71. Nuremberg: Verlag für moderne Kunst, 2007.

Latour, Bruno. *Aramis; or, The Love of Technology*. Translated by Catherine Porter. Cambridge, MA: Harvard University Press, 1996.

Latour, Bruno. "On Technical Mediation: Philosophy, Sociology, Genealogy." *Common Knowledge* 3, no. 2 (1994): 29–64.

Latour, Bruno. *Pandora's Hope: Essays on the Reality of Science Studies*. Cambridge, MA: Harvard University Press, 1999.

Latour, Bruno. *Reassembling the Social: An Introduction to Actor-Network-Theory*. Oxford: Oxford University Press, 2005.

Latour, Bruno. "Technology Is Society Made Durable." In *A Sociology of Monsters: Essays on Power, Technology and Domination*, edited by John Law, 103–131. London: Routledge, 1991.

Latour, Bruno. "Visualisation and Cognition: Drawing Things Together." In *Representation in Scientific Practice*, edited by Michael Lynch and Steve Woolgar, 19–68. Cambridge, MA: MIT Press, 1990.

Latour, Bruno. *We Have Never Been Modern*. Translated by Catherine Porter. Cambridge, MA: Harvard University Press, 1993.

Law, John. "Note on the Theory of the Actor Network: Ordering, Strategy and Heterogeneity." *Systems Practice* 5, no. 4 (1992): 379–393.

Leboff, David und Tim Demuth. *No Need to Ask! Early Maps of London's Underground Railways*. Harrow Weld, UK: Capital Transport, 1999.

Leibniz, Gottfried Wilhelm. "Characteristica Geometrica. Analysis Geometrica propria. Calculus situs." In *Mathematische Schriften V*, edited by Carl Immanuel Gerhardt, 141–219. Hildesheim: Olms, 1971. Translated as "On Analysis Situs." In *Philosophical Papers and Letters: A Selection*. Edited and translated by Leroy E. Loemker. 2nd ed., 254–258. Dordrecht: D. Reidel, 1969.

Lenoir, Timothy. "Helmholtz and the Materialities of Communication." *Osiris* 9 (1994): 184–207.

Leroi-Gourhan, André. *Gesture and Speech*. Translated by Anna Bostock Berger. Cambridge, MA: MIT Press, 1993.

Leroi-Gourhan, André. *Milieu et techniques*. Paris: Éditions Albin Michel, 1945.

Lévi-Strauss, Claude. *Structural Anthropology*. Vol. 1. Translated by Claire Jacobson and Brooke Grundfest Schoepf. New York: Basic Books, 1963.

Lévi-Strauss, Claude. *Structural Anthropology*. Vol. 2. Translated by Monique Layton. New York: Basic Books, 1976.

Levy, Dana, Lea Sneider, and Frank Gibney. *Kanban: The Art of the Japanese Shop Sign*. San Francisco: Chronicle Books, 1983.

Licklider, Joseph Carl Robnett. *Libraries of the Future*. 5th ed. Cambridge, MA: MIT Press, 1971.

Licklider, Joseph Carl Robnett, and Robert William Taylor. "The Computer as a Communication Device." *Science and Technology* 76 (1968): 21–31.

Lin, Tan. "Two or Three Sentences about Mark Lombardi." In *Mark Lombardi: Preparatory Drawings, 1994–2000*, 8–22. New York: Pierogi, 2003.

Lindner, Kurt. *Geschichte des deutschen Weidwerks*. Vol. 1, *Die Jagd der Vorzeit*. Berlin: De Gruyter, 1937.

Lindner, Kurt. *Geschichte des deutschen Weidwerks*. Vol. 2, *Die Jagd im frühen Mittelalter*. Berlin: De Gruyter, 1940.

Lipartito, Kenneth. "When Women Were Switches: Technology, Work, and Gender in the Telephone Industry, 1890–1920." *The American Historical Review* 99, no. 4 (1994): 1075–1111.

Lips, Julius. *Fallensysteme der Naturvölker*. Leipzig: Ernst Wiegandt Verlagsbuchhandlung, 1927.

Lombardi, Mark. *100 Notes—100 Thoughts, No. 071*, introduction by Carolyn Christov-Bakargiev, exh. cat. Kassel, documenta 13. Ostfildern: Hatje Cantz, 2012.

London County Council. Special Committee on Electricity Supply, ed. *Report on London Electricity Supply by Merz and McLellan*. London: The Council, 1914.

London Passenger Transport Board, ed. *First Annual Report and Statement of Accounts and Statistics for the Year Ended 30 June 1934*. London: LPTB, 1934.

Loraux, Nicole. "Le lien de la division." *Le Cahier (Collège international de philosophie)*, no. 4 (1987): 101–124.

Lovink, Geert. *The Principle of Notworking: Concepts in Critical Internet Culture*. Amsterdam: Hogeschool van Amsterdam, 2005.

Lucas, Édouard. *Récréations mathématiques: Qui perd gagne, les dominos, les marelles . . .* Vol. 2. Paris: A. Blanchard, 1892.

Luhmann, Niklas. *Ecological Communication*. Translated by John Bednarz, Jr. Chicago: University of Chicago Press, 1989.

Luhmann, Niklas. "Gleichzeitigkeit und Synchronisation." In *Soziologische Aufklärung 5: Konstruktivistische Perspektiven*, 2nd ed., 95–130. Wiesbaden: Westdeutscher Verlag, 1993.

Luhmann, Niklas. "Kausalität im Süden." *Soziale Systeme* 1 (1995): 7–28. http://www.soziale-systeme.ch/leseproben/luhmann.htm.

Luhmann, Niklas. *Theory of Society*. Vol. 1. Translated by Rhodes Barrett. Stanford, CA: Stanford University Press, 2012.

Luhmann, Niklas. *The Reality of the Mass Media*. Translated by Kathleen Cross. Stanford, CA: Stanford University Press, 2000.

Lundberg, George A. "Social Attraction-Patterns in a Rural Village: A Preliminary Report." *Sociometry* 1, nos. 1–2 (1937): 77–80.

Lundberg, George A., and Mary Steele. "Social Attraction-Patterns in a Village." *Sociometry* 1, nos. 3–4 (1938): 375–419.

Macho, Thomas. "Körper der Zukunft: Vom Vor- und Nachleben der Bilder." In *Bilderfragen: Die Bildwissenschaften im Aufbruch*, edited by Hans Belting, 181–194. Paderborn: Wilhelm Fink, 2007.

Macho, Thomas. "Second-Order Animals: Cultural Techniques of Identity and Identification." Translated by Michael Wutz. *Theory, Culture & Society* 30, no. 6 (2013): 30–47.

Macho, Thomas. "Stimmen ohne Körper: Anmerkungen zur Technikgeschichte der Stimme." In *Stimme: Annäherung an ein Phänomen*, edited by Doris Kolesch and Sybille Krämer, 130–146. Frankfurt am Main: Suhrkamp, 2006.

Mackenzie, Adrian. *Wirelessness: Radical Empiricism in Network Cultures*. Cambridge, MA: MIT Press, 2010.

Magee, John Francis, and David Morris Boodman. *Production Planning and Inventory Control*. 2nd ed. New York: McGraw-Hill, 1967.

Mahr, Bernd. "Cargo: Zum Verhältnis von Bild und Modell." In *Visuelle Modelle*, edited by Ingeborg Reichle, Steffen Siegel, and Achim Spelten, 17–40. Paderborn: Wilhelm Fink, 2008.

Mahr, Bernd. "Modellieren: Beobachtungen und Gedanken zur Geschichte des Modellbegriffs." In *Bild—Schrift—Zahl*, edited by Sybille Krämer and Horst Bredekamp, 59–86. Paderborn: Wilhelm Fink, 2003.

Malcolm, Donald G., J. H. Rosenbaum, C. E. Clark, and W. Fazar. "Application of a Technique for Research and Development Program Evaluation." *Operations Research* 7, no. 5 (1959): 646–669.

Malpighi, Marcello. "De anatome plantarum." In *Opera omnia: figuris elegantissimis in aes incisis illustrata; zwei Bände in einem Band*. Vol. 1, 1–78. Hildesheim: Olms, 1975.

Malpighi, Marcello. "De pulmonibus." In *Opera omnia: figuris elegantissimis in aes incisis illustrata; zwei Bände in einem Band*. Vol. 2, 133–145. Hildesheim: Olms, 1975.

Malpighi, Marcello. *The Correspondence of Marcello Malpighi. 1658–1669*. Edited by Howard B. Adelmann. Vol. 1. Ithaca, NY: Cornell University Press, 1975.

Marcuse, Herbert. *Eros and Civilization: A Philosophical Inquiry into Freud*. Boston: Beacon Press, 1974. First published in 1955.

Marill, Thomas, and Lawrence G. Roberts. "Toward a Cooperative Network of Time-Shared Computers." *Proceedings of AFIPS Fall Joint Computer Conference*, Vol. 29, *Nov. 7–10 San Francisco, California*, 425–431. Washington, DC: Spartan Books, 1966.

Marschall, Brigitte. *"Ich bin der Mythe": Von der Stegreifbühne zum Psychodrama Jakob Levy Morenos*. Vienna: Böhlau, 1988.

Marschall, Brigitte. "Jakob Levy Morenos Theaterkonzept: Die Zeit-Räume des Lebens als Szenenraum der Begegnung." *Zeitschrift für Psychodrama und Soziometrie* 4, no. 2 (2005): 229–243.

Martin, James. *Future Developments in Telecommunications*. Englewood Cliffs, NJ: Prentice-Hall, 1971.

Martin, Michèle: *Hello Central? Gender, Technology, and Culture in the Formation of the Telephone System*. Montreal: Gill University Press, 1991.

Marvin, Carolyn. *When Old Technologies Were New: Thinking about Electric Communication in the Late Nineteenth Century*. New York: Oxford University Press, 1988.

Mauss, Marcel. *A General Theory of Magic*. Translated by Robert Brain. London: Routledge, 2005.

Mauss, Marcel. "Techniques of the Body." *Economy and Society* 2, no. 1 (1973): 70–88.

Maye, Harun. "Was ist eine Kulturtechnik?" *Zeitschrift für Medien- und Kulturforschung: Schwerpunkt Kulturtechnik* 1, no. 1 (2010): 121–135.

McCarty, Christopher, and José Luis Molina. "Innovative Use of Social Network Analysis in Cultural Anthropology." In *Knoten und Kanten 2.0. Soziale Netzwerkanalyse in Medienforschung und Kulturanthropologie*, edited by Markus Gamper, Linda Reschke, and Michael Schönhuth, 217–230. Bielefeld: transcript, 2012.

McCulloch, Warren S. "A Heterarchy of Values Determined by the Topology of Nervous Nets." *Bulletin of Mathematical Biophysics* 7, no. 2 (1945): 89–93.

McCulloch, Warren S., and Walter Pitts. "A Logical Calculus of the Ideas Immanent in Nervous Activity." *Bulletin of Mathematical Biophysics* 5, no. 4 (1943): 115–133.

McGrath, Patrick. *Spider*. New York: Vintage, 1990.

McKelvey, Fenwick, and Kevin Driscoll. "ARPANET and Its Boundary Devices: Modems, IMPs, and the Inter-Structuralism of Infrastructures." *Internet Histories* 3, no. 1 (2019): 31–50.

McKenzie, Alex. *RFC 413: Traffic Statistics (October 1972)*. Tech. rep. BBN, 1972. https://www.rfc-editor.org/rfc/rfc413.txt.

McKenzie, Alex. *RFC 422: Traffic Statistics (November 1972)*. Tech. rep. BBN, 1972. https://www.rfc-editor.org/rfc/rfc422.txt.

McLaren, Kenneth Gordon, and Eric Leonard Buesnel. *Network Analysis in Project Management: An Introductory Manual Based on Unilever Experience*. London: Cassell, 1969.

McLuhan, Herbert Marshall. *The Mechanical Bride: Folklore of Industrial Man*. New York: Vanguard, 1951.

McMillan, Charles J. *The Japanese Industrial System*. 3rd ed. Berlin: De Gruyter, 1996.

Meinel, Christoph. "Molecules and Croquet Balls." In *Models: The Third Dimension of Science*, edited by Soraya de Chadarevian and Nick Hopwood, 242–275. Stanford, CA: Stanford University Press, 2004.

Meister, Carolin. "Einleitung." In *Randgänge der Zeichnung*, edited by Werner Busch, Oliver Jehle, and Carolin Meister, 7–10. Paderborn: Wilhelm Fink. 2007.

Melley, Timothy. *Empire of Conspiracy: The Culture of Paranoia in Postwar America*. Ithaca, NY: Cornell University Press, 2000.

Mercklé, Pierre. *Sociologie des réseaux sociaux*. Paris: La Découverte, 2004.

Mersch, Dieter. "Visual Arguments: The Role of Images in Sciences and Mathematics." Translated by Rett Rossi. In *Science Images and Popular Images of the Sciences*, edited by Bernd-Rüdiger Hüppauf and Peter Weingart, 181–198. New York: Routledge, 2011.

Meteling, Arno. "The Parallax View. Verschwörungstheorie zur Einführung." *Transkriptionen* 9 (2008): 15–18.

Meyer Jr., Edwin E. *RFC 46: ARPA Network Protocol Notes*. Tech. rep. MIT, 1970. https://www.rfc-editor.org/rfc/rfc46.txt.

Meyer Jr., Edwin W. *RFC 49: Conversations with Steve Crocker (UCLA)*. Tech. rep. MIT Project MAC, 1970. https://www.rfc-editor.org/rfc/rfc49.txt.

Meyer Jr., Edwin W. *RFC 82: Network Meeting Notes*. Tech. rep. MIT Project MAC, 1970. https://www.rfc-editor.org/rfc/rfc82.txt.

Michalski, Katarzyna, and Sergiusz Michalski. *Spider*. London: Reaktion, 2010.

Millman, Sidney, ed. *A History of Engineering and Science in the Bell System*. Vol. 5, *Communication Sciences (1925–1980)*. Indianapolis, IN: AT&T Bell Laboratories, 1984.

Mills, Mara. "Testing Hearing with Speech." In *Testing Hearing: The Making of Modern Aurality*, edited by Viktoria Tkaczyk, Mara Mills, and Alexandra Hui, 23–48. Oxford: Oxford University Press, 2020.

Mitchell, J. Clyde, ed. *Social Networks in Urban Situations: Analyses of Personal Relationships in Central African Towns*. Manchester: Manchester University Press, 1969.

Mizruchi, Mark S. *The American Corporate Network, 1904–1974*. Beverly Hills, CA: Sage, 1982.

Moder, Joseph John, and Cecil R. Philipps. *Project Management with CPM and PERT*. 4th ed. New York: Reinhold, 1967.

Möller-Wiering, Susanne. "Maschenstoffe." In *Reallexikon der Germanischen Altertumskunde*. Vol. 19, *Luchs—Metrum*. Edited by Johannes Hoops, 377–383. Berlin: De Gruyter, 2001.

Montel, Nathalie. "Le chantier du Canal de Suez (1859–1869): Techniques et techniciens de travaux publics du XIXème siècle." PhD diss., Université de Lille III, 1995.

Moore, Edward F., and Claude Elwood Shannon. *Reliable Circuits Using Crummy Relays*. Tech. rep. Bell Telephone Laboratories, 1954.

Moreno, Jacob Levy, and Helen Jennings. *Die Grundlagen der Soziometrie: Wege zur Neuordnung der Gesellschaft*. 2nd ed. Cologne: Westdeutscher Verlag, 1967. First published 1954.

Moreno, Jacob Levy, and Helen Jennings. *Who Shall Survive? A New Approach to the Problem of Human Interrelations*. Washington, DC: Nervous and Mental Disease Publishing, 1934.

Moreno, Jacob Levy, and Helen Jennings. *Who Shall Survive? Foundations of Sociometry, Group Psychotherapy and Sociodrama*. 2nd ed. Beacon, NY: Beacon House, 1953.

Moreno, Jonathan D. *Impromptu Man: J. L. Moreno and the Origins of Psychodrama, Encounter Culture, and the Social Network*. New York: Bellevue Literary Press, 2014.

Moses, Claire Goldberg, and Leslie Wahl Rabine. *Feminism, Socialism, and French Romanticism*. Bloomington: Indiana University Press, 1993.

Mueller, Milton. "The Switchboard Problem: Scale, Signaling, and Organization in Manual Telephone Switching." *Technology and Culture* 30, no. 3 (1989): 534–560.

Mullaney, Thomas S., Benjamin Peters, Mar Hicks, and Kavita Philip, eds. *Your Computer Is on Fire*. Cambridge, MA: MIT Press, 2021.

Münkler, Herfried, and Martina Münkler. *Lexikon der Renaissance*. Munich: C. H. Beck, 2000.

Munster, Anna. *An Aesthesia of Networks: Conjunctive Experience in Art and Technology*. Cambridge, MA: MIT Press, 2013.

Musso, Pierre. *Télécommunications et philosophie des réseaux: La posterité paradoxale de Saint-Simon*. Paris: Presses Universitaires de France, 1997.

Nanz, Tobias. *Grenzverkehr: Eine Mediengeschichte der Diplomatie*. Zurich: diaphanes, 2010.

Network Working Group. *RFC 1958: Architectural Principles of the Internet*. Tech. rep. Internet Architecture Board, 1996. https://www.rfc-editor.org/rfc/rfc1958.txt.

Neubert, Christoph. "Onto-Logistik: Kommunikation und Steuerung im Internet der Dinge." In *Agenten und Agenturen: Archiv für Mediengeschichte*, edited by Lorenz Engell, Joseph Vogl, and Bernhard Siegert, 119–133. Weimar: Universitätsverlag, 2008.

Neubert, Christoph. "The End of the Line: Zu Theorie und Geschichte der Selbststeuerung in der modernen Logistik." In *Unsichtbare Hände: Automatismen in Medien-, Technik- und Diskursgeschichte*, edited by Hannelore Bublitz, Irina Kaldrack, Theo Röhle, and Hartmut Winkler, 191–214. Paderborn: Wilhelm Fink, 2011.

Neubert, Christoph, and Gabriele Schabacher, eds. *Verkehrsgeschichte und Kulturwissenschaft. Analysen an der Schnittstelle von Technik, Kultur und Medien*. Bielefeld: transcript, 2012.

Neurath, Otto. "Bildstatistik und Wirtschaftsplan." In *Gesammelte bildpädagogische Schriften*, edited by Rudolf Haller and Robin Kinross, 177–179. Vienna: Hölder-Pichler-Tempsky, 1991.

Newkirk, John, Mike Kraley, Jon Postel, and Steve Crocker. *RFC 55: A Prototypical Implementation of the NCP*. Tech. rep. Harvard and UCLA, 1970. https://www.rfc-editor.org/rfc/rfc55.txt.

New York Times. "Emotions Mapped by New Geography: Charts Seek to Portray the Psychological Currents of Human Relationships." *New York Times*, April 3, 1933, 17.

Nichols, Sarah. "Pollux's Spears." *Grey Room* 71 (2018): 141–155.

Niderberger, Leonz. *Der Suezkanal: Seine Geschichte, Lage und Bedeutung für den Weltverkehr und im Weltkrieg*. Limburg an der Lahn: Steffen, 1916.

Nietzsche, Friedrich. "On the Genealogy of Morality." Translated by Adrian del Caro. In *The Complete Works of Friedrich Nietzsche*. Vol. 8, *Beyond Good and Evil / On the Geneaology of Morality*, 207–422. Stanford, CA: Stanford University Press, 2014.

Nikolow, Sybilla. "Kurven, Diagramme, Zahlen- und Mengenbilder. Die Wiener Methode der Bildstatistik als statistische Bildform." In *Bildwelten des Wissens* 3, no. 1 (2005): 20–33.

Nordsieck, Fritz. *Die schaubildliche Erfassung der Betriebsorganisation*. Stuttgart: Poeschel, 1932.

Oettermann, Stephan. *The Panorama: History of a Mass Medium*. Translated by Deborah Lucas Schneider. New York: Zone Books, 1997.

Offner, Jean-Marc. "Are There Such Things as Small Networks?" In *The Governance of Large Technical Systems*, edited by Olivier Coutard, 217–238. London: Routledge, 1999.

Ohno, Taiichi. *Toyota Production System: Beyond Large-Scale Production*. Boca Raton, FL: CRC, 1988. First published in Japanese 1978 by Diamond (Tokyo).

Ornstein, Severo M., Frank E. Heart, William. M. Crowther, Hawley R. Rising, S. B. Russel, and A. Michel. "The Terminal IMP for the ARPA computer network." In *AFIPS Conference Proceedings*. Vol. 40, *Spring Joint Computer Conference, May 16–18, 1972. Atlantic City, New Jersey*, 243–254. Montvale, NJ: AFIPS Press, 1972.

Osterhammel, Jürgen. *The Transformation of the World: A Global History of the Nineteenth Century*. Translated by Patrick Camiller. Princeton, NJ: Princeton University Press, 2015.

Otis, Laura. *Networking: Communicating with Bodies and Machines in the Nineteenth Century*. Ann Arbor: University of Michigan Press, 2001.

Otto, Isabell. *Prozess und Zeitordnung: Temporalität unter den Bedingungen digitaler Vernetzung*. Konstanz: Konstanz University Press, 2020.

Ovenden, Mark. *Metro Maps of the World*. 2nd ed. Harrow Weld, UK: Capital Transport, 2004.

Owen, Kenneth. *RFC 828: Data Communications: IFIP's International Network of Experts*. Tech. rep. IFIP, 1982. https://www.rfc-editor.org/rfc/rfc828.txt.

Oxford English Dictionary, online, Oxford University Press. S.v. "protocol, n," Accessed January 22, 2022. https://www.oed.com/view/Entry/153243?rskey=3NlvLN&result=1&isAdvanced=false.

Paloque-Bergès, Camille, and Valérie Schafer. "Arpanet (1969–2019)." *Internet Histories* 3, no. 1 (2019): 1–14.

Panofsky, Erwin. *Perspective as Symbolic Form*. Translated by Christopher S. Wood. New York: Zone Books, 1997.

Parsons, Talcott. *The Structure of Social Action*. Vol. 1, *Marshall, Pareto, Durkheim*. New York: Free Press, 1968.

Passepartout, ed. *Weltnetzwerke—Weltspiele: Jules Vernes "In 80 Tagen um die Welt."* Konstanz: Konstanz University Press, 2013.

Pauer, Erich. "Der Technologietransfer nach Japan: Strukturen und Strategien." In *Technologietransfer Deutschland–Japan*, edited by Erich Pauer, 48–72. Munich: iudicium, 1992.

Pavlov, Ivan. "Natural Science and the Brain." In *Selected Works*. Edited by J. Gibbons. Translated by S. Belsky, 206–219. Moscow: Foreign Languages Publishing House, 1955.

Peirce, Charles Sanders. "Of Reasoning in General." In *The Essential Peirce: Selected Philosophical Writings*. Vol. 2 (1893–1913). Edited by Peirce Edition Project, 11–26. Bloomington: Indiana University Press, 1998.

Peirce, Charles Sanders. "What Is a Sign?" In *The Essential Peirce. Selected Philosophical Writings*. Vol. 2. Edited by Peirce Edition Project. Bloomington: Indiana University Press, 1998, 4–10.

Peters, Ben. *How Not to Network a Nation: The Uneasy History of the Soviet Internet*. Cambridge, MA: MIT Press, 2016.

Peters, John Durham. *Speaking into the Air: A History of the Idea of Communication*. Chicago: University of Chicago Press, 2000.

Pfeifer, Wolfgang, ed. *Etymologisches Wörterbuch des Deutschen*. 7th ed. Munich: dtv, 2004.

Pflüger, Jörg. "Writing, Building, Growing: Leitvorstellungen der Programmiergeschichte." In *Geschichten der Informatik*, edited by Hans-Dieter Hellige, 275–320. Berlin: Springer, 2004.

Philip, Kavita. "The Internet Will Be Decolonized." In *Your Computer Is on Fire*, edited by Thomas S. Mullaney, Benjamin Peters, Mar Hicks, and Kavita Philip, 91–115. Cambridge, MA: MIT Press, 2021.

Pias, Claus, ed. *Cybernetics/Kybernetik. The Macy Conferences 1946–1953*. Vol. 1. *Transactions/Protokolle*. Zurich: diaphanes, 2003.

Pias, Claus, ed. *Cybernetics/Kybernetik. The Macy Conferences 1946–1953*. Vol. 2. *Essays & Dokumente*. Zurich: diaphanes, 2004.

Pias, Claus. "Elektronenhirn und verbotene Zone: Zur kybernetischen Ökonomie des Digitalen." In *Analog/Digital—Opposition oder Kontinuum?* Edited by Jens Schröter and Alexander Böhnke, 295–309. Bielefeld: transcript, 2004.

Pias, Claus, and Sebastian Vehlken. "Einleitung." In *Think Tanks: Die Beratung der Gesellschaft*, edited by Thomas Brandstetter, Claus Pias, and Sebastian Vehlken, 7–15. Zurich: diaphanes, 2010.

Picon, Antoine. *Les saint-simoniens: Raison, imaginaire et utopie*. Paris: Belin, 2002.

Pikovsky, Arkady, Michael Rosenblum, and Jürgen Kurths. *Synchronization: A Universal Concept in Nonlinear Sciences*. Cambridge: Cambridge University Press, 2001.

Pilbeam, Pamela. *Saint-Simonians in Nineteenth-Century France: From Free Love to Algeria*. New York: Palgrave Macmillan, 2013.

Piquet, Caroline. *Histoire du canal de Suez*. Paris: Perrin, 2009.

Pircher, Wolfgang. "Im Schatten der Kybernetik: Rückkopplung im operativen Einsatz; 'operational research.'" In *Die Transformation des Humanen: Beiträge zur Kulturgeschichte der Kybernetik*, edited by Michael Hagner and Erich Hörl, 348–376. Frankfurt am Main: Suhrkamp, 2008.

Poe, Edgar Allan. "The Purloined Letter." In *Poetry and Tales*, 680–698. New York: Library of America, 1984. First published 1844.

Polanyi, Michael. *The Tacit Dimension*. Chicago: University of Chicago Press, 2009.

Polianski, Igor J. *Die Kunst, die Natur vorzustellen: Die Ästhetisierung der Pflanzenkunde um 1800 und Goethes Gründung des botanischen Gartens zu Jena im Spannungsfeld kunsttheoretischer und botanischer Diskussionen der Zeit*. Jena: Lehrstuhl für Kunstgeschichte und Kunsthistorisches Seminar mit Kustodie Friedrich-Schiller-Universität, 2004.

Pollux [Georges Baehler]. *Trusts in der Schweiz? Die schweizerische Politik im Schlepptau der Hochfinanz*. Zurich: Verein für wirtschaftliche Studien, 1945.

Pollux [Georges Baehler]. *Wer leitete Deutschland? Die eigentlichen Kriegsverbrecher: 50 Drahtzieher hinter den Kulissen*. Zurich: Verein für wirtschaftliche Studien, 1945.

Postel, Jon. *RFC 77: Network Meeting Report*. Tech. rep. UCLA, 1970. https://www.rfc-editor.org/rfc/rfc77.txt.

Postel, Jon. *RFC 694: Protocol Information*. Tech. rep. Stanford Research Institute, 1975. https://www.rfc-editor.org/rfc/rfc694.txt.

Postel, Jon, and Steve Crocker. *RFC 48: A Possible Protocol Plateau*. Tech. rep. UCLA, 1970. https://www.rfc-editor.org/rfc/rfc48.txt.

Pottage, Alain, and Brad Sherman. *Figures of Invention. A History of Modern Patent Law*. Oxford: Oxford University Press, 2010.

Potthast, Jörg. *Die Bodenhaftung der Netzwerkgesellschaft: Eine Ethnografie von Pannen an Grossflughäfen*. Bielefeld: transcript, 2007.

Pouzin, Louis, ed. *The Cyclades Computer Network: Towards Layered Network Architecture*. Amsterdam: North Holland, 1982.

Powell, Walter P. "Neither Market nor Hierarchy: Network Forms of Organization." *Research in Organizational Behavior* 12 (1990): 295–336.

Prioreschi, Plinio. "Galenicae Quaestiones Disputatae Duae: Rete Mirabile and Pulmonary Circulation." *Vesalius* 2, no. 2 (1996): 67–78.

Proust, Marcel. *The Guermantes Way*. Translated by Charles Kenneth Scott-Moncrieff and Dennis Joseph Enright. New York: Vintage, 1996. Originally published as *Le Côté de Guermantes*. 2 vols. (Paris: Gallimard: 1920–1921).

Pynchon, Thomas. *Gravity's Rainbow*. London: Vintage, 2000. First published 1973 by Viking (New York).

Pynchon, Thomas. *The Crying of Lot 49*. London: Vintage, 2000. First published 1966 by Lippincott (Philadelphia).

Quack, Martin. "Modelle in der Chemie." In *Modelle des Denkens: Streitgespräch in der wissenschaftlichen Sitzung der Versammlung der Berlin-Brandenburgischen Akademie der Wissenschaften am 12. Dezember 2003*, 21–33. Berlin: BBAW, 2005.

Radcliffe-Brown, Alfred Reginald. *Structure and Function in Primitive Society: Essays and Addresses*. 6th ed. Glencoe, IL: Free Press, 1952.

Radin, Paul. "Zur Netztechnik der südamerikanischen Indianer." *Zeitschrift für Ethnologie* 36 (1906): 926–938.

Railway Gazette and Railway News. "Centralised Control of Railway Traffic." *Railway Gazette and Railway News*, January 22, 1932, 106–107, 115.

Railway Gazette and Railway News. "Public Notes on Railways." *Railway Gazette and Railway News*, January 8, 1932, 44–45.

Rancière, Jacques. *The Names of History: On the Poetics of Knowledge*. Translated by Hassan Melehy. Minneapolis: University of Minnesota Press, 1994.

Rand, Graham K., and Luis Valadares Tavares. "Network Planning." In *Encyclopedia of Operations Research and Management Science*, edited by Saul I. Gass, 561–564. Boston: Kluwer, 2001.

Rathenau, Walter. "Die Resurrection Co." Translated by Louis Kaplan. *New German Critique*, no. 62 (1994): 63–69. Originally published as "Die Resurrection Co." In *Gesammelte Schriften IV. Reflexionen und Aufsätze*, 337–349. Berlin: S. Fischer, 1925. First published 1898.

Ratzel, Friedrich. *Politische Geographie*. 3rd ed. Munich: Oldenbourg, 1923. First published 1897.

Raymond, Eric S. *The New Hacker's Dictionary*. 2nd ed. Cambridge, MA: MIT Press, 1997.

Raymond, Joad, and Noah Moxham, eds. *News Networks in Early Modern Europe*. Leiden: Brill, 2016.

Régnier, Philippe, ed. *Le livre nouveau des Saint-Simoniens: Manuscrits d'Émile Barrault, Michel Chevalier, Charles Duveyrier, Prosper Enfantin, Charles Lambert, Léon Simon et Thomas-Ismayl Urbain (1832–1833)*. Tusson: Lérot, 1991.

Régnier, Philippe. "Le mythe oriental des saint-simoniens." In *Les Saint-Simoniens et l'Orient: Vers la modernité*, edited by Magali Morsy, 29–49. Aix en Provence: Édisud, 1989.

Régnier, Philippe. *Les Saint-Simoniens en Egypte 1833–1851*. Cairo: Banque de l'Union Européene, 1989.

Reich, Robert B. *The Work of Nations: Preparing Ourselves for 21st-Century Capitalism*. New York: Alfred A. Knopf, 1991.

Reinecke, Michael. *Galen und Vesal: Ein Vergleich der anatomisch-physiologischen Schriften*. Münster: LIT, 1997.

Renn, Jürgen, and Peter Damerow. "Mentale Modelle als kognitive Instrumente der Transformation von technischem Wissen." In *Übersetzung und Transformation*, edited by Hartmut Böhme, Christof Rapp, and Wolfgang Rösler, 311–331. Berlin: De Gruyter, 2007.

Reynolds, Joyce K., and Jon Postel. *RFC 1000: The Request for Comments Reference Guide*. Tech. rep. ISI, 1987. https://www.rfc-editor.org/rfc/rfc1000.txt.

Rheinberger, Hans-Jörg. *Toward a History of Epistemic Things: Synthesizing Proteins in the Test Tube*. Stanford, CA: Stanford University Press, 1997.

Ricardo, David. "Three Letters to the Morning Chronicle on the Bullion Report, 1810." In *The Works and Correspondence of David Ricardo*. Vol. 3, *Pamphlets and Papers 1809–1811*. Edited by Piero Sraffa, 129–153. Cambridge: Cambridge University Press, 1951.

Riot-Sarcey, Michèle. *Histoire du feminisme*. Paris: La Decouverte, 2002.

Roberts, Larry. "The ARPANET and Computer Networks." In *A History of Personal Workstations*, edited by Adele Goldberg, 141–171. New York: ACM Press, 1988.

Roberts, Lawrence G. "The Evolution of Packet Switching." *Proceedings of the IEEE* 66, no. 11 (1978): 1307–1313.

Roberts, Lawrence G., and Barry D. Wessler. "Computer Network Development to Achieve Resource Sharing." In *AFIPS Conference Proceedings*. Vol. 36, *Spring Joint Computers Conference. May 5–7, 1970. Atlantic City, New Jersey*, 543–549. New York: ACM Press, 1970.

Rocke, Alan J. *Image and Reality: Kekulé, Kopp, and the Scientific Imagination*. Chicago: University of Chicago Press, 2010.

Roisman, Joseph. *The Rhetoric of Conspiracy in Ancient Athens*. Berkeley: University of California Press, 2006.

Ronell, Avital. *The Telephone Book: Technology, Schizophrenia, Electric Speech*. Lincoln: University of Nebraska Press, 1989

Rose, Douglas. *The London Underground: A Diagrammatic History*. 7th ed. Harrow Weld, UK: Capital Transport, 2000.

Rosenberg, Emily, ed. *A World Connecting, 1870–1945*. Cambridge, MA: Belknap Press of Harvard University Press, 2012.

Rosenkranz, Friedrich. *Netzwerktechnik und wirtschaftliche Anwendung: Mit besonderer Berücksichtigung der Netzplantechnik*. Meisenheim am Glan: Anton Hain, 1968.

Roskothen, Johannes. *Verkehr: Zu einer poetischen Theorie der Moderne*. Paderborn: Wilhelm Fink, 2003.

Rossler, Gustav. "Galerie neuer Dingbegriffe: Hybriden, Quasi-Objekte, Grenzobjekte, epistemische Dinge." In *Bruno Latours Kollektive: Kontroversen zur Entgrenzung des Sozialen*, edited by Georg Kneer, Markus Schroer, and Erhard Schüttpelz, 76–107. Frankfurt am Main: Suhrkamp, 2008.

Roudinesco, Élisabeth, and Michel Plon, eds. *Dictionnaire de la psychanalyse*. Paris: Fayard, 1997.

Roy, Bernard, ed. *Les problèmes d'ordonnancement*. Paris: Dunod, 1964.

Ruchatz, Jens. "Das Telefon: Ein sprechender Telegraf." In *Einführung in die Geschichte der Medien*, edited by Albert Kümmel, Leander Scholz, and Eckhard Schumacher, 125–149. Paderborn: UTB | Fink, 2004.

Russell, Andrew L. *Open Standards and the Digital Age: History, Ideology, and Networks*. Cambridge: Cambridge University Press, 2014.

Russell, Andrew L. "Rough Consensus and Running Code and the Internet-OSI Standards War." *IEEE Annals of the History of Computing* 28, no. 3 (2006): 48–61.

Sainte-Laguë, André. "Les réseaux (ou graphes)." *Mémorial des Sciences Mathematiques* 18 (1926): 1–62.

Saltzer, Jerome H., David Patrick Reed, and David Dana Clark. "End-To-End Arguments in System Design." *ACM Transactions on Computer Systems* 2, no. 4 (1984): 277–288.

Salus, Peter H. *Casting the Net: From ARPANET to INTERNET and Beyond . . .* Reading, MA: Addison-Wesley, 1995.

Salus, Peter H., ed. *The ARPANET Sourcebook: The Unpublished Foundations of the Internet*. Charlottesville, VA: Peer-to-Peer-Communications, 2008.

Sampson, Tony D., and Jussi Parikka. "Learning from Network Dysfunctionality: Accidents, Enterprise, and Small Worlds of Infection." In *A Companion to New Media Dynamics*, edited by John Hartley, Jean Burgess, and Axel Bruns, 450–460. Oxford: Blackwell, 2013.

Sansovino, Francesco. *Venetia citta nobilissima et singolare, Descritta in XIIII. Libri*. Venice: Iacomo Sansovino, 1581.

Saraceno, Tomás. *Cloud Cities*. Edited by Marion Ackermann, Daniel Birnbaum, Udo Kittelmann, and Hans Ulrich Obrist. Berlin: Distanz, 2011. Exhibition catalog.

Sassen, Saskia. *Territory, Authority, Rights. From Medieval to Global Assemblages*. 4th ed. Princeton, NJ: Princeton University Press, 2008.

Schabacher, Gabriele. "Im Zwischenraum der Lösungen: Reparaturarbeit und Workarounds." *ilinx, Berliner Beiträge zur Kulturwissenschaft* 4 (2017): xiii–xxviii.

Schabacher, Gabriele. "Raum-Zeit-Regime: Logistikgeschichte als Wissenszirkulation zwischen Medien, Verkehr und Ökonomie." In *Agenten und Agenturen. Archiv für Mediengeschichte*, edited by Lorenz Engell, Joseph Vogl, and Bernhard Siegert, 135–148. Weimar: Universitätsverlag, 2008.

Schäffner, Wolfgang. "Electric Graphs: Charles Sanders Peirce und die Medien." In *Electric Laokoon. Zeichen und Medien, von der Lochkarte zur Grammatologie*, edited by Michael Franz, Wolfgang Schäffner, Bernhard Siegert, and Robert Stockhammer, 313–326. Berlin: Akademie, 2007.

Scheftelowitz, Isidor. *Das Schlingen- und Netzmotiv im Glauben und Brauch der Völker*. Giessen: Alfred Töpelmann, 1912.

Schivelbusch, Wolfgang. *The Railway Journey: The Industrialization of Time and Space in the Nineteenth Century*. Oakland: University of California Press, 2014.

Schlimm, Anette. *Ordnungen des Verkehrs: Arbeit an der Moderne; Deutsche und britische Verkehrsexpertise im 20. Jahrhundert.* Bielefeld: transcript, 2011.

Schmidt, Gunnar. *Ästhetik des Fadens: Zur Medialisierung eines Materials in der Avantgardekunst.* Bielefeld: transcript, 2007.

Schnaithmann, Christine. "'Fifty Ways to Cut Expenses': Praktiken der Effizienzsteigerung in amerikanischen Management-Handbüchern, 1910–1920," *ilinx, Berliner Beiträge zur Kulturwissenschaft* 3 (2013): 39–56.

Schneider, Birgit. *Textiles Prozessieren: Eine Mediengeschichte der Lochkartenweberei.* Zurich: diaphanes, 2007.

Schröter, Jens. *Das Netz und die virtuelle Realität: Zur Selbstprogrammierung der Gesellschaft durch die universelle Maschine.* Bielefeld: transcript, 2004.

Schultz, Oliver Lerone. "Marshall McLuhan—Medien als Infrastrukturen und Archetypen." In *Medientheorien. Eine philosophische Einführung,* edited by Alice Lagaay, 31–68. Frankfurt am Main: Campus, 2004.

Schumaker, Lyn. *Africanizing Anthropology: Fieldwork, Networks, and the Making of Cultural Knowledge in Central Africa.* Durham, NC: Duke University Press, 2001.

Schüttpelz, Erhard. "Der Punkt des Archimedes: Einige Schwierigkeiten des Denkens in Operationsketten." In *Bruno Latours Kollektive: Kontroversen zur Entgrenzung des Sozialen,* edited by Georg Kneer, Markus Schroer, and Erhard Schüttpelz, 234–258. Frankfurt am Main: Suhrkamp, 2008.

Schüttpelz, Erhard. "Die ältesten in den neuesten Medien. Folklore und Massenkommunikation um 1950." *Navigationen* 6, no. 1 (2006): 33–46.

Schüttpelz, Erhard. "Die medienanthropologische Kehre der Kulturtechniken." In *Archiv für Mediengeschichte. Kulturgeschichte als Mediengeschichte (oder vice versa)?* Edited by Lorenz Engell, Joseph Vogl, and Bernhard Siegert, 87–110. Weimar: Universitätsverlag, 2006.

Schüttpelz, Erhard. "Ein absoluter Begriff: Zur Genealogie und Karriere des Netzwerkkonzepts." In *Vernetzte Steuerung: Soziale Prozesse im Zeitalter technischer Netzwerke,* edited by Stefan Kaufmann, 25–46. Zurich: Chronos, 2007.

Schüttpelz, Erhard. "Elemente einer Akteur-Medien-Theorie." In *Akteur-Medien-Theorie,* edited by Tristan Thielmann and Erhard Schüttpelz, 9–67. Bielefeld: transcript, 2013.

Schüttpelz, Erhard. "Frage nach der Frage, auf die das Medium eine Antwort ist." In *Signale der Störung,* edited by Albert Kümmel and Erhard Schüttpelz, 15–29. Paderborn: Wilhelm Fink, 2003.

Schüttpelz, Erhard. "'Get the message through': From the Channel of Communication to the Message of the Medium, 1945–1960." In *Media, Culture, and Mediality: New Insights into the Current State of Research,* edited by Ludwig Jäger, Erika Linz, and Irmela Schneider, 109–138. Bielefeld: transcript, 2010.

Schüttpelz, Erhard. "The Irreducibility of Technical Skills: Before and After Science." In *Materiality of Cooperation,* edited by Sebastian Giessmann, Tobias Röhl, Ronja Trischler, and Martin Zillinger, 307–325. Wiesbaden: Springer VS, 2023.

Schwenk, Sigrid. *Zur Terminologie des Vogelfangs im Deutschen: Eine sprachliche Untersuchung aufgrund der deutschen didaktischen Literatur des 14. bis 19. Jahrhunderts.* Clausthal-Zellerfeld: Bönecke-Druck, 1967.

Sedgwick, Eve Kosofsky. "Paranoid Reading and Reparative Reading; or, You're so Paranoid, You Probably Think This Essay Is about You." In *Touching Feeling: Affect, Pedagogy, Performativity*, 123–151. Durham, NC: Duke University Press, 2003.

Seesslen, Georg. *Die Matrix entschlüsselt.* Berlin: Bertz, 2003.

Seitter, Walter. *Die Physik der Medien: Materialien, Apparate, Präsentierungen.* Weimar: Verlag und Datenbank für Geisteswissenschaften, 2002.

Seligo, Arthur. "Die Fischerei in den Fliessen, Seen und Strandgewässern Mitteleuropas." In *Handbuch der Binnenfischerei Mitteleuropas*. Vol. V, edited by Reinhard Demoll and Hermann Nikolaus Maier, 1–422. Stuttgart: E. Schweizerbart'sche Verlagsbuchhandlung, 1926.

Selz, Gebhard. *Untersuchungen zur Götterwelt des altsumerischen Stadtstaates von Lagaš.* Philadelphia: University of Pennsylvania Museum, 1995.

Sennett, Richard. *Flesh and Stone: The Body and the City in Western Civilization.* New York: W. W. Norton, 1994.

Serres, Michel. *Esthétiques sur Carpaccio.* Paris : Hermann, 1975.

Serres, Michel. *Hermès ; ou, La communication.* Paris: Éditions de minuit, 1968.

Serres, Michel. "Introduction." In *A History of Scientific Thought: Elements of a History of Science*, edited by Michel Serres, 1–16. Oxford : Blackwell, 1995.

Serres, Michel. *La naissance de la physique dans le texte de Lucrèce : Fleuves et turbulences.* Paris: Éditions de minuit, 1977.

Serres, Michel. *The Natural Contract.* Translated by Elizabeth MacArthur and William Paulson. Ann Arbor: University of Michigan Press, 1995.

Serres, Michel. *The Parasite.* Translated by Lawrence R. Schehr. Baltimore: Johns Hopkins University Press, 1982.

Serres, Michel, and Bruno Latour. *Conversations on Science, Culture, and Time.* Translated by Roxanne Lapidus. Ann Arbor: University of Michigan Press, 1995.

Severin, Ingrid. "Technische Vernetzungen und ihre Auswirkungen auf zeitgenössische Kunst." In *Technik ohne Grenzen*, edited by Ingo Braun and Bernward Joerges, 212–250, 386–409. Frankfurt am Main: Suhrkamp, 1994.

Shannon, Claude Elwood. "A Mathematical Theory of Communication." *The Bell System Technical Journal* 27, no. 3 (1948): 379–423.

Shiomi, Haruhito und Kazuo Wada, eds. *Fordism Transformed. The Development of Production Methods in the Automobile Industry.* Oxford: Oxford University Press, 1995.

Shulman, Seth. *The Telephone Gambit. Chasing Alexander Graham Bell's Secret.* New York: W. W. Norton, 2009.

Siegel, Steffen. *Tabula: Figuren der Ordnung um 1600.* Berlin: Akademie, 2009.

Siegel, Steffen und Petra Weigel, eds. *Die Werkstatt des Kartographen: Materialien und Praktiken visueller Welterzeugung*. Paderborn: Wilhelm Fink, 2011.

Siegert, Bernhard. *Cultural Techniques: Grids, Filters, Doors, and Other Articulations of the Real*. Translated by Geoffrey Winthrop-Young. New York: Fordham University Press, 2015.

Siegert, Bernhard. "Currents and Currency: Elektrizität, Ökonomie und Ideenumlauf um 1800." In *Netzwerke: Eine Kulturtechnik der Moderne*, edited by Hartmut Böhme, Jürgen Barkhoff, and Jeanne Riou, 53–68. Cologne: Böhlau, 2004.

Siegert, Bernhard. "Das Amt des Gehorchens: Hysterie der Telefonistinnen oder Wiederkehr des Ohres, 1874–1913." In *Armaturen der Sinne: Literarische und technische Medien, 1870 bis 1920*, edited by Jochen Hörisch and Michael Wetzel, 83–106. Munich: Wilhelm Fink, 1990.

Siegert, Bernhard. "Die trinität des gastgebers." In *Cyberspace: Gemeinschaften, virtuelle Kolonien, Öffentlichkeiten*, edited by Manfred Fassler and Wulf R. Halbach, 281–294. Munich: Wilhelm Fink, 1994.

Siegert, Bernhard. "Repräsentationen diskursiver Räume: Einleitung." In *Topographien der Literatur: Deutsche Literatur im transnationalen Kontext*, edited by Hartmut Böhme, 3–11. Stuttgart: J. B. Metzler, 2005.

Siegert, Bernhard. "Switchboards and Sex: The Nut(t) Case." In *Inscribing Science: Scientific Texts and the Materiality of Communication*, edited by Timothy Lenoir, 78–90. Stanford, CA: Stanford University Press, 1998.

Siegert, Paul Ferdinand. *Die Geschichte der E-Mail: Erfolg und Krise eines Massenmediums*. Bielefeld: transcript, 2008.

Siegrist, Hansmartin. "Screening the Phone: Wie das Telefon ins Kino kam." In *Dis Connecting Media: Technik, Praxis und Ästhetik des Telefons: Vom Festnetz zum Handy*, edited by Ulla Autenrieth, Andreas Blättler, Regine Buschauer, and Doris Gassert, 11–52. Basel: Merian, 2011.

Simmel, Georg. "Space and the Spatial Ordering of Society." In *Sociology: Inquiries into the Construction of Social Forms*, edited and translated by Anthony J. Blasi, Anton K. Jacobs, and Mathew Kanjirathinkal, 549–626. Leiden: Brill, 2009.

Simmel, Georg. "The Philosophy of Landscape." Translated by Josef Bleicher. *Theory, Culture & Society* 24, nos. 7–8 (2007): 20–29.

Smith, P. C. "Curious Patents in Mechanical Switching." *Bell Laboratories Record* 7 (1929): 265–269.

Smith, Roberta. "Mark Lombardi, 48, an Artist Who Was Inspired by Scandals." *New York Times*, March 25, 2000. http://web.archive.org/web/20080215173912/http://www.pierogi2000.com/memorial/lombardiobit.html.

Stäheli, Urs. "Entnetzt Euch! Praktiken und Ästhetiken der Anschlusslosigkeit." In *Kunst und Arbeit: Zum Verhältnis von Ästhetik und Arbeitsanthropologie vom 18. Jahrhundert bis zur Gegenwart*, edited by Anja Lemke and Alexander Weinstock, 77–97. Paderborn: Wilhelm Fink, 2014.

Stäheli, Urs. *Soziologie der Entnetzung*. Berlin: Suhrkamp, 2021.

Star, Susan Leigh. "This Is Not a Boundary Object: Reflections on the Origin of a Concept." *Science, Technology, & Human Values* 35, no. 5 (2010): 601–617.

Starosielski, Nicole. *The Undersea Network*. Durham, NC: Duke University Press, 2015.

Stegbauer, Christian, and Roger Häussling, eds. *Handbuch Netzwerkforschung*. Wiesbaden: VS Verlag für Sozialwissenschaften, 2010.

Šteger, Aleš. *Buch der Dinge*. Translated by Urška P. Černe and Matthias Göritz. Frankfurt am Main: Suhrkamp, 2006.

Steinhauer, Fabian. *Vom Scheiden: Geschichte und Theorie einer juristischen Kulturtechnik*. Berlin: Duncker & Humblodt, 2014.

Stephenson, Neal. *Snow Crash*. London: Penguin, 1993.

Sterne, Jonathan. *The Audible Past: Cultural Origins of Sound Reproduction*. 3rd ed. Durham, NC: Duke University Press, 2006.

Stjernfelt, Frederik. *Diagrammatology: An Investigation on the Borderlines of Phenomenology, Ontology, and Semiotics*. Dordrecht: Springer, 2007.

Stockhammer, Robert. *Kartierung der Erde: Macht und Lust in Karten und Literatur*. Paderborn: Wilhelm Fink, 2007.

Stokman, Frans N., Rolf Ziegler, and John Scott, eds. *Networks of Corporate Power: A Comparative Analysis of Ten Countries*. Cambridge: Polity Press, 1985.

Strathern, Marilyn. "Cutting the Network." *Journal of the Royal Anthropological Institute* 2, no. 3 (1996): 517–535.

Strathern, Marilyn. *Partial Connections*. Rev. ed. Walnut Creek, CA: Altamira, 2004.

Straube, Frank. *E-Logistik: Ganzheitliches Logistikmanagement*. Berlin: Springer, 2004.

Strowger, Almon Brown. "Automatic Telephone Exchange." US Patent 447 918, filed March 12, 1889, and issued March 10, 1891.

Sugimoro, Y., K. Kusunoki, F. Cho, and S. Uchikawa. "Toyota Production System and Kanban System of Materialization of Just-in-Time and Respect-for-Human System." *International Journal of Production Research* 15, no. 6 (1977): 553–564.

Sydow, Jörg, and Arnold Windeler. "Steuerung von und in Netzwerken: Perspektiven, Konzepte, vor allem aber offene Fragen." In *Steuerung von Netzwerken: Konzepte und Praktiken*, edited by Jörg Sydow and Arnold Windeler, 1–24. Wiesbaden: Westdeutscher Verlag, 2000.

Sylvester, James Joseph. "On an Application of the New Atomic Theory to the Graphical Representation of the Invariants and Covariants of Binary Quantics." *American Journal of Mathematics* 1 (1878): 64–104.

Taha, Nadine. "Patent in Action: Das US-amerikanische Patent aus der Perspektive der Science and Technology Studies." *Zeitschrift für Medienwissenschaft* 6 (2012): 36–47.

Takeda, Hitoshi. *The Synchronized Production System: Going beyond Just-in-Time through Kaizen*. Translated by Greta Konradt. London: Kogan Page, 2006.

Tarde, Gabriel. *The Laws of Imitation*. Translated by Elsie Clews Parsons. New York: Henry Holt, 1903.

Thacker, Andrew. "Uncompleted Life: The Modernist Underground." In *The Railway and Modernity. Time, Space and the Machine Ensemble*, edited by Matthew Beaumont und Michael Freeman, 101–124. Bern: Peter Lang, 2007.

Thacker, Eugene. *Biomedia*. Minneapolis: University of Minnesota Press, 2004.

Thacker, Eugene. "Networks, Swarms, Multitudes. Part I." *ctheory.net*. May 18, 2004. https://journals.uvic.ca/index.php/ctheory/article/view/14542.

Thielmann, Tristan, and Erhard Schüttpelz, eds. *Akteur-Medien-Theorie*. Bielefeld: transcript, 2013.

Thomas, John Pattinson. *Handling London's Underground Traffic*. London: London's Underground, 1928.

Tolksdorf, Robert. "Mark Lombardi Networks: Towards the complete Mark Lombardi digital." https://lombardinetworks.net.

Töpfer, Georg. "Linien, Bäume, Netze—und die Gegenstände der Biologie." In *Netzwerke: Beiträge zur 13. Jahrestagung der DGGTB in Neuburg an der Donau 2004*, edited by Michael Kaasch, Joachim Kaasch, and Volker Wissemann, 74–94. Berlin: VWB, 2006.

Transport for London. "Thank you for Getting Ahead of the Games." August 4, 2012. https://web.archive.org/web/20120808132057/http://www.getaheadofthegames.com/news/1364.html.

Tresch, John. *The Romantic Machine: Utopian Science and Technology after Napoleon*. Chicago: University of Chicago Press, 2012.

Troske, Ludwig. *Die Londoner Untergrundbahnen: Sonder-Abdruck aus der Zeitschrift des Vereines deutscher Ingenieure 1891 und 1892*. Berlin: Julius Springer, 1892.

Tufte, Edward. *Design of Causal Diagrams: Barr Art Chart, Lombardi Diagrams, Evolutionary Trees, Feynman Diagrams, Timelines*, 2003. https://www.edwardtufte.com/bboard/q-and-a-fetch-msg?msg_id=0000yO&topic_id=1&topic=.

Turner, Fred. *From Counterculture to Cyberculture: Stewart Brand, the Whole Earth Network and the Rise of Digital Utopianism*. Chicago: University of Chicago Press, 2006.

Unterluggauer, Mariann. "Wie Europa beinahe das Internet erfand." *MIT Technology Review* 1 (2012): 52–55. https://www.heise.de/hintergrund/Wie-Europa-beinahe-das-Internet-erfand-1398231.html.

Vandermonde, Théophile. "Remarques sur les problèmes de situation." *Mémoires de l'Académie Royale 1771* (1774): 566–574.

van der Vleuten, Erik. "Understanding Network Societies. Two Decades of Large Technical System Studies." In *Networking Europe: Transnational Infrastructures and the Shaping of Europe, 1850–2000*, edited by Erik van der Vleuten and Arne Kaiser, 279–314. Sagamore Beach, MA: Science History Publications, 2006.

van der Vleuten, Erik, and Arne Kaijser, eds. *Networking Europe: Transnational Infrastructures and the Shaping of Europe, 1850–2000*. Sagamore Beach, MA: Science History Publications, 2006.

van de Velde, Carl. "The Painted Decoration of Florin's House." In *Netherlandish Mannerism: Papers Given at a Symposium in Nationalmuseum Stockholm, September 21–22, 1984*, edited by Görel Cavalli-Björkman, 127–134. Stockholm: Nationalmuseum, 1985.

van Gennep, Arnold. *The Rites of Passage*. Translated by Monika B. Vizedom and Gabrielle L. Caffee. Chicago: University of Chicago Press, 1960.

van Laak, Dirk. "Der Begriff Infrastruktur und was er vor seiner Erfindung besagte." *Archiv für Begriffsgeschichte* 41 (1999): 280–299.

van Laak, Dirk. *Imperiale Infrastruktur: Deutsche Planungen für eine Erschließung Afrikas, 1880 bis 1960*. Paderborn: Schöningh, 2004.

van Schewick, Barbara. *Internet Architecture and Innovation*. Cambridge, MA: MIT Press, 2010.

Vehlken, Sebastian. *Zootechnologies: A Media History of Swarm Research*. Translated by Valentine A. Pakis. Amsterdam: Amsterdam University Press, 2019.

Velminski, Wladimir. "Nachwort." In *Leonhard Euler, Die Geburt der Graphentheorie: Ausgewählte Schriften von der Topologie zum Sudoku*, edited by Wladimir Velminski, 151–206. Berlin: Kadmos, 2008.

Vertesi, Janet. "Mind the Gap: The London Underground Map and Users' Representation of Urban Space." *Social Studies of Science* 38, no. 1 (2008): 7–33.

Vesalius, Andreas. *De Humani Corporis Fabrica Libri septem*. Basel: Johann Oporinus, 1543.

Virilio, Paul. *The Aesthetics of Disappearance*. 2nd ed. Translated by Philip Beitchman. Los Angeles: Semiotext(e), 2009.

Vismann, Cornelia. "Cultural Techniques and Sovereignty." Translated by Ilinca Iurascu. *Theory, Culture & Society* 30, no. 6 (2013): 83–93.

Vismann, Cornelia, and Markus Krajewski. "Computer Juridisms." *Grey Room* 29 (2008): 99–109.

Vogl, Joseph. "Mittler und Lenker. Goethes *Wahlverwandtschaften*." In *Poetologien des Wissens um 1800*, edited by Joseph Vogl, 145–161. Munich: Wilhelm Fink, 1999.

Vogl, Joseph. "Regierung und Regelkreis. Historisches Vorspiel." In *Cybernetics/Kybernetik. The Macy Conferences 1946–1953*. Vol. 2, edited by Claus Pias, 67–79. Zurich: diaphanes, 2004.

Vogl, Joseph. *Über das Zaudern*. Zurich: diaphanes, 2007.

von der Osten-Sacken, Elisabeth. "Netz." In *Reallexikon der Assyriologie und Vorderasiatischen Archäologie*. Vol. 9, edited by Otto Dietz Edzard, 235–242. Berlin: De Gruyter, 2001.

von Foerster, Heinz. "Memory without Record." In *Learning, Remembering, and Forgetting*. Vol. 1, *The Anatomy of Memory*, edited by Daniel P. Kimble, 388–433. Palo Alto, CA: Stanford University Press, 1965.

von Foerster, Heinz, and Bernhard Pörksen. *Wahrheit ist die Erfindung eines Lügners: Gespräche für Skeptiker*. 4th ed. Heidelberg: Carl Auer, 2001.

von Hagen, Kirsten. "Wahlverwandtschaften: Spielformen des Telefons im Film." In *Intermedialität Analog/Digital: Theorien, Methoden, Analysen*, edited by Joachim Paech and Jens Schröter, 333–343. Paderborn: Wilhelm Fink, 2008.

von Stephan, Heinrich. *Das Telephon betreffend. Brief des Generalpostmeisters Heinrich v. Stephan an Seine Durchlaucht den Fürsten Reichskanzler Otto v. Bismarck. Vorschlag für die Einführung der ersten Telephon-Verbindungen Berlin 9.11.1877*. Edited by Uwe Otto. Berlin: Berliner Handpresse, 1984.

Wagner, Kirsten. *Datenräume, Informationslandschaften, Wissensstädte: Zur Verräumlichung des Wissens und Denkens in der Computermoderne*. Freiburg: Rombach, 2006.

Walden, David C. "Looking Back at the ARPANET Effort Decades Later." In *The ARPANET Sourcebook: The Unpublished Foundations of the Internet*, edited by Peter Salus, 360–362. Charlottesville, VA: Peer-to-Peer-Communications, 2008.

Walden, David C., and Alexander A. McKenzie. "The Evolution of Host-to-Host Protocol Technology." *Computer* 12, no. 9 (1979): 29–38.

Waldrop, M. Mitchell. *The Dream Machine: J. C. R. Licklider and the Revolution That Made Computing Personal*. New York: Viking, 2001.

Warncke, Carsten-Peter. *Symbol, Emblem, Allegorie: Die zweite Sprache der Bilder*. Cologne: Deubner, 2005.

Warnke, Martin. *Political Landscape: The Art History of Nature*. Translated by David McLintock. Cambridge, MA: Harvard University Press, 1995.

Warnke, Martin. *Theorien des Internet zur Einführung*. Hamburg: Junius, 2011.

Watts, Duncan. *Six Degrees: The Science of a Connected Age*. New York: W. W. Norton, 2003.

Weber, Samuel M. *Targets of Opportunity: On the Militarization of Thinking*. New York: Fordham University Press, 2005.

Weigel, Sigrid. *Body- and Image-Space: Re-Reading Walter Benjamin*. Translated by Georgina Paul with Rachel McNicholl and Jeremy Gaines. London: Routledge, 1996.

Werber, Niels. *Die Geopolitik der Literatur*. Munich: Hanser, 2007.

Werber, Niels. "Netzwerkgesellschaft: Zur Kommunikationsgeschichte von 'technoiden' Selbstbeschreibungsformeln." In *Kulturgeschichte als Mediengeschichte (oder vice versa?). Archiv für Mediengeschichte*, edited by Lorenz Engell, Joseph Vogl, and Bernhard Siegert, 179–191. Weimar: Universitätsverlag, 2006.

Werner, Gabriele, ed. *Systemische Räume: Bildwelten des Wissens* 5, no. 1. Berlin: Akademie, 2007.

Widlok, Thomas. "Kulturtechniken: Ethnographisch fremd und anthropologisch fremd; Eine Kritik an ökologisch-phänomenologischen und kognitiv-modularisierenden Ansätzen." In *Universitätsforschungen zur prähistorischen Archäologie* 264: *Fremdheit—Perspektiven auf das Andere*, edited by Tobias L. Kienlin, 41–59. Bonn: Verlag Dr. Rudolf Habelt, 2015.

Wiener, Norbert. *Cybernetics; or, Control and Communication in the Animal and the Machine*. 2nd ed. Cambridge, MA: MIT Press, 1965.

Wigley, Mark. "Network Fever." *Grey Room* 4 (2001): 82–122.

Wigley, Mark. "The Architectural Brain." In *Network Practices: New Strategies in Architecture and Design*, edited by Anthony Burke and Therese Tierney, 30–53. New York: Princeton Architectural Press, 2007.

Wikipedia. S.v. "Complex network." Last modified December 8, 2021, 04:29, https://en.wikipedia.org/wiki/Complex_network.

Wikipedia. S.v. "Lilian Dring." Last modified November 13, 2020, 18:44, https://en.wikipedia.org/wiki/Lilian_Dring.

Wikipedia. S.v. "SS France 1960." Last modified February 7, 2022, 00:19, https://en.wikipedia.org/wiki/SS_France_(1960).

Wikipedia. S.v. "Telephone newspaper." Last modified January 5, 2022, 14:50, https://en.wikipedia.org/wiki/Telephone_newspaper.

Winkler, Hartmut. *Docuverse: Zur Medientheorie der Computer*. Munich: Boer, 1997.

Winthrop-Young, Geoffrey. "Cultural Techniques: Preliminary Remarks." *Theory, Culture & Society* 30, no. 6 (2013): 3–19.

Winthrop-Young, Geoffrey, Ilinca Iurascu, and Jussi Parikka, eds., special issue "Cultural Techniques," *Theory, Culture & Society* 30, no. 6 (2013).

Wirth, Uwe. "Vorüberlegungen zu einer Logik der Kulturforschung." In *Kulturwissenschaft: Eine Auswahl grundlegender Texte*, edited by Uwe Wirth, 9–67. Frankfurt am Main: Suhrkamp, 2008.

Wirth, Uwe. "Kultur als Pfropfung, Pfropfung als Kulturmodell: Prolegomena zu einer allgemeinen Greffologie (2.0)." In *Impfen, Pfropfen, Transplantieren*, edited by Uwe Wirth, 9–27. Berlin: Kadmos, 2011.

Wittgenstein, Ludwig. *Vermischte Bemerkungen / Culture and Value*. Edited by Georg Henrik von Wright with Heikki Nyman. Translated by Peter Winch. Chicago: Chicago University Press, 1980.

Wittmann, Barbara, and Christoph Hoffmann. "Knowledge in the Making: Drawing and Writing as Research Techniques." *Science in Context* 26, no. 2 (2013): 203–213.

Wu, Tim. *The Master Switch: The Rise and Fall of Information Empires*. New York: Alfred A. Knopf, 2011.

Yates, JoAnne. *Control through Communication: The Rise of System in American Management*. Baltimore: Johns Hopkins University Press, 1989.

Yates, JoAnne, and Craig Murphy. *Engineering Rules: Global Standard Setting since 1880*. Baltimore: Johns Hopkins University Press, 2019.

Young, James. "Malpighi's 'De Pulmonibus.'" *Proceedings of the Royal Society of Medicine* 23, no. 1 (1929): 1–11.

Zedler, Johann Heinrich, ed. "Netz." In *Johann Heinrich Zedlers Grosses vollständiges Universal-Lexicon aller Wissenschafften und Künste*. Vol. 23, *N–Net*. Halle: Zedler, 1740, cols. 1993–2006.

FILMS

Cameron, James, dir. *Terminator*. 1984; Image Entertainment, 1997. DVD. 107 min.

Cameron, James, dir. *Terminator 2: Judgment Day*. 1991; Columbia Tri-Star Home Video, 1997. DVD. 137 min.

Chvany, Peter, dir. *Computer Networks: The Heralds of Resource Sharing*. 1972; Archive.org, 2022. Online-Video. 30 min. https://archive.org/details/ComputerNetworks_TheHeraldsOfResourceSharing.

Condon, Bill, dir. *Kinsey*. 2004; 20th Century Fox Home Entertainment, 2005. DVD. 118 min.

Coppola, Francis Ford, dir. *The Conversation*. 1974; Paramonut Pictures, 2000. DVD. 113 min.

Coppola, Francis Ford, dir. *The Godfather*. 1972; Paramount Pictures, 2008. DVD. 177 min.

Forde, Walter, dir. *Bulldog Jack*. 1935; Gaumont British Picture Corporation, 1994. VHS. 72 min.

Gaghan, Stephen, dir. *Syriana*. 2005; Warner Home Video, 2006. DVD. 128 min.

Goulding, Edmund, dir. *Grand Hotel*. 1932; Turner Entertainment Co. and Warner Home Video, 2004. DVD. 108 min.

Harlin, Renny, dir. *Die Hard 2*. 1990; 20th Century Fox, 2005. DVD. 124 min.

Herzog, Werner, dir. *Lo and Behold: Reveries of the Connected World*. 2016; Magnolia Pictures, 2016. DVD. 98 min.

Kubrick, Stanley, dir. *2001: A Space Odyssey*. 1968; Warner Home Video, 2011. DVD. 142 min.

Maida, Marco J. D., dir. *Jacob Levy Moreno: His Life and His Muses*. 1997; Psychotherapy.net, 2003. Video. 40 min.

Marclay, Christian. *Telephones*. 1995; Video installation. Single-channel video, black-and-white and color, with sound, 7.30 min, dimensions variable. https://vimeo.com/176259496.

McGinty Nichol, Joseph, dir. *Terminator Salvation*. 2009; Sony Pictures Releasing, 2009. DVD. 115 min.

McTiernan, John, dir. *Die Hard*. 1988; 20th Century Fox, 2001. DVD. 132 min.

McTiernan, John, dir. *Die Hard with a Vengeance*. 1995; 20th Century Fox, 2005. DVD. 128 min.

Melville, Jean-Pierre, dir. *Le Samouraï*. 1967; S. N. Prodis, 2001. DVD. 103 min.

Moore, John, dir. *A Good Day to Die Hard*. 2013; 20th Century Fox, 2013. DVD. 97 min.

Mostow, Jonathan, dir. *Terminator 3: Rise of the Machines*. 2003; Colombia Tri-Star Home Video, 2003. DVD. 109 min.

Pakula, Alan J., dir. *Parallax View*. 1974; Paramount Pictures, 2004. DVD. 102 min.

Pollack, Sydney, dir. *Three Days of the Condor*. 1975; Paramount Pictures produced by Stanley Schneider, 2001. DVD. 117 min.

Pollack, Sydney, dir. *The Interpreter*. 2005: Universal Pictures, 2009. DVD. 128 min.

Roddenberry, Gene [Eugene Wesley], prod. *Star Trek: The Original Series (season 1–3)*. 1966–1969; CBS Paramount Television, 2008. DVD, TV Series, 79 episodes.

Sargent, Joseph, dir. *Colossus: The Forbin Project*. 1970; Universal Pictures, 2004. DVD. 100 min.

Scorsese, Martin, dir. *Mean Streets*. 1973; Warner Home Video, 2005. DVD. 112 min.

The Right and Wrong Way to Board an Electric Train. 1920, Gaumont; London Transport Museum, 2007. Video. Film Collection, no. 125.

Underground Group Signals Section, dir. *Signalling on the Ealing and Shepherd's Bush Railway*. 1921; TfL from the London Transport Museum Collection, 2007. Digitized 35 mm Video. 14 min. Film Collection, no. 121.

Wachowski Lilly [Andy], and Lana [Laurence] Wachowski, dir. *The Matrix*. 1999; Warner Home Video, 1999. DVD. 136 min.

Wegener, Mareike, dir. *Mark Lombardi: Kunst und Konspiration*. 2012; RealFiction, 2013. DVD. 80 min.

Wiseman, Len, dir. *Live Free or Die Hard*. 2007; 20th Century Fox, 2007. DVD. 129 min.

INDEX

Page numbers followed by an *f* denote a figure.

Infrastructures Series

Edited by Paul N. Edwards and Janet Vertesi

Paul N. Edwards, *A Vast Machine: Computer Models, Climate Data, and the Politics of Global Warming*

Lawrence M. Busch, *Standards: Recipes for Reality*

Lisa Gitelman, ed., *"Raw Data" Is an Oxymoron*

Finn Brunton, *Spam: A Shadow History of the Internet*

Nil Disco and Eda Kranakis, eds., *Cosmopolitan Commons: Sharing Resources and Risks across Borders*

Casper Bruun Jensen and Brit Ross Winthereik, *Monitoring Movements in Development Aid: Recursive Partnerships and Infrastructures*

James Leach and Lee Wilson, eds., *Subversion, Conversion, Development: Cross-Cultural Knowledge Exchange and the Politics of Design*

Olga Kuchinskaya, *The Politics of Invisibility: Public Knowledge about Radiation Health Effects after Chernobyl*

Ashley Carse, *Beyond the Big Ditch: Politics, Ecology, and Infrastructure at the Panama Canal*

Alexander Klose, translated by Charles Marcrum II, *The Container Principle: How a Box Changes the Way We Think*

Eric T. Meyer and Ralph Schroeder, *Knowledge Machines: Digital Transformations of the Sciences and Humanities*

Geoffrey C. Bowker, Stefan Timmermans, Adele E. Clarke, and Ellen Balka, eds., *Boundary Objects and Beyond: Working with Leigh Star*

Clifford Siskin, *System: The Shaping of Modern Knowledge*

Lawrence Busch, *Knowledge for Sale: The Neoliberal Takeover of Higher Education*

Bill Maurer and Lana Swartz, *Paid: Tales of Dongles, Checks, and Other Money Stuff*

Katayoun Shafiee, *Machineries of Oil: An Infrastructural History of BP in Iran*

Megan Finn, *Documenting Aftermath: Information Infrastructures in the Wake of Disasters*

Ann M. Pendleton-Jullian and John Seely Brown, *Design Unbound: Designing for Emergence in a White Water World*, Volume 1: *Designing for Emergence*

Ann M. Pendleton-Jullian and John Seely Brown, *Design Unbound: Designing for Emergence in a White Water World*, Volume 2: *Ecologies of Change*

Jordan Frith, *A Billion Little Pieces: RFID and Infrastructures of Identification*

Morgan G. Ames, *The Charisma Machine: The Life, Death, and Legacy of One Laptop per Child*

Ryan Ellis, *Letters, Power Lines, and Other Dangerous Things: The Politics of Infrastructure Security*

Mario Biagioli and Alexandra Lippman, eds, *Gaming the Metrics: Misconduct and Manipulation in Academic Research*

Malcolm McCullough, *Downtime on the Microgrid: Architecture, Electricity, and Smart City Islands*

Emmanuel Didier, translated by Priya Vari Sen, *America by the Numbers: Quantification, Democracy, and the Birth of National Statistics*

Andrés Luque-Ayala and Simon Marvin, *Urban Operating Systems; Producing the Computational City*

Michael Truscello, *Infrastructural Brutalism: Art and the Necropolitics of Infrastructure*

Christopher R. Henke and Benjamin Sims, *Repairing Infrastructures: The Maintenance of Materiality and Power*

Stefan Höhne, *New York City Subway: The Invention of the Urban Passenger*

Timothy Moss, *Conduits of Berlin: Remaking the City through Infrastructure, 1920–2020*

Claude Rosental, translated by Catherine Porter, *The Demonstration Society*

Blake Atwood, *Underground: The Secret Life of Videocassettes in Iran*

Huub Dijstelbloem, *Borders as Infrastructure: The Technopolitics of Border Control*

Dylan Mulvin, *Proxies: Standards and their Media*

Eric Monteiro, *Digital Oil: Machineries of Knowing*

Klaus Hoeyer, *Data Paradoxes: The Politics of Intensified Data Sourcing in Healthcare*

Dirk van Laak, *Lifelines of Our Society: A Global History of Infrastructure*

Jonathan Silver, *The Infrastructural South: Techno-Environments of the Third Wave of Urbanization*

Christine E. Evans and Lars Lundgren, *No Heavenly Bodies: A History of Satellite Communications Infrastructure*

Sebastian Giessmann, *The Connectivity of Things: Network Cultures since 1832*